Springer Proceedings in Mathematics & Statistics

Volume 395

This book series features volumes composed of selected contributions from workshops and conferences in all areas of current research in mathematics and statistics, including data science, operations research and optimization. In addition to an overall evaluation of the interest, scientific quality, and timeliness of each proposal at the hands of the publisher, individual contributions are all refereed to the high quality standards of leading journals in the field. Thus, this series provides the research community with well-edited, authoritative reports on developments in the most exciting areas of mathematical and statistical research today.

Melvyn B. Nathanson
Editor

Combinatorial and Additive Number Theory V

CANT, New York, USA, 2021

Springer

Editor
Melvyn B. Nathanson
Department of Mathematics
Lehman College
Bronx, NY, USA

ISSN 2194-1009 ISSN 2194-1017 (electronic)
Springer Proceedings in Mathematics & Statistics
ISBN 978-3-031-10795-5 ISBN 978-3-031-10796-2 (eBook)
https://doi.org/10.1007/978-3-031-10796-2

Mathematics Subject Classification: 11B13, 11B05, 11B34, 11B75, 11P99, 20M14

This Springer imprint is published by the registered company Springer Nature Switzerland AG
The registered company address is: Gewerbestrasse 11, 6330 Cham, Switzerland

Preface: Mathematics in the Time of COVID

Workshops on Combinatorial and Additive Number Theory (CANT) have been organized at the CUNY Graduate Center in New York every year since 2003. We would usually meet for 4 days in May, in the week immediately preceding Memorial Day. These in-person meetings had become a fixed point in the number theory calendar. CANT 2020 and CANT 2021 were different. In 2020, because of the COVID pandemic, it was impossible to meet in person at the CUNY Graduate Center, and CANT 2020 was held on Zoom. In 2021, the COVID pandemic still had not abated, and CANT 2021, the 19th annual meeting, again took place on Zoom. A mega-meeting, with over 100 talks, it ran for 5 days, from Monday, May 24, to Friday, May 28, 2021, and was a great success. Zoom attendance was high and the videos of the talks, uploaded to YouTube, continue to attract viewers.

In the coming years, historians and sociologists of science will investigate how COVID affected research in mathematics. How much does mathematics depend on international travel and face-to-face interactions? How essential are the large grants that pay for them? The pandemic affected everything, but it is likely that the future will show that research in mathematics and, probably, in most other sciences did not decline under COVID, and, in fact, benefitted in many ways.

There are five transformative inventions (all Internet-enabled) that saved science. All are recent inventions, though only one was created during and, in part, because of the COVID pandemic.

The first, which needs no introduction, is Zoom. It made online conferencing easy, efficient, and cheap. Face-to-face discussions on Zoom may convey less substantive and emotional information than in-person meetings, but the difference is probably slight.

Wikipedia and arXiv.org are online creations of enormous value to the mathematics research community. The arXiv provides free worldwide access to a vast library of preprints in all fields of mathematics and is a viable alternate to the admittedly pleasurable activity of browsing through the new journals shelf in a library. The website arXiv was created as a preprint server for articles in physics but has grown to include mathematics and some other sciences.

The amazingly broad range of mathematics articles on Wikipedia provides an effective alternative to meeting someone in the department lounge to satisfy your curiosity and learn about topics in mathematics that might be far from your areas of specialization. Wikipedia articles vary in quality and reliability, but the average article is informative and contains useful references.

The fourth factor is YouTube.com. A treasure trove of videos of mathematics lectures by distinguished researchers has been uploaded to YouTube and can be watched by anyone anywhere for free. This is an amazing resource.

The fifth and final Internet invention that importantly contributes to science is the website researchseminars.org. It was created as mathseminars.org by a group of mathematicians at MIT and launched on April 10, 2020. A few months later it was renamed researchseminars.org and extended to include other sciences. The website provides a master list of seminars and conferences (with direct videoconference links) taking place at universities and institutes around the world. Now, sitting at home with a computer and the internet, one can participate in an unimaginably broad and rich array of research seminars.

The access to information provided by Zoom, Wikipedia, arXiv, YouTube, and researchseminars.org would have been unimaginable a few years ago. COVID forced the international research community to adopt these technologies much faster and more intensely than would otherwise have happened. The cataclysmic change in the scientific environment caused by the pandemic created rapid evolutionary adaptation. In this sense, COVID benefitted science.

I hope that the online CANT conferences, with lectures listed on research-seminars.org and presented on Zoom, and with videos of the talks available on YouTube.com, contribute to the international mathematics research enterprise.

I am grateful to Springer and mathematics editor Dahlia Fisch for making possible the publication of the proceedings of the CANT 2021 workshop.

Steven J. Miller and Kevin O'Bryant provided technical support for CANT 2021. Steve Miller also enabled funding from *Journal of Number Theory* (Elsevier) in support of the conference. I thank the following students who assisted with the meeting: Sunita Bhattacharya (UMass Amherst), Aidan Dunkelberg (Williams College), Teresa Dunn (UC Davis), Phuc Lam (University of Rochester), Clayton Mizgerd (Williams College), and Chenyang Sun (Williams College).

Previous volumes in this series are [1], [2], [3], and [4].

Short Hills, NY, USA
June 2022

Melvyn B. Nathanson

References

1. M. B. Nathanson, editor, *Combinatorial and Additive Number Theory—CANT 2011 and 2012*, Springer Proc. Math. Stat., vol. 101, Springer, New York, 2014.
2. M. B. Nathanson, editor, *Combinatorial and Additive Number Theory II--CANT 2015 and 2016*, Springer Proc. Math. Stat., vol. 220, Springer, New York, 2017.

3. M. B. Nathanson, editor, *Combinatorial and Additive Number Theory III--CANT 2017 and 2018*, Springer Proc. Math. Stat., vol. 297, Springer, New York, 2020.
4. M. B. Nathanson, editor, *Combinatorial and Additive Number Theory IV--CANT 2019 and 2020*, Springer Proc. Math. Stat., vol. 347, Springer, New York, 2021.

Contents

On the Number of Dot Product Chains in Finite Fields and Rings

Vincent Blevins, David Crosby, Ethan Lynch, and Steven Senger

1 Introduction

Background

In 1946, in [9], Paul Erdős raised a question that eludes mathematicians to this day: In a large finite point set in the plane, how many pairs of points can be separated by the same distance? This is often referred to as the unit distance problem. While optimal bounds are not yet known, there has been much activity on this and related problems. See, for example, [4, 10, 17].

A more general family of questions is: Given a large subset of an ambient set with some structure (such as a vector space or module), how many instances of some class of point configurations can be present? Let q be some power of an odd prime, p, and consider d-dimensional vector spaces over finite fields, $\mathbb{F}_q^d$, or the d-rank free modules over $\mathbb{Z}_q^d$, instead of the plane. There has been much work studying different types of analogs of the unit distance problem in these settings, as can be seen in [13], by Alex Iosevich and Misha Rudnev.

While Erdős asked about pairs of points determining a fixed distance, one can investigate k-tuples of points determining other functions, such as dot products. Again, there is an abundance of work on such generalizations. See [5, 8, 12, 14, 19] and the references contained therein.

V. Blevins · D. Crosby · E. Lynch · S. Senger (✉)
Missouri State University, Springfield, MO, USA
e-mail: stevensenger@gmail.com

M. B. Nathanson (ed.), *Combinatorial and Additive Number Theory V*,
Springer Proceedings in Mathematics & Statistics 395,
https://doi.org/10.1007/978-3-031-10796-2_1

Notation

We now give a precise definition of our main object of study. We follow the convention of similar definitions from those given in [2, 16].

Definition 1 Given a sequence of field (or ring) elements $(\alpha_1, \ldots, \alpha_k)$, a **dot product k-chain** is a $(k+1)$-tuple of distinct points $(x_1, \ldots, x_{k+1})$ satisfying $x_j \cdot x_{j+1} = \alpha_j$ for every $1 \leq j \leq k$.

In [1], Daniel Barker and the fourth listed author gave bounds on the number of dot product 2-chains in the plane. These have been expanded and generalized recently in [11, 15] This approach was adapted to the settings of $\mathbb{F}_q^d$ and $\mathbb{Z}_q^d$ by Dave Covert and the fourth listed author in [6].

In this note, we study k-chains in these settings. Given a subset of $E \subseteq \mathbb{F}_q^d$ $\left(\text{or } \mathbb{Z}_q^d\right)$, and a k-tuple of elements $\alpha = (\alpha_1, \ldots, \alpha_k) \in \mathbb{F}_q^k$ (or $\mathbb{Z}_q^k$), let $\Pi_\alpha(E)$ denote the set of dot product k-chains corresponding to α whose members are points in E. Throughout this paper, we assume that k is like a constant compared to the size of any subset E. To ease exposition, we use the asymptotic symbols $X \lesssim Y$ if $X = O(Y)$, and $X \approx Y$ when $X = \Theta(Y)$. Moreover, we write $X \gtrapprox Y$ when for every $\epsilon > 0$, there exists a constant $C_\epsilon > 0$ such that $X \gtrsim C_\epsilon q^\epsilon Y$.

Main Results

Our first new result is on 3-chains in $\mathbb{F}_q^d$.

Theorem 1 *Given $E \subseteq \mathbb{F}_q^d$, with q a power of an odd prime, and $\alpha \in \mathbb{F}_q^3$, if $|E| \gtrapprox q^{\frac{d+3}{2}}$ (or $|E| \gtrapprox q^{\frac{d+2}{2}}$ if one of the α_j is nonzero, or $|E| \gtrapprox q^{\frac{d+1}{2}}$ if the α_j are nonzero), then*

$$|\Pi_\alpha(E)| = (1 + o(1))\frac{|E|^4}{q^3}.$$

The method of proof is to count the number of dot product 3-chains with a character sum, which we break up into a main term and several remainder terms. The main idea is to estimate the remainder terms. Using this same line of reasoning, but with less strict hypotheses, we get the following results which are more general, but weaker.

Theorem 2 *Given $E \subseteq \mathbb{Z}_q^d$, with $q = p^\ell$ a power of an odd prime p, $k \in \mathbb{N}$, and a k-tuple of units, $\alpha \in \mathbb{Z}_q^k$, if $|E| \gtrapprox q^{\frac{d(2\ell-1)+1}{2\ell} + \frac{k-2}{2}}$, then*

$$|\Pi_\alpha(E)| = (1 + o(1))\frac{|E|^{k+1}}{q^k}.$$

Theorem 3 *Given $E \subseteq \mathbb{F}_q^d$, with q a power of an odd prime and $\alpha \in \mathbb{F}_q^k$, if $|E| \gtrsim q^{\frac{d+k}{2}}$ (or $|E| \gtrsim q^{\frac{d+k-1}{2}}$ if the components of α are nonzero), then*

$$|\Pi_\alpha(E)| = (1 + o(1))\frac{|E|^{k+1}}{q^k}.$$

While all of the results so far have stated that particular types of dot product k-chains occur with regularity within large enough subsets, we now turn to upper bounds on the number of a given type of dot product k-chain in a subset of $\mathbb{F}_q^2$ in terms of the size of the subset.

Theorem 4 *Given $E \subseteq \mathbb{F}_q^2$, with q a power of an odd prime, and $\alpha \in \mathbb{F}_q^k$, with all components nonzero,*

$$|\Pi_\alpha(E)| \lesssim |E|^{\left\lceil \frac{2(k+1)}{3} \right\rceil}.$$

These bounds hold without strict size constraints on the size of E found in the estimates above. However, if the size of E is sufficiently large to apply Theorems 1 or 3, then Theorem 4 will yield weaker bounds. This result is a straightforward corollary of an estimate in [6] (Theorem 3 in that paper), though that result has a mistake as stated. The finite fields result is true, but there is a fatal flaw in the way that incidences of points and lines were counted. This is detailed in Sect. 5.

Sharpness and Relevance of Hypotheses

While these results are almost certainly far from sharp, we provide some constructions that demonstrate some of what is known. These also help indicate why size conditions on E make sense as hypotheses, as comparatively small sets can exhibit behavior far from what would be expected from a random subset. We also show how the zero dot product has distinct behavior, to demonstrate why we have the relevant hypotheses on the α_j being units or at least nonzero.

Remark 1 Consider the set

$$E := (\{(x, 0) : x \in \mathbb{F}_q\}) \cup (\{(0, y) : y \in \mathbb{F}_q\}) \subseteq \mathbb{F}_q^2.$$

Clearly, this set has $\gtrsim q^{k+1}$ dot product k-chains of dot product zero, obtained by alternately selecting points from each of the subsets listed in the union. Here, $|E| \approx q$.

While the previous remark shows that zero dot products can exhibit markedly different behavior, we also have related examples for nonzero dot products.

Remark 2 Consider the set

$$E := (\{(x, 0, \alpha) : x \in \mathbb{F}_q\}) \cup (\{(0, y, 1) : y \in \mathbb{F}_q\}) \subseteq \mathbb{F}_q^3.$$

Clearly, this set has $\gtrsim q^{k+1}$ dot product k-chains of dot product α, obtained by alternately selecting points from each of the subsets listed in the union. Again, we have $|E| \approx q$.

Clearly, both of these examples can be modified to fit into $\mathbb{F}_q^d$ or $\mathbb{Z}_q^d$ for any d large enough to admit an embedding. This indicates that the occurrence of more dot product k chains of a given type is an artifact of "lower dimensional" subsets. If E is a large enough subset of the ambient space (or module), then these occurrences are outweighed by the behavior of the rest of the set E.

2 Bounds on 3-Chains in $\mathbb{F}_q^d$

Let $\chi(\alpha)$ to denote the canonical additive character of $\mathbb{F}_q$. The plan will be to count the number of 3-chains using a character sum. We will then split this sum into a main term and several error terms. The key idea will be getting nontrivial bounds on the error terms based on the size of the set E.

For 3-chains in $\mathbb{F}_q^d$:

$$\left|\Pi_{\alpha_1,\alpha_2,\alpha_3}(E)\right| = |\{(x_1, x_2, x_3, x_4) \in E^4 : x_j \cdot x_{j+1} = \alpha_j\}|$$

$$= q^{-3} \sum_{s_j} \sum_{x_j} \chi(s_j(x_j \cdot x_{j+1} - \alpha_j))$$

$$= M + R_1 + R_2 + R_3.$$

Here, M is the case where all auxiliary variables are zero, and each R_n is the case where n auxiliary variables are nonzero. Moreover, $R_{\{n_i,\dots,n_j\}}$ is the case where the set of specific auxiliary variables, $n_i, \dots, n_j$ are nonzero.

$$M = q^{-3} \sum_{s_j=0} \sum_{x_j \in E} \chi(s_j(x_j \cdot x_{j+1} - \alpha_j))$$

$$= q^{-3} \sum_{x_j \in E} \chi(0 \cdot (x_j \cdot x_{j+1} - \alpha_j))$$

$$= q^{-3} \sum_{x_j \in E} \chi(0)$$

$$= q^{-3}|E|^4.$$

The hope is that we have M be the dominant term in the sum so that it can serve as our estimate for $|\Pi_{\alpha_1,\alpha_2,\alpha_3}(E)|$, while the other terms can be bounded.

$$R_1 = q^{-3} \sum_{\substack{s_2=s_3=0\\ s_1\in\mathbb{F}_q^*}} \sum_{x_j\in E} \chi(s_j(x_j\cdot x_{j+1}-\alpha_j)) + q^{-3} \sum_{\substack{s_1=s_3=0\\ s_2\in\mathbb{F}_q^*}} \sum_{x_j\in E} \chi(s_j(x_j\cdot x_{j+1}-\alpha_j)) +$$

$$q^{-3} \sum_{\substack{s_1=s_2=0\\ s_3\in\mathbb{F}_q^*}} \sum_{x_j\in E} \chi(s_j(x_j\cdot x_{j+1}-\alpha_j))$$

$$= R_{\{1\}} + R_{\{2\}} + R_{\{3\}}.$$

We now introduce the following result due to Derrick Hart, Alex Iosevich, Doo Won Koh, and Misha Rudnev, in [12], to help us get a handle on quantities like $R_{\{1\}}$.

Lemma 1 (Eq. 2.5 from [12]) *For any set $E \subseteq \mathbb{F}_q^d$, we have the bound*

$$\left|\sum_{s\neq 0} \sum_{x,y\in E} \chi(s(x\cdot y-\gamma))\right| \leq |E| q^{\frac{d+1}{2}} \lambda(\gamma), \tag{1}$$

where $\lambda(\gamma) = 1$ for $\gamma \in \mathbb{F}_q^*$ and $\lambda(0) = \sqrt{q}$.

$$\begin{aligned} R_{\{1\}} &= q^{-3} \sum_{\substack{s_2=s_3=0\\ s_1\in\mathbb{F}_q^*}} \sum_{x_j\in E} \chi(s_j(x_j\cdot x_{j+1}-\alpha_j)). \\ &= q^{-3}|E|^2 \sum_{s_1\in\mathbb{F}_q^*} \sum_{x_1,x_2\in E} \chi(s_j(x_j\cdot x_{j+1}-\alpha_j)). \end{aligned}$$

By Lemma 1, we have

$$\begin{aligned} R_{\{1\}}| \leq& q^{-3}|E|^2 \cdot |E| q^{\frac{d+1}{2}} \lambda(\alpha_1) \\ &= |E|^3 q^{\frac{d-5}{2}} \lambda(\alpha_1). \end{aligned}$$

By similar arguments, we also get that

$$\left|R_{\{2\}}\right| \leq |E|^3 q^{\frac{d-5}{2}} \lambda(\alpha_2), \text{ and}$$

$$\left|R_{\{3\}}\right| \leq |E|^3 q^{\frac{d-5}{2}} \lambda(\alpha_3).$$

So,

$$
\begin{aligned}
|R_1| &= \left|R_{\{1\}} + R_{\{2\}} + R_{\{3\}}\right| \\
&\leq |E|^3 q^{\frac{d-5}{2}} (\lambda(\alpha_1) + \lambda(\alpha_2) + \lambda(\alpha_3)).
\end{aligned}
$$

Now let's look at the case with two nonzero auxiliary variables. This case in particular is special because we have two sub-cases. The first is when the two dot products share a point, and the second is when they share no points in common.

$$R_2 = R_{\{1,2\}} + R_{\{1,3\}} + R_{\{2,3\}}$$

We'll start with $R_{\{1,2\}}$ and $R_{\{2,3\}}$ which are similar in that the auxiliary variables involved are consecutive.

$$
\begin{aligned}
R_{\{1,2\}} &= q^{-3}|E| \sum_{s_1,s_2 \in \mathbb{F}_q^*} \sum_{x_1,x_2,x_3 \in E} \chi(s_1(x_1 \cdot x_2 - \alpha_1))\chi(s_2(x_2 \cdot x_3 - \alpha_2)) \\
&= q^{-3}|E| \sum_{s_1,s_2 \in \mathbb{F}_q^*} \sum_{x_1,x_2,x_3 \in E} \chi(s_1(x_1 \cdot x_2 - \alpha_1))\chi(s_2(x_2 \cdot x_3 - \alpha_2)) \\
&= q^{-3}|E| \cdot T_1(E),
\end{aligned}
$$

where we define

$$T_i(E) := \sum_{s_i,s_{i+1} \in \mathbb{F}_q^*} \left(\sum_{\substack{x_i,x_{i+1}, \\ x_{i+2} \in E}} \chi(s_i(x_i \cdot x_{i+1} - \alpha_i))\chi(s_{i+1}(x_{i+1} \cdot x_{i+2} - \alpha_{i+1})) \right)$$

We state an adaptation of one of the key estimates in [6], due to Dave Covert and the fourth listed author as a lemma.

Lemma 2 (Estimate of III from [6])

$$|T_i(E)| \lesssim q^{d+1}|E|\lambda(\alpha_i)\lambda(\alpha_{i+1}).$$

By the above argument and Lemma 2, we get

$$\left|R_{\{1,2\}}\right| \leq q^{-3}|E|\,|T_1(E)| \leq q^{-3}|E| \cdot q^{d+1}|E|\lambda(\alpha_1)\lambda(\alpha_2),$$

which yields

$$\left|R_{\{1,2\}}\right| \leq q^{d-2}|E|^2\lambda(\alpha_1)\lambda(\alpha_2). \tag{2}$$

Similarly, we can see that

$$\left|R_{\{2,3\}}\right| \leq q^{d-2}|E|^2\lambda(\alpha_2)\lambda(\alpha_3).$$

Next, we'll look at $R_{\{1,3\}}$.

$$R_{\{1,3\}} = q^{-3} \sum_{s_1, s_3 \in \mathbb{F}_q^*} \sum_{x_j \in E} \chi(s_1(x_1 \cdot x_2 - \alpha_1))\chi(s_3(x_3 \cdot x_4 - \alpha_3)).$$

By taking absolute values and appealing to Cauchy–Schwarz, we get

$$\left|R_{\{1,3\}}\right| \leq q^{-3} \left| \sum_{x_j \in E} \sum_{s_1 \in \mathbb{F}_q^*} \chi(2s_1(x_1 \cdot x_2 - \alpha_1)) \right|^{\frac{1}{2}} \left| \sum_{x_j \in E} \sum_{s_3 \in \mathbb{F}_q^*} \chi(2s_3(x_3 \cdot x_4 - \alpha_3)) \right|^{\frac{1}{2}},$$

where we write $2s_j$ to mean $s_j + s_j$. Since s_1 and s_3 are ranging over $\mathbb{F}_q^*$, and q is odd, we know that $2s_j \neq 0$. So the sums over s_1 and s_3 will just be over a permutation of the elements of $\mathbb{F}_q^*$. Therefore, by a change of variables, $t_j = 2s_j$, the above quantity can be written

$$= q^{-3} \left| \sum_{x_j \in E} \sum_{t_1 \in \mathbb{F}_q^*} \chi(t_1(x_1 \cdot x_2 - \alpha_1)) \right|^{\frac{1}{2}} \left| \sum_{x_j \in E} \sum_{t_3 \in \mathbb{F}_q^*} \chi(t_3(x_3 \cdot x_4 - \alpha_3)) \right|^{\frac{1}{2}}.$$

Since x_3 and x_4 are left out of the first sum, and x_1 and x_2 are left out of the second, we can write this as

$$= q^{-3} \left| |E|^2 \sum_{x_1, x_2 \in E} \sum_{t_1 \in \mathbb{F}_q^*} \chi(t_1(x_1 \cdot x_2 - \alpha_1)) \right|^{\frac{1}{2}} \left| |E|^2 \sum_{x_3, x_4 \in E} \sum_{t_3 \in \mathbb{F}_q^*} \chi(t_3(x_3 \cdot x_4 - \alpha_3)) \right|^{\frac{1}{2}}$$

$$= q^{-3}|E|^2 \left| \sum_{x_1, x_2 \in E} \sum_{t_1 \in \mathbb{F}_q^*} \chi(t_1(x_1 \cdot x_2 - \alpha_1)) \right|^{\frac{1}{2}} \left| \sum_{x_3, x_4 \in E} \sum_{t_3 \in \mathbb{F}_q^*} \chi(t_3(x_3 \cdot x_4 - \alpha_3)) \right|^{\frac{1}{2}}.$$

Applying Lemma 1 twice as in our bound of R_1, we get that this is bounded above by

$$\leq q^{-3}|E|^3 \left(|E| q^{\frac{d+1}{2}} \lambda(\alpha_1)\right)^{\frac{1}{2}} \left(|E| q^{\frac{d+1}{2}} \lambda(\alpha_3)\right)^{\frac{1}{2}}$$

$$\leq q^{\frac{d-5}{2}} |E|^3 \sqrt{\lambda(\alpha_1)\lambda(\alpha_3)}.$$

Combining the above bounds on the components of R_2 yields

$$|R_2| \leq q^{\frac{d-5}{2}} |E|^3 \sqrt{\lambda(\alpha_1)\lambda(\alpha_2) + \lambda(\alpha_1)\lambda(\alpha_3) + \lambda(\alpha_2)\lambda(\alpha_3)}.$$

Now we can start bounding R_3. The strategy here will be to add in some terms with auxiliary variables equal to zero and break the new sum up into pieces that we can estimate separately. We recall $R_3 = R_{\{1,2,3\}}$, so we write

$$R_3 = q^{-3} \sum_{s_1,s_2,s_3 \in \mathbb{F}_q^*} \sum_{x_j \in E} \chi(s_1(x_1 \cdot x_2 - \alpha_1))\chi(s_2(x_2 \cdot x_3 - \alpha_2))\chi(s_3(x_3 \cdot x_4 - \alpha_3))$$

We define the secondary portion R_3' is made up of terms similar to those of R_3, but with $s_3 = 0$. We follow this process unless α_3 is the only nonzero α_j, in which case, we follow the same procedure as below, but reversing the roles of α_1 and α_3, and the corresponding variables. This allows us to have fewer lambda factors contributing $\sqrt{q}$ to our final bound.

$$R_3' := q^{-3} \sum_{s_1,s_2 \in \mathbb{F}_q^*} \sum_{x_j \in E} \chi(s_1(x_1 \cdot x_2 - \alpha_1))\chi(s_2(x_2 \cdot x_3 - \alpha_2))\chi(0(x_3 \cdot x_4 - \alpha_3)).$$

We can apply the triangle inequality and recall that $\chi(0) = 1$ to get

$$|R_3'| = q^{-3}|E| \left| \sum_{s_1,s_2 \in \mathbb{F}_q^*} \sum_{x_j \in E} \chi(s_1(x_1 \cdot x_2 - \alpha_1))\chi(s_2(x_2 \cdot x_3 - \alpha_2)) \right|.$$

Notice that the inner sum is exactly $T_1(E)$, so we can appeal to Lemma 2 to get that

$$|R_3'| \leq q^{d-2}|E|^2\lambda(\alpha_1)\lambda(\alpha_2). \tag{3}$$

By the triangle inequality, we get that

$$|R_3| = |R_3 + R_3' - R_3'| \leq |R_3 + R_3'| + |R_3'|$$

The first term in this sum involves $R_3 + R_3'$, which we write as

$$= q^{-3} \sum_{s_1,s_2 \in \mathbb{F}_q^*} \sum_{x_j \in E} \chi(s_1(x_1 \cdot x_2 - \alpha_1))\chi(s_2(x_2 \cdot x_3 - \alpha_2)) \sum_{s_3 \in \mathbb{F}_q} \chi(s_3(x_3 \cdot x_4 - \alpha_3)).$$

Here, we can re-associate and get the sum in two parts, the sum of terms where $x_3 \cdot x_4 = \alpha_3$ and the terms where $x_3 \cdot x_4 \neq \alpha_3$. In the case where $x_3 \cdot x_4 = \alpha_3$, the inner sum collapses to q, as we get $\chi(0) = 1$ for each element of $\mathbb{F}_q$. In the case where $x_3 \cdot x_4 \neq \alpha_3$, we get zero by orthogonality. Altogether, we get

$$R_3 + R_3' = q \cdot q^{-3} \sum_{s_1,s_2 \in \mathbb{F}_q^*} \sum_{x_j \in E} \chi(s_1(x_1 \cdot x_2 - \alpha_1))\chi(s_2(x_2 \cdot x_3 - \alpha_2)) +$$

$$0 \cdot q^{-3} \sum_{s_1, s_2 \in \mathbb{F}_q^*} \sum_{x_j \in E} \chi(s_1(x_1 \cdot x_2 - \alpha_1))\chi(s_2(x_2 \cdot x_3 - \alpha_2)).$$

We can ignore the second sum as it is multiplied by zero. Now we take absolute values on both sides to get

$$\left|R_3 + R_3'\right| = q \left| q^{-3} \sum_{s_1, s_2 \in \mathbb{F}_q^*} \sum_{x_j \in E} \chi(s_1(x_1 \cdot x_2 - \alpha_1))\chi(s_2(x_2 \cdot x_3 - \alpha_2)) \right|$$

$$= q \cdot \left|R_{\{1,2\}}\right| \le q^{d-1}|E|^2\lambda(\alpha_1)\lambda(\alpha_2),$$

where we applied (2) in the last step. Comparing the estimates of $|R_1|$, $|R_2|$, and $|R_3|$ to M yields the desired result.

3 Bounds on k-Chains in $\mathbb{Z}_q^d$

In this proof, we provide an asymptotic bound for the number of dot product k-chains of units in a sufficiently large finite subset of $\mathbb{Z}_q^d$ (the d-rank free module over $\mathbb{Z}_q$), where d is a positive integer and $q = p^\ell$ for some odd prime p and positive integer ℓ. Given a subset E of $\mathbb{Z}_q^d$ and a k-tuple of units $\alpha = (\alpha_1, \ldots, \alpha_k)$ in $\mathbb{Z}_q^\times$, denote the set of k-chains in E by $\Pi_\alpha(E)$. The asymptotic bound for $\Pi_\alpha(E)$ is as follows:

Theorem 5 *Let $E \subseteq \mathbb{Z}_q^d$ where $q = p^\ell$ a power of an odd prime p, and let $\alpha = (\alpha_1, \alpha_2, \ldots, \alpha_k)$ be a k-tuple of units in $\mathbb{Z}_q$ where $k \ge 2$. Then we have*

$$|\Pi_\alpha(E)| = \frac{|E|^{k+1}}{q^k}(1 + o(1))$$

provided $|E| \gtrapprox q^{\frac{d(2\ell-1)+1}{2\ell} + \frac{k-2}{2}}$.

The proof of this bound relies on previous results which we restate here for the reader's convenience. We first define a useful quantity. For $E \subseteq \mathbb{Z}_q^d$, and α a unit in $\mathbb{Z}_q$, let

$$S_{E,\alpha}(x) := \sum_{s \in \mathbb{Z}_q \setminus \{0\}} \sum_{y \in E} \chi(s(x \cdot y - \alpha))$$

. The following estimate is due to Dave Covert, Alex Iosevich, and Jonathan Pakianathan, in [7].

Lemma 3 (from [7]) *Suppose that $E \subseteq \mathbb{Z}_q^d$, where $q = p^\ell$ is a power of an odd prime. Let $\gamma \in \mathbb{Z}_q^\times$ be a unit. Then we have the following upper bound:*

$$\sum_{x\in E} S_{E,\alpha}(x) \le 2|E|q^{(\frac{d-1}{2})(2-\frac{1}{\ell})+1}.$$

Next, notice that by the definition of $S_{E,\alpha}(x)$ and Cauchy–Schwarz, we get

$$\left|\sum_{x\in\mathbb{Z}_q^d} S_{E,\alpha}(x)S_{E,\beta}(x)\right| \le \left(\sum_{x\in\mathbb{Z}_q^d} \left|S_{E,\alpha}(x)\right|^2\right)^{\frac{1}{2}} \cdot \left(\sum_{x\in\mathbb{Z}_q^d} \left|S_{E,\beta}(x)\right|^2\right)^{\frac{1}{2}}. \tag{4}$$

We now record a technical estimate from [6] $\left(\text{the bound of } III \text{ in } \mathbb{Z}_q^d\right)$, which we state here as a lemma.

Lemma 4 (from [6]) *Suppose that $E \subseteq \mathbb{Z}_q^d$, where $q = p^\ell$ is a power of an odd prime. Let $\alpha, \beta \in \mathbb{Z}_q^\times$ be units. We have the following upper bounds:*

$$\left(\sum_{x\in\mathbb{Z}_q^d} \left|S_{E,\alpha}(x)\right|^2\right)^{\frac{1}{2}} \left(\sum_{x\in\mathbb{Z}_q^d} \left|S_{E,\beta}(x)\right|^2\right)^{\frac{1}{2}} \le 2|E|q^{\frac{d(2\ell-1)}{\ell}+\frac{1}{\ell}}. \tag{5}$$

Proof (*Proof of Main Result*) Let $q = p^\ell$ be a power of an odd prime, and consider a subset E of $\mathbb{Z}_q^d$ where d is a positive integer. If $\alpha = (\alpha_1, \ldots, \alpha_k)$ is a k-tuple of units (for an integer $k > 2$), then we have by orthogonality

$$|\Pi_\alpha(E)| = q^{-k} \sum_{\substack{s_j\in\mathbb{Z}_q \\ 1\le j\le k}} \sum_{\substack{x_j\in E \\ 1\le j\le k+1}} \prod_{j=1}^{k} \chi(s_j(x_j \cdot x_{j+1} - \alpha_j)).$$

As above, we proceed by decomposing this sum into others based on which s_i are zero. Given a binary k-tuple $\mathbf{j}$, let $R_\mathbf{j}$ be the component sum where $s_i = 0$ if $\mathbf{j}(\mathrm{i}) = 0$ and $s_i \neq 0$ if $\mathbf{j}(\mathrm{i}) = 1$. Denote the sum of the $R_\mathbf{j}$, where exactly n entries of $\mathbf{j}$ are nonzero by R_n. First, consider the sum R_0, where every $s_i = 0$. Because $\chi(0) = 1$, we get

$$\begin{aligned} R_0 &= q^{-k} \sum_{\substack{s_j=0 \\ 1\le j\le k}} \sum_{\substack{x_j\in E \\ 1\le j\le k+1}} \prod_{j=1}^{k} \chi(s_j(x_j \cdot x_{j+1} - \alpha_j)) \\ &= q^{-k}|E|^{k+1}. \end{aligned}$$

Now let $\mathbf{j}$ be a binary k-tuple with the ith entry equal to 1 and the others zero. After simplifying, it follows from Lemma 3 that

$$|R_{\mathbf{j}}| = q^{-k} \sum_{\substack{s_j=0\\1\le j\le k\\j\neq i}} \sum_{\substack{x_j\in E\\1\le j\le k+1}} \prod_{j=1}^{k} \chi(s_j(x_j\cdot x_{j+1}-\alpha_j)) \sum_{s_i\neq 0} S_{E,\alpha_i}(x_i)$$

$$\le q^{-k} \sum_{\substack{s_j=0\\1\le j\le k\\j\neq i}} \sum_{\substack{x_j\in E\\1\le j\le k+1}} \prod_{j=1}^{k} \chi(s_j(x_j\cdot x_{j+1}-\alpha_j)) \left(2|E|q^{(\frac{d-1}{2})(2-\frac{1}{\ell})+1}\right)$$

$$= 2q^{-k}|E|^k q^{(\frac{d-1}{2})(2-\frac{1}{\ell})+1}.$$

As there are $\binom{k}{2} = k$ such tuples, it follows from the triangle inequality that

$$|R_1| \le 2kq^{-k}|E|^k q^{(\frac{d-1}{2})(2-\frac{1}{\ell})+1}. \tag{6}$$

Next, consider a sum $R_{\mathbf{j}}$ where exactly two entries of $\mathbf{j}$ are nonzero. If the nonzero entries of $\mathbf{j}$ are consecutive, say entries numbered i and $i+1$, then, reasoning as before, we can apply the triangle inequality, (4), and Lemma 4, and so

$$|R_{\mathbf{j}}| \le q^{-k} \sum_{\substack{s_j=0\\1\le j\le k\\j\neq i,i+1}} \sum_{\substack{x_j\in E\\1\le j\le k+1}} \prod_{j=1}^{k} \chi(s_j(x_j\cdot x_{j+1}-\alpha_j)) \sum_{s_i,s_{i+1}\neq 0} S_{E,\alpha_i}(x_i)S_{E,\alpha_i}(x_i)$$

$$\le q^{-k} \sum_{\substack{s_j=0\\1\le j\le k\\j\neq i,i+1}} \sum_{\substack{x_j\in E\\1\le j\le k+1}} \prod_{j=1}^{k} \chi(s_j(x_j\cdot x_{j+1}-\alpha_j)) \left| \sum_{s_i,s_{i+1}\neq 0} S_{E,\alpha_i}(x_i)S_{E,\alpha_i}(x_i) \right|$$

$$\le q^{-k} \sum_{\substack{s_j=0\\1\le j\le k\\j\neq i,i+1}} \sum_{\substack{x_j\in E\\1\le j\le k+1}} \prod_{j=1}^{k} \chi(s_j(x_j\cdot x_{j+1}-\alpha_j)) \times$$

$$\left| \left(\sum_{x\in\mathbb{Z}_q^d} |S_{E,\alpha}(x)|^2 \right)^{\frac{1}{2}} \left(\sum_{x\in\mathbb{Z}_q^d} |S_{E,\beta}(x)|^2 \right)^{\frac{1}{2}} \right|$$

$$\le q^{-k} \sum_{\substack{s_j=0\\1\le j\le k\\j\neq i,i+1}} \sum_{\substack{x_j\in E\\1\le j\le k+1}} \prod_{j=1}^{k} \chi(s_j(x_j\cdot x_{j+1}-\alpha_j)) \cdot \left(2|E|q^{\frac{d(2\ell-1)}{\ell}+\frac{1}{\ell}}\right)$$

$$\le q^{-k}|E|^{k-2}\left(2|E|q^{\frac{d(2\ell-1)}{\ell}+\frac{1}{\ell}}\right)$$

$$= 2q^{-k}|E|^{k-1}q^{\frac{d(2\ell-1)}{\ell}+\frac{1}{\ell}}.$$

Putting this all together yields (in the case of consecutive nonzero entries)

$$|R_{\mathbf{j}}| \leq 2q^{\frac{d(2\ell-1)}{\ell}+\frac{1}{\ell}-k}|E|^{k-1}. \tag{7}$$

However, if the nonzero entries are not consecutive, say $\mathbf{j}(u), \mathbf{j}(v) \neq 0$, where $1 \leq u, v \leq k$ and $u \neq v+1, v-1$, then we need to bound the exponent of $|E|$ appearing in $R_{\mathbf{j}}$ after simplifying. Let $1 < i \leq k$ be an integer and take $s_0 = 0$ by convention. Then if s_i and s_{i-1} are both zero, it follows for fixed $x_{i-1}, x_{i+1} \in E$ that

$$\begin{aligned}\sum_{s_i=0}\sum_{x_i\in E} &\chi(s_{i-1}(x_{i-1}\cdot x_i-\alpha_{i-1}))\chi(s_i(x_i\cdot x_{i+1}-\alpha_i))\\ &=\sum_{s_i=0}\sum_{x_i\in E}\chi(s_i(x_i\cdot x_{i+1}-\alpha_i))=|E|,\end{aligned} \tag{8}$$

and furthermore, the preceding sum can be factored out of $R_{\mathbf{j}}$. Hence, the contribution to $R_{\mathbf{j}}$ due to these auxiliary variables is $|E|$ raised to the power of the number of integers $1 \leq i \leq k$ such that s_i and s_{i-1} are zero.

Since the nonzero entries of $\mathbf{j}$ are not consecutive in this case, there will be exactly four distinct x_i appearing in $R_{\mathbf{j}}$, namely x_u, x_{u+1}, x_v, and x_{v+1}. Hence, by reasoning as in (8), the contribution to $R_{\mathbf{j}}$ by the other $k-3$ variables x_i and the similarly indexed auxiliary variables s_i will be $|E|^{k-3}$, giving

$$R_{\mathbf{j}} = q^{-k}|E|^{k-3}\left(\sum_{x_{u+1}\in E} S_{E,\alpha_u}(x_u)\right)\left(\sum_{x_v\in E} S_{E,\alpha_v}(x_v)\right).$$

By applying the triangle inequality, applying Cauchy–Schwarz, and dominating the sum over E by the sum over $\mathbb{Z}_q^d$, we get

$$|R_{\mathbf{j}}| \leq q^{-k}|E|^{k-3}\sum_{x_{u+1}\in E}\left|S_{E,\alpha_u}(x_u)\right|\sum_{x_v\in E}\left|S_{E,\alpha_v}(x_v)\right|$$

$$|R_{\mathbf{j}}| \leq q^{-k}|E|^{k-3}\left(\sum_{x_{u+1}\in E}\left|S_{E,\alpha_u}(x_u)\right|^2\right)^{\frac{1}{2}}\left(\sum_{x_v\in E}\left|S_{E,\alpha_v}(x_v)\right|^2\right)^{\frac{1}{2}}$$

$$|R_{\mathbf{j}}| \leq q^{-k}|E|^{k-3}\left(\sum_{x_{u+1}\in \mathbb{Z}_q^d}\left|S_{E,\alpha_u}(x_u)\right|^2\right)^{\frac{1}{2}}\left(\sum_{x_v\in \mathbb{Z}_q^d}\left|S_{E,\alpha_v}(x_v)\right|^2\right)^{\frac{1}{2}}$$

We apply Lemma 4 to the product of sums in the last inequality to get,

$$|R_{\mathbf{j}}| \leq 2q^{-k}|E|^{k-2}|E|q^{\frac{d(2\ell-1)}{\ell}+\frac{1}{\ell}} = 2q^{-k}|E|^{k-1}q^{\frac{d(2\ell-1)}{\ell}+\frac{1}{\ell}}. \tag{9}$$

Combining (8) and (9) gives us

$$|R_2| \leq 2\binom{k}{2} q^{-k}|E|^{k-1} q^{\frac{d(2\ell-1)}{\ell}+\frac{1}{\ell}}. \tag{10}$$

Lastly, consider $R_{\mathbf{j}}$ where $n \geq 3$ entries of $\mathbf{j}$ are nonzero. To this end, we record a known result often used to bound quadratic forms. See [18].

Lemma 5 *Let m and n be positive integers. Then for each double sequence $\{c_{jk} : 1 \leq j \leq m, 1 \leq k \leq n\}$ and pair of sequences $\{z_j : 1 \leq j \leq m\}$ $\{y_k : 1 \leq k \leq n\}$ of complex numbers, we have the bound*

$$\left| \sum_{j=1}^{m} \sum_{k=1}^{n} c_{jk} z_j y_k \right| \leq \sqrt{RC} \left(\sum_{j=1}^{m} |z_j|^2 \right)^{\frac{1}{2}} \left(\sum_{k=1}^{n} |y_k|^2 \right)^{\frac{1}{2}},$$

where R and C are, respectively, the row and column sum maxima defined by

$$R = \max_j \sum_{k=1}^{n} |c_{jk}| \text{ and } C = \max_k \sum_{j=1}^{n} |c_{jk}|.$$

Let u and v be the smallest and largest integers, respectively, such that $\mathbf{j}(u), \mathbf{j}(v) = 1$. The idea will be to change the order of summation of $R_{\mathbf{j}}$ so that it resembles the form of the double sum in Lemma 4. Repeated applications of the triangle inequality will then give upper bounds for the quantities corresponding to R and C in Lemma 5, allowing for its application. Finally, the resulting upper bound sum will be bounded via Lemma 4. Specifically, define the sequences

$$U_{x_{u+1}} := \left\{ S_{E,\alpha_u}(x_{u+1}) : x_{u+1} \in E \right\}$$
$$V_{x_v} := \left\{ S_{E,\alpha_v}(x_v) : x_v \in E \right\}$$

corresponding to the sequences $\{z_j\}$ and $\{y_k\}$ in the lemma. The terms of the double sequence corresponding to $\{c_{jk}\}$ will be the remaining part of $R_{\mathbf{j}}$ after the sums $S_{E,\alpha_u}(x_{u+1})$ and $S_{E,\alpha_{v+1}}(x_v)$ are factored out. We call this double sequence $W_{u+1,v}$. Note that this factoring out is possible because x_u and x_{v+1} each appear in only one χ expression in $R_{\mathbf{j}}$.

Now if m is the number of distinct variables x_i appearing in the summand of $R_{\mathbf{j}}$, it follows after repeated applications of the triangle inequality and the fact that complex exponentials have absolute value 1 that

$$R = \max_{x_{u+1}} \sum_{x_v \in E} |W_{x_{u+1},x_v}| \leq q^{n-2}|E|^{m-3}$$

as all but three of the index variables $x_i \in E$ and all but two of the index variables $s_i \in \mathbb{Z}_q \setminus \{0\}$ vary in $\sum_{x_v \in E} |W_{x_{u+1},x_v}|$. The column maximum C can be bounded in the same fashion. Thus, all that remains before we can apply Lemma 5 is to bound the exponent of $|E|$ that appears upon simplifying $R_{\mathbf{j}}$. We will do this with the following simple counting argument.

Lemma 6 *Let* $\mathbf{j}$ *be a binary* k*-tuple, and suppose that* m *distinct variables* x_i *appear in the associated sum* $R_{\mathbf{j}}$. *Then the number of integers* $1 \leq i \leq k$ *such that* $s_i = s_{i-1} = 0$ *is bounded above by* $k - m + 1$.

Proof Let n be the number of nonzero entries. Define $z = |\{i : s_i = s_{i-1} = 0\}|$ and $z' = |\{i : s_i = 0, s_{i-1} \neq 0\}|$. As exactly n entries of $\mathbf{j}$ are nonzero, it is clear that $z + z' = k - n$. Now for each i such that $s_i \neq 0$, consider the expression $\chi(s_i(x_i \cdot x_{i+1} - \alpha_i))$. If $s_{i-1} \neq 0$ as well, then only x_{i+1} appears for the first time (assuming the χ expressions are written in order by the indices of their auxiliary variables) in $R_{\mathbf{j}}$ in this expression. However, if $s_{i-1} = 0$, then both x_i and x_{i+1} appear for the first time in this expression. Thus, $m = n + a$ where a is the number of integers i such that $s_i \neq 0$ and $s_{i-1} = 0$.

Next, let i be an integer such that $s_i \neq 0$ and $s_{i-1} = 0$ but i is not the smallest integer such that $\mathbf{j}(i) = 1$. If no such i exists, then the nonzero entries of $\mathbf{j}$ must be consecutive, and so $m = n + 1$ and $z \leq k - n = k - m - 1$. Otherwise, there is a largest integer b with $1 \leq b < i - 1$ such that $s_b \neq 0$ and $s_c = 0$ for every integer c satisfying $b < c < i$. Hence, s_{b+1} contributes to the quantity z'. It follows then that $z' \geq a - 1$, and therefore

$$z = k - n - z' \leq k - n - a + 1 = k - m + 1.$$

We have by Lemma 6 that we can estimate the exponent of $|E|$ out front, then continue by applying Lemma 5, bound the sums over E by the sums over $\mathbb{Z}_q^d$, and apply Lemma 4.

$$|R_{\mathbf{j}}| \leq q^{-k}|E|^{k-m+1} \times$$

$$\left| \sum_{x_{u+1} \in E} \sum_{x_v \in E} \chi(s_u(x_u \cdot x_{u+1} - \alpha_u))\chi(s_v(x_v \cdot x_{v+1} - \alpha_v))W_{x_{u+1},x_v} \right|$$

$$\leq q^{-k}|E|^{k-m+1}q^{n-2}|E|^{m-3} \left(\sum_{x_{u+1} \in E} |S_{E,\alpha_u}(x_{u+1})|^2 \right)^{\frac{1}{2}} \left(\sum_{x_v \in E} |S_{E,\alpha_v}(x_v)|^2 \right)^{\frac{1}{2}}$$

$$\leq q^{n-k-2}|E|^{k-2} \left(\sum_{x_{u+1} \in \mathbb{Z}_q^d} |S_{E,\alpha_u}(x_{u+1})|^2 \right)^{\frac{1}{2}} \left(\sum_{x_v \in \mathbb{Z}_q^d} |S_{E,\alpha_v}(x_v)|^2 \right)^{\frac{1}{2}}$$

$$\leq 2q^{n-k-2}|E|^{k-1}q^{\frac{d(2\ell-1)}{\ell}+\frac{1}{\ell}}.$$

Thus, for n nonzero entries, we get

$$|R_n| \leq 2\binom{k}{n} q^{n-k-2}|E|^{k-1} q^{\frac{d(2\ell-1)}{\ell}+\frac{1}{\ell}}. \tag{11}$$

This completes the proof since

$$|\Pi_\alpha(E)| = \frac{|E|^{k+1}}{q^k} + \sum_{i=1}^{k} R_i,$$

and we have that (6), (10), and (11) together show that

$$\left|\sum_{i=1}^{k} R_i\right| \lesssim q^{-k}|E|^k q^{\left(\frac{d-1}{2}\right)(2-\frac{1}{\ell})+1} + \sum_{n=2}^{k} q^{n-k-2}|E|^{k-1} q^{\frac{d(2\ell-1)}{\ell}+\frac{1}{\ell}} \lesssim |E|^{k+1} q^{-k}$$

provided $|E| \gtrsim q^{\frac{k-2}{2}+\frac{d(2\ell-1)+1}{2\ell}}$.

4 Bounds on k-chains in $\mathbb{F}_q^d$

A similar result holds for k-chains of dot products in vector spaces over finite fields of odd order.

Theorem 6 *Let $E \subseteq \mathbb{F}_q^d$ where q is a power of an odd prime, and let $\alpha = (\alpha_1, \alpha_2, \ldots, \alpha_k)$ be a k-tuple of elements in $\mathbb{F}_p$, where $k \geq 2$. Then we have*

$$|\Pi_\alpha(E)| = \frac{|E|^{k+1}}{q^k}(1 + o(1))$$

provided $|E| \gtrapprox q^{\frac{d+k}{2}}$ or $|E| \gtrsim q^{\frac{d+k-1}{2}}$ if the components of α are nonzero.

Proof The proof of this result is mostly analogous to that of the previous theorem. Note that Lemmas 5 and 6 still hold in this case. Now fix integers $d, k \geq 1$ and a power of an odd prime q and consider a subset E of $\mathbb{F}_q^d$. If $\alpha = (\alpha_1, \alpha_2, \ldots, \alpha_k)$ is a k-tuple of elements in $\mathbb{F}_q$, then we have by orthogonality

$$|\Pi_\alpha(E)| = q^{-k} \sum_{\substack{s_j \in \mathbb{F}_q \\ 1 \leq j \leq k}} \sum_{\substack{x_j \in E \\ 1 \leq j \leq k+1}} \prod_{j=1}^{k} \chi(s_j(x_j \cdot x_{j+1} - \alpha_j)).$$

As in the previous proof, we decompose this sum into different remainder terms based on which s_i are nonzero. Let R_n ($n \geq 0$) have the meaning analogous to the similarly named quantity from the previous proof. It is easy to see that

$$|R_0| = q^{-k}|E|^{k+1}.$$

Next, consider a binary k-tuple $\mathbf{j}$ with exactly one entry equal to 1. The remainder sum $R_\mathbf{j}$ is of the form of the sum in Lemma 1, and hence

$$|R_\mathbf{j}| \leq q^{-k}|E|^{k-1}|E|q^{\frac{d+1}{2}}\lambda(\alpha_i) = q^{-k}|E|^k q^{\frac{d+1}{2}}\lambda(\alpha_i)$$

where $\mathbf{j}(i) \neq 0$. Thus,

$$|R_1| \leq \sum_{i=1}^{k} q^{-k}|E|^k q^{\frac{d+1}{2}}\lambda(\alpha_i)$$

The upper bound for $|R_2|$ is obtained in an analogous way to the method of the preceding proof but with the application of Lemma 2. As there are $\binom{k}{2}$ remainder sums of this type, it follows that

$$|R_2| \leq \sum_{1\leq i\neq j\leq k} q^{d-k+1}|E|^{k-1}\lambda(\alpha_i)\lambda(\alpha_j)$$

Lastly, consider $R_\mathbf{j}$ where exactly $3 \leq n \leq k$ entries of $\mathbf{j}$ are nonzero. Suppose that there are m distinct x_i appearing in $R_\mathbf{j}$, and let u and v be, respectively, the smallest and largest integers such that $\mathbf{j}(u), \mathbf{j}(v) \neq 0$. Following the reasoning in the proof of Theorem 5 and applying Lemma 2, we have

$$|R_\mathbf{j}| \leq q^{d+n-k-1}|E|^{k-1}\lambda(\alpha_u)\lambda(\alpha_v)$$

It follows then, after dominating the sum of the lambda factors, that

$$|R_n| \leq \sum_{1\leq i\neq j\leq k} q^{d+n-k-1}|E|^{k-1}\lambda(\alpha_i)\lambda(\alpha_j).$$

The result now follows as we have shown that

$$|\Pi_\alpha(E)| = |E|^{k+1}q^{-k} + \sum_{i=1}^{k} R_i,$$

where

$$\left|\sum_{i=1}^{k} R_i\right| \lesssim \sum_{i=1}^{k} q^{-k}|E|^k q^{\frac{d+1}{2}}\lambda(\alpha_i) + \sum_{1\le i\ne j\le k} q^{d-k+1}|E|^{k-1}\lambda(\alpha_i)\lambda(\alpha_j)+$$

$$\sum_{n=3}^{k}\sum_{1\le i\ne j\le k} q^{d+n-k-1}|E|^{k-1}\lambda(\alpha_i)\lambda(\alpha_j),$$

and the claimed result holds.

5 Small Set Results

We now turn to the estimates that apply to smaller subsets of $\mathbb{F}_q^2$. We begin by stating a result from [6] (the corrected form of Theorem 3 from that paper). We then derive Theorem 4 as a corollary. Finally, we describe the error in [6].

Theorem 7 (Theorem 3 from [6]) *Given $E \subseteq \mathbb{F}_q^2$, with q a power of an odd prime, and $\alpha, \beta \in \mathbb{F}_q^*$,*

$$\left|\Pi_{(\alpha,\beta)}(E)\right| \lesssim |E|^2.$$

Proof of Theorem 4

We start by assuming that $(k+1)$ is a multiple of three. Notice that for any $(\alpha_1, \ldots, \alpha_k) \in \mathbb{F}_q^k$, with all α_j nonzero, every dot product k-chain of this type must have a triple of points (x_1, x_2, x_3) forming a dot product 2-chain of type (α_1, α_2), followed by a triple of points (x_4, x_5, x_6) forming a dot product 2-chain of type (α_4, α_5), and so on, for a total of $(k+1)/3$ triples of points. Theorem 7 tells us that there are no more than $|E|^2$ choices of triples of each type. So when $(k+1)$ is a multiple of three, the total number of dot product k-chains can be no more than

$$\lesssim \left(|E|^2\right)^{\frac{k+1}{3}} = |E|^{\frac{2(k+1)}{3}}.$$

Now, if $(k+1)$ is one more than a multiple of three, we reason as above to get that there are no more than

$$\lesssim |E|^{\frac{2k}{3}}$$

occurrences of dot product $(k-1)$-chains of the type $(\alpha_1, \ldots, \alpha_{k-1})$, accounting for all choices of the relevant k-tuples $(x_1, \ldots, x_k)$. Then there are no more than $|E|$ choices for the point x_{k+1}, giving us a total upper bound of no more than

$$\lesssim |E|^{\frac{2k}{3}+1}.$$

Notice that in this case, we are not using the fact that the $x_k \cdot x_{k+1} = \alpha_k$ at all.

Finally, if $(k+1)$ is two more than a multiple of 3, we again reason as before to get that there are no more than

$$\lesssim |E|^{\frac{2(k-1)}{3}}$$

occurrences of dot product $(k-2)$-chains of the type $(\alpha_1, \ldots, \alpha_{k-2})$, and now have $|E|$ choices for x_k, as well as $|E|$ choices for x_{k+1}, for a total upper bound of no more than

$$\lesssim |E|^{\frac{2(k-1)}{3}+2}.$$

Combining these yields the claimed upper bound on dot product k-chains of

$$\lesssim |E|^{\left\lceil \frac{2(k+1)}{3} \right\rceil}.$$

Erratum

We now detail the error in Theorem 3 from [6]. In that paper, it was claimed that an analog of Theorem 7 above held for $\mathbb{Z}_q^d$. However, there is a mistake in the proof. First, we present a construction showing that the claim is false. Then we describe how the method employed fails.

Proposition 1 *Given $q = p^\ell$, for a natural number $\ell \geq 2$ and an odd prime p, and units $\alpha, \beta \in \mathbb{Z}_q$, there exists a subset of $E \subseteq \mathbb{Z}_q^2$ with*

$$\left|\Pi_{(\alpha,\beta)}(E)\right| = |E|^3.$$

Proof Given distinct units α and β, both different from 1, we define $E \subseteq Z_q^2$, to be the union of three sets: $E = X \cup Y \cup Z$ where,

$$X = \left\{(ap, \alpha) : a \in Z_p\right\},\ Y = \left\{\left(bp^{\ell-1}, 1\right) : b \in Z_p\right\},\ Z = \left\{(cp, \beta) : c \in Z_p\right\}.$$

By definition, we have $|E| = |X| + |Y| + |Z| = 3p$. Now notice that for any $x \in X$, and any $y \in Y$, we have that for some $a, b \in \mathbb{Z}_p$,

$$x \cdot y = (ap)\left(bp^{\ell-1}\right) + (\alpha)(1) = abp^\ell + \alpha = \alpha.$$

Similarly, for any $y \in Y$, and any $z \in Z$, we will have $y \cdot z = \beta$. Therefore we can compute explicitly,

$$|\Pi_{\alpha,\beta}(E)| \geq |\{(x, y, z) \in X \times Y \times Z : x \cdot y = \alpha, x \cdot z = \beta\} = p^3 = |E|^3.$$

The issue lies in how incidences were counted in $\mathbb{Z}_q^d$. Given a unit $\alpha \in \mathbb{Z}_q$, where $q = p^\ell$, for a natural number $\ell \geq 2$ and an odd prime p, and an element $v \in \mathbb{Z}_q^2$, we define

$$L\alpha(x) := \left\{ y \in \mathbb{Z}_q^2 : x \cdot y = \alpha \right\}.$$

Looking through the proof of Theorem 3 in [6], we find the incorrect claim that if w and v are distinct elements in $\mathbb{Z}_q^2$, with $|L_\alpha(v) \cap L_\beta(w)| > 1$, then $L_\alpha(v) = L_\beta(w)$. While the analogous result holds for $v, w \in \mathbb{F}_q^2$, we give an explicit counterexample for this statement in Z_9^2. Let $v = (3, 2)$ and $w = (3, 4)$. Notice that

$$L_2(v) = \{(0, 1), (1, 4), (2, 7), (3, 1), (4, 4), (5, 7), (6, 1), (7, 4), (8, 7)\}, \text{ and}$$

$$L_4(w) = \{(0, 1), (1, 7), (2, 4), (3, 1), (4, 7), (5, 4), (6, 1), (7, 7), (8, 4)\},$$

yet $|L_2(v) \cap L_4(w)| = |\{(0, 1), (3, 1), (6, 1)\}| = 3 > 1$, but we can see that $L_2(v) \neq L_4(w)$, contradicting the erroneous claim.

References

1. D. Barker and S. Senger, *Upper bounds on pairs of dot products*, Journal of Combinatorial Mathematics and Combinatorial Computing, Volume 103, November, 2017, pp. 211–224.
2. M. Bennett, A. Iosevich, and K. Taylor, *Finite chains inside thin subsets of $\mathbb{R}^d$*, Analysis and PDE, volume 9, no. 3, (2016).
3. J. Bourgain, N. Katz, and T. Tao, *A sum-product estimate in finite fields, and applications,* Geom. Funct. Anal. **14** (2004), pp. 27–57.
4. P. Brass, W. Moser, and J. Pach, *Research Problems in Discrete Geometry,* Springer (2000), 499 pp.
5. J. Chapman, B. Erdoğan, D. Hart, A. Iosevich, D. Koh, *Pinned distance sets, k-simplices, Wolff's exponent in finite fields and sum-product estimates*, Math. Z. 271 (2012), no. 1–2, 63–93.
6. D. Covert and S. Senger, *Pairs of dot products in finite fields and rings*, Nathanson M. (eds) Combinatorial and Additive Number Theory II. CANT 2015, CANT 2016. Springer Proceedings in Mathematics & Statistics, vol 220. Springer, Cham. (Appeared 14 Jan 2018)
7. D. Covert, A. Iosevich, J. Pakianathan, *Geometric configurations in the ring of integers modulo p^ℓ*, Indiana University Mathematics Journal, 61 (2012), 1949–1969.
8. D. Covert, D. Hart, A. Iosevich, S. Senger, and I. Uriarte-Tuero, *An analog of the Furstenberg-Katznelson-Weiss theorem on triangles in sets of positive density in finite field geometries,* Discrete Math. 311, no. 6, 423–430, (2011).
9. P. Erdős, *On sets of distances of n points*, Amer. Math. Monthly **53** (1946) 248–250.
10. L. Guth and N. H. Katz, *On the Erdős distinct distance problem in the plane,* Annals of Math., Pages 155–190, Volume 181 (2015), Issue 1.
11. S. Gunter, E. Palsson, B. Rhodes, and S. Senger, *Bounds on point configurations determined by distances and dot products,* Nathanson M. (eds) Combinatorial and Additive Number Theory IV. CANT 2019, CANT 2020. Springer Proceedings in Mathematics & Statistics, vol 347. Springer, Cham.
12. D. Hart, A. Iosevich D. Koh, M. Rudnev, *Averages over hyperplanes, sum-product theory in vector spaces over finite fields and the Erdős-Falconer distance conjecture*, Trans. Amer. Math. Soc. 363 (2011), no. 6, 3255–3275.

13. A. Iosevich and M. Rudnev, *Erdős-Falconer distance problem in vector spaces over finite fields*, Trans. Amer. Math. Soc., 359 (2007), no. 12, 6127–6142.
14. A. Iosevich and S. Senger, *Orthogonal systems in vector spaces over finite fields,* Electronic J. of Combinatorics, Volume 15, December (2008).
15. S. Kilmer, C. Marshall, and S. Senger, *Dot product chains,* (submitted) arXiv:2006.11467 (2020).
16. E. Palsson, S. Senger, and A. Sheffer, *On the number of discrete chains,* (to appear in the Proceedings of the American Mathematical Society).
17. J. Spencer, E. Szemerédi, and W. T. Trotter, *Unit distances in the Euclidean plane* B. Bollobás, editor, "Graph Theory and Combinatorics", pages 293–303, Academic Press, New York, NY, (1984).
18. J. M. Steele, *The Cauchy-Schwarz Master Class ICM Edition: An Introduction to the Art of Mathematical Inequalities*, Cambridge University Press, (2010).
19. T. V. Pham and L. A. Vinh *Orthogonal Systems in Vector Spaces over Finite Rings*, Electronic J. of Combinatorics, Volume 19, Issue 2 (2012).

Completeness of Positive Linear Recurrence Sequences

Elżbieta Bołdyriew, John Haviland, Phúc Lâm, John Lentfer, Steven J. Miller, and Fernando Trejos Surez

MSC 2010: 11P99 primary · 11K99 secondary

1 Introduction

Few sequences of integers have attracted as much study as the Fibonacci numbers. One of their many interesting properties is that they can be used to construct a unique decomposition for any positive integer. Zeckendorf proved that every positive integer can be written uniquely as a sum of nonconsecutive Fibonacci numbers, when indexing Fibonacci numbers $\{1, 2, 3, 5, \dots\}$; this unique decomposition is called the *Zeckendorf decomposition* [10]. This result, of unique decompositions, has

The authors were partially supported by NSF grants DMS1265673 and DMS1561945, the Claire Booth Luce Foundation, and Carnegie Mellon University. We thank the students from the Math 21–499 Spring '16 research class at Carnegie Mellon and the participants from CANT 2016 and 2017 and the 18th Fibonacci Conference, especially Russell Hendel, for many helpful conversations.

E. Bołdyriew · J. Haviland · P. Lâm · J. Lentfer · S. J. Miller (✉) · F. T. Surez
Williams College, Yale University, and the University of Rochester, Rochester, USA
e-mail: sjm1@williams.edu

E. Bołdyriew
e-mail: eboldyriew@colgate.edu

J. Haviland
e-mail: havijw@umich.edu

P. Lâm
e-mail: plam6@u.rochester.edu

J. Lentfer
e-mail: jlentfer@hmc.edu

F. T. Surez
e-mail: fernando.trejos@yale.edu

M. B. Nathanson (ed.), *Combinatorial and Additive Number Theory V*,
Springer Proceedings in Mathematics & Statistics 395,
https://doi.org/10.1007/978-3-031-10796-2_2

been generalized to a much larger class of linear recurrence relations; the following definitions are from [9].

Definition 1.1 We say a sequence $\{H_n\}_{n=1}^{\infty}$ of positive integers is a **Positive Linear Recurrence Sequence (PLRS)** if the following properties hold:

1. *Recurrence relation:* There are nonnegative integers $L, c_1, \ldots, c_L$ such that

$$H_{n+1} \;=\; c_1 H_n + \cdots + c_L H_{n+1-L}, \tag{1}$$

   with L, c_1, and c_L positive.
2. *Initial conditions:* $H_1 = 1$, and for $1 \leq n < L$ we have

$$H_{n+1} \;=\; c_1 H_n + c_2 H_{n-1} + \cdots + c_n H_1 + 1. \tag{2}$$

Definition 1.2 (*Legal decompositions*) We call a decomposition $\sum_{i=1}^{m} a_i H_{m+1-i}$ of a positive integer N (and the sequence $\{a_i\}_{i=1}^{m}$) legal if $a_1 > 0$, the other $a_i \geq 0$, and one of the following two conditions holds:

1. We have $m < L$ and $a_i = c_i$ for $1 \leq i \leq m$.
2. There exists $s \in \{1, \ldots, L\}$ such that

$$a_1 \;=\; c_1,\; a_2 \;=\; c_2,\; \ldots,\; a_{s-1} \;=\; c_{s-1} \text{ and } a_s < c_s, \tag{3}$$

   $a_{s+1}, \ldots, a_{s+\ell} \;=\; 0$ for some $\ell \geq 0$, and $\{b_i\}_{i=1}^{m-s-\ell}$ (with $b_i = a_{s+\ell+i}$) is legal or empty.

The following theorem is due to [5], and stated in this form in [9].

Theorem 1.3 (Generalized Zeckendorf's Theorem for PLRS) *Let* $\{H_n\}_{n=1}^{\infty}$ *be a* Positive Linear Recurrence Sequence. *Then there is a unique legal decomposition for each positive integer* $N \geq 0$.

Next, we introduce *completeness*, as defined by [7].

Definition 1.4 An arbitrary sequence of positive integers $\{f_i\}_{i=1}^{\infty}$ is **complete** if and only if every positive integer n can be represented in the form $n = \sum_{i=1}^{\infty} \alpha_i f_i$, where $\alpha_i \in \{0, 1\}$. A sequence that fails to be complete is **incomplete**.

In other words, a sequence of positive integers is complete if and only if each positive integer can be written as a sum of unique terms of the sequence.

Example 1.5 The Fibonacci sequence, indexed from $\{1, 2, \ldots\}$, is complete. This follows from Zeckendorf's theorem, which is a stronger statement. Completeness does not require that the decompositions be unique, and that they use nonconsecutive terms.

After seeing this example, it is natural to ask if Theorem 1.3 implies that all PLRSs are complete. Previous work in numeration systems by Gewurz and Merola [6] has shown that specific classes of recurrences as defined by Fraenkel [4] are complete under their greedy expression. However, we cannot generalize this result to all PLRS's. For legal decompositions, the decomposition rule can permit sequence terms to be used more than once. This is not allowed for completeness decompositions, where each unique term from the sequence can be used at most once.

Example 1.6 The PLRS $H_{n+1} = H_n + 3H_{n-1}$ has terms $\{1, 2, 5, 11, \ldots\}$. The unique *legal* decomposition for 9 is $1 \cdot 5 + 2 \cdot 2$, where the term 2 is used twice. However, no *complete* decomposition for 9 exists. Adding all terms from the sequence less than 9 is $1 + 2 + 5 = 8$, and to include 11 or any subsequent term surpasses 9.

We also make use of the following criterion for completeness of a sequence, due to Brown [3].

Theorem 1.7 (Brown's Criterion) *If a_n is a nondecreasing sequence, then a_n is complete if and only if $a_1 = 1$ and for all $n > 1$,*

$$a_{n+1} \leq 1 + \sum_{i=1}^{n} a_i. \tag{4}$$

An immediate corollary is the following sufficient, though not necessary condition for completeness, which we call the doubling criterion. The proof is left to the appendix, as Corollary 6.2.

Corollary 1.8 (Doubling Criterion) *If a_n is a nondecreasing sequence such that $a_n \leq 2a_{n-1}$ for all $n \geq 2$, then a_n is complete.*

Remark 1.9 By considering the special case when $a_n = 2a_{n-1}$, this immediately implies that the doubling sequence itself $\{1, 2, 4, 8, \ldots\}$ is complete.

In this paper, we characterize many types of PLRS by whether they are complete or not complete.

Notation 1.10 *We use the notation $[c_1, \ldots, c_L]$, which is the collection of all L coefficients, to represent the PLRS $H_{n+1} = c_1 H_n + \cdots + c_L H_{n+1-L}$.*

A simple case to consider is when all coefficients in $[c_1, \ldots, c_L]$ are positive. The following result, proved in Sect. 2, completely characterizes these sequences are either complete or incomplete.

Theorem 1.11 *If $\{H_n\}$ is a PLRS generated by all positive coefficients $[c_1, \ldots, c_L]$, then sequence is complete if and only if the coefficients are $[\underbrace{1, \ldots, 1}_{L}]$ or $[\underbrace{1, \ldots, 1}_{L-1}, 2]$ for $L \geq 1$.*

The situation becomes much more complicated when we consider all PLRSs, in particular, those that have at least one 0 as a coefficient. In order to be able to make progress on determining completeness of these PLRSs, we develop several tools. The following three theorems are results that allow certain modifications of the coefficients $[c_1, \dots, c_L]$ that generate a PLRS that is known to be complete or incomplete, and preserve completeness or incompleteness. They are proved in Sect. 2.

Theorem 1.12 *Consider sequences* $\{G_n\} = [c_1, \dots, c_L]$ *and* $\{H_n\} = [c_1, , \dots, c_L, c_{L+1}]$*, where* c_{L+1} *is any positive integer. If* $\{G_n\}$ *is incomplete, then* $\{H_n\}$ *is incomplete as well.*

Theorem 1.13 *Consider sequences*

$$\{G_n\} = [c_1, \dots, c_{L-1}, c_L] \text{ and } \{H_n\} = [c_1, \dots, c_{L-1}, k_L],$$

where $1 \leq k_L \leq c_L$*. If* $\{G_n\}$ *is complete, then* $\{H_n\}$ *is also complete.*

Theorem 1.14 *Consider sequences*

$$\{G_n\} = [c_1, \dots, c_{L-1}, c_L] \text{ and } \{H_n\} = [c_1, \dots, c_{L-1} + c_L].$$

If $\{G_n\}$ *is incomplete, then* $\{H_n\}$ *is also incomplete.*

The next two theorems are results that classify two families of PLRSs as complete or incomplete. They are shown in Sect. 3.

Theorem 1.15 *The sequence generated by* $[1, \underbrace{0, \dots, 0}_{k}, N]$ *is complete if and only if* $1 \leq N \leq \lceil (k+2)(k+3)/4 \rceil$*, where* $\lceil \cdot \rceil$ *is the ceiling function.*

Theorem 1.16 *The sequence generated by* $[1, 1, \underbrace{0, \dots, 0}_{k}, N]$ *is complete if and only if* $1 \leq N \leq \lfloor (F_{k+6} - k - 5)/4 \rfloor$*, where* F_n *are the Fibonacci numbers with* $F_1 = 1$*,* $F_2 = 2$ *and* $\lfloor \cdot \rfloor$ *is the floor function.*

We have a partial extension of these theorems to when there are g initial ones followed by k zeroes in the collection of coefficients.

Theorem 1.17 *Consider a PLRS generated by coefficients* $[\underbrace{1, \dots, 1}_{g}, \underbrace{0, \dots, 0}_{k}, N]$*, with* $g, k \geq 1$*.*

1. *For* $g \geq k + \lceil \log_2 k \rceil$*, the sequence is complete if and only if* $1 \leq N \leq 2^{k+1} - 1$*.*
2. *For* $k \leq g \leq k + \lceil \log_2 k \rceil$*, the sequence is complete if and only if* $1 \leq N \leq 2^{k+1} - \lceil k/2^{g-k} \rceil$*.*

Finally, in Sect. 4, we introduce some results and conjectures on completeness based on the principal roots of a PLRS. We determine some criteria for completeness based on the size of the principal root and find that there is a certain indeterminate region where the principal root does not reveal any information.

2 Modifying Sequences

A basic question to ask is how far we can tweak the coefficients used to generate a sequence, yet preserve its completeness. The modifying process turns out to be well behaved and heavily dependent on the location of coefficients that are changed. Before we start looking into implementing any changes to our sequences, we first need to understand the maximal complete sequence.

Maximal Complete Sequence

We introduce the maximal complete sequence, which serves an important role. First, we look at all complete sequences with only positive coefficients.

Proof (*Proof of Theorem* 1.11.) Assume that $\{H_n\}$ is complete. By the definition of a PLRS and by Brown's criterion, we have

$$c_1 H_{L-1} + c_2 H_{L-2} + \cdots + c_{L-1} H_1 + 1 = H_L \leq 1 + H_1 + H_2 + \cdots + H_{L-1}. \quad (1)$$

Since $c_i \geq 1$ for $1 \leq i \leq L$, this implies that $c_i = 1$ for $1 \leq i < L$. By the definition of a PLRS,

$$H_{L+1} = c_1 H_L + c_2 H_{L-1} + \cdots + c_L H_1 = H_L + H_{L-1} + \cdots + H_2 + c_L H_1. \quad (2)$$

Combining this with Brown's criterion gives

$$\begin{aligned} H_{L+1} = H_L + H_{L-1} + \cdots + c_L H_1 &\leq 1 + H_1 + H_2 + \cdots + H_{L-1} \\ c_L H_1 &\leq 1 + H_1 = 2. \end{aligned} \quad (3)$$

Hence $c_L \leq 2$, which completes the forward direction of the proof.

We know that if the coefficients are just [2], then the sequence is complete by Remark 1.9. So, now assume that $c_1 = \cdots = c_{L-1} = 1$ and $1 \leq c_L \leq 2$. We argue by strong induction on n that H_n satisfies Brown's criterion. We can show this explicitly for $1 \leq n < L$. First, if $n = 1$, then $H_n = 1$, as desired. Next, if $1 \leq n < L$, then

$$H_{n+1} = c_1 H_n + \cdots + c_n H_1 + 1 = H_n + \cdots + H_1 + 1, \quad (4)$$

so these terms satisfy Brown's criterion. Now assume that for some $n \geq L$, for all $n' < n$,

$$H_{n'+1} \leq H_{n'} + \cdots + H_1 + 1. \quad (5)$$

It follows that

$$
\begin{aligned}
H_{n+2} &= H_{n+1} + \cdots + H_{n+2-L} + c_L H_{n+1-L} \\
&\le H_{n+1} + \cdots + H_{n+2-L} + 2H_{n+1-L} \\
&\le H_{n+1} + \cdots + H_{n+2-L} + H_{n+1-L} + (H_{n-L} + \cdots + H_1 + 1), \qquad (6)
\end{aligned}
$$

where the inductive hypothesis was applied to H_{n+1-L} to obtain (6). This completes the induction.

Now that we have found some complete sequences, it turns out that the sequence generated by the coefficient [2], i.e., $\{2^{n-1}\}$, is the maximal complete sequence.

Lemma 2.1 *The complete sequence with largest span in summands is* $\{2^{n-1}\}$.

Proof Suppose there exists a complete sequence $\{H_n\}$ with the largest span in summands. As a complete sequence must satisfy Brown's criterion, it suffices to take $H_{n+1} = 1 + \sum_{i=1}^{n} H_i$. Hence,

$$
H_{n+1} = 1 + \sum_{1}^{n} H_i = 1 + \sum_{1}^{n-1} H_i + H_n = 2H_n. \qquad (7)
$$

By the initial conditions for a PLRS, $H_1 = 1$ and $H_2 = 2$. Thus, $H_n = 2H_{n-1} = 2^{n-1}$.

Remark 2.2 Thus $\{H_k\} = \{2^{k-1}\}$ is an inclusive upper bound for any complete sequence.

As it turns out, this sequence can be generated by multiple collections of coefficients.

Corollary 2.3 *A PLRS with coefficients* $[\underbrace{1, \ldots, 1}_{L-1}, 2]$ *generates the sequence* $H_n = 2^{n-1}$.

Proof Consider the sequence $\{H_n\}$ generated by $[\underbrace{1, \ldots, 1}_{L-1}, 2]$. We proceed by induction on L. Note $H_1 = 1 = 2^{1-1}$ by the definition of the PLRS. Now, suppose $H_k = 2^{k-1}$ for $k \in \{1, \ldots, n\}$. For $n < L$, note

$$
\begin{aligned}
H_{n+1} &= c_1 H_n + c_2 H_{n-1} + \cdots + c_n H_1 + 1 \\
&= H_n + H_{n-1} + \cdots + H_1 + 1 \\
&= 2^{n-1} + 2^{n-2} + \cdots + 1 + 1 = 2^n. \qquad (8)
\end{aligned}
$$

Hence, the claim holds for all $n < L$. Now, for $n \ge L$, note

$$
\begin{aligned}
H_{n+1} &= c_1 H_n + c_2 H_{n-1} + \cdots + c_L H_{n+1-L} \\
&= H_n + H_{n-1} + \cdots + 2H_{n+1-L} \\
&= 2^{n-1} + 2^{n-2} + \cdots + 2^{n-L+1} + 2 \cdot 2^{n-L} = 2^n.
\end{aligned} \tag{9}
$$

Thus, by induction, the claim holds for all $n, L \in \mathbb{N}$.

Modifications of Sequences of Arbitrary Coefficients

Modifying coefficients in order to preserve completeness proves to be a balancing act. Sometimes increasing a coefficient causes an incomplete sequence to become complete, while other times, increasing a coefficient causes a complete sequence to become incomplete. For example, $[1, 0, 0, 0, 0, 0, 15]$ is incomplete; increasing the second coefficient to 1, i.e., $[1, 1, 0, 0, 0, 0, 15]$ is complete. Further increasing it to 2, i.e., $[1, 2, 0, 0, 0, 0, 15]$ is again incomplete. To study how such modifications preserve completeness or incompleteness, we add a new definition to our toolbox.

Definition 2.4 For a sequence $\{H_n\}$, we define its ***n*th Brown's gap**

$$B_{H,n} := 1 + \sum_{i=1}^{n-1} H_i - H_n. \tag{10}$$

Thus, from Brown's criterion, $\{H_n\}$ is complete if and only if $B_{H,n} \geq 0$ for all $n \in \mathbb{N}$.

Our next questions is: What happens if we append one more coefficient to $[c_1, \ldots, c_L]$? It turns out that if our sequence is already incomplete, appending any new coefficients will never make it complete. This is Theorem 1.12, which we are ready to prove using Brown's gap.

Proof (*Proof of Theorem* 1.12.) By Brown's criterion, it is clear that $\{G_n\}$ is incomplete if and only if there exists n such that $B_{G,n} < 0$. We claim that for all m, $B_{H,m} \leq B_{G,m}$. If true, our lemma is proven: suppose $B_{G,n} < 0$ for some n, we would see $B_{H,n} \leq B_{G,n} < 0$, implying $\{H_n\}$ is incomplete as well.

We proceed by induction. Clearly, $B_{H,k} = B_{G,k}$ for $1 \leq k \leq L$. Further, for $k = L$, we see

$$B_{G,L+1} - B_{H,L+1} = 1 + \sum_{i=1}^{L} G_i - G_{L+1} - \left(1 + \sum_{i=1}^{L} H_i - H_{L+1}\right) = H_{L+1} - G_{L+1} = 1 > 0. \tag{11}$$

Now, let $m \geq 2$ be arbitrary, and suppose

$$B_{H,\ L+m-1} \leq B_{G,\ L+m-1}. \tag{12}$$

We wish to show that $B_{H,\ L+m} \leq B_{G,\ L+m}$. Note that

$$B_{H,\,L+m} - B_{H,\,L+m-1} = 2H_{L+m-1} - H_{L+m}. \tag{13}$$

Similarly,

$$B_{G,\,L+m} - B_{G,\,L+m-1} = 2G_{L+m-1} - G_{L+m}. \tag{14}$$

We use Lemma 7.1, which states that for all $k \geq 2$,

$$H_{L+k} - G_{L+k} \;\geq\; 2\left(H_{L+k-1} - G_{L+k-1}\right). \tag{15}$$

Applying it to Eqs. (13) and (14), we see that $B_{H,\,L+m} - B_{H,\,L+m-1} \leq B_{G,\,L+m} - B_{G,\,L+m-1}$. Summing this inequality to both sides of inequality (12), we arrive at $B_{H,L+m} \leq B_{G,L+m}$, as desired.

Now, we turn our attention to the behavior when we decrease the last coefficient for any complete sequence. In Theorem 1.13, we find that decreasing the last coefficient for any complete sequence preserves completeness.

Proof (*Proof of Theorem* 1.13.) Given that $\{G_n\}$ is complete, suppose for the sake of contradiction that there exists an incomplete $\{H_n\}$. Thus, let m be the least such that

$$H_m > 1 + \sum_{i=1}^{m-1} H_i. \tag{16}$$

Simultaneously, as $\{G_n\}$ is complete, by Brown's criterion,

$$G_m \leq 1 + \sum_{i=1}^{m-1} G_i. \tag{17}$$

First, suppose $m \leq L$. However, for all $n \leq L$, $G_n = H_n$, hence

$$H_m = G_m \leq 1 + \sum_{i=1}^{m-1} G_i = 1 + \sum_{i=1}^{m-1} H_i, \tag{18}$$

which contradicts (16). Now, suppose $m > L$. Therefore,

$$G_m \leq 1 + \sum_{i=1}^{m-1} G_i = 1 + \sum_{i=1}^{L} G_i + \sum_{i=L+1}^{m-1} G_i = 1 + \sum_{i=1}^{L} H_i + \sum_{i=L+1}^{m-1} G_i.$$

This implies,

$$1 + \sum_{i=1}^{L} H_i \geq G_m - \sum_{i=L+1}^{m-1} G_i. \tag{19}$$

Now, we know that

$$H_m > 1 + \sum_{i=1}^{m-1} H_i = 1 + \sum_{i=1}^{L} H_i + \sum_{i=L+1}^{m-1} H_i \geq G_m - \sum_{i=L+1}^{m-1} G_i + \sum_{i=L+1}^{m-1} H_i, \quad (20)$$

and thus

$$H_m - \sum_{i=L+1}^{m-1} H_i > G_m - \sum_{i=L+1}^{m-1} G_i. \quad (21)$$

We claim that the opposite of (21) is true, arguing by induction on m. For $m = L + 1$, we obtain $G_{L+1} \geq H_{L+1}$ as $k_L \leq c_L$. Now, assume that

$$G_m - \sum_{i=L+1}^{m-1} G_i \geq H_m - \sum_{i=L+1}^{m-1} H_i \quad (22)$$

is true for a positive integer m. Using the inductive hypothesis, it then follows that

$$G_{m+1} - \sum_{i=L+1}^{m} G_i = G_{m+1} - \sum_{i=L+1}^{m-1} G_i - G_m \geq G_{m+1} - 2G_m + H_m - \sum_{i=L+1}^{m-1} H_i. \quad (23)$$

Finally, we use Lemma 7.2, proved in Appendix B, which states that for all $k \in \mathbb{N}$, $H_{L+k+1} - 2H_{L+k} \leq G_{L+k+1} - 2G_{L+k}$. Note

$$G_{m+1} - 2G_m + H_m - \sum_{i=L+1}^{m-1} H_i \geq H_{m+1} - 2H_m + H_m - \sum_{i=L+1}^{m-1} H_i = H_{m+1} - \sum_{i=L+1}^{m} H_i, \quad (24)$$

which does contradict (21) for all $m > L$. Therefore, for all $m \in \mathbb{N}$, we have contradicted (16). Hence, $\{H_n\}$ must be complete as well.

The result above is crucial in our characterization of *families* of complete sequences in Sect. 3; finding one complete sequence allows us to decrease the last coefficient to find more. Next, we prove two lemmas that together prove Theorem 1.14.

Lemma 2.5 *Let $\{G_n\}$ be the sequence defined by $[c_1, \ldots, c_L]$, and let $\{H_n\}$ be the sequence defined by $[c_1, \ldots, c_{L-1} + 1,\ c_L - 1]$. If $\{G_n\}$ is incomplete, then $\{H_n\}$ must be incomplete as well.*

Proof We claim that for all m, $B_{H,m} \leq B_{G,m}$. This lemma is proven using similar reasoning as for Lemma 1.12. We proceed by induction. Clearly, $B_{H,k} = B_{G,k}$ for $1 \leq k \leq L - 1$. Further, for $k = L$, we see

$$B_{G,L} - B_{H,L} = 1 + \sum_{i=1}^{L-1} G_i - G_L - \left(1 + \sum_{i=1}^{L-1} H_i - H_L\right) = H_L - G_L = 1 > 0. \tag{25}$$

Now, let $m \geq 0$ be arbitrary, and suppose

$$B_{H,\ L+m} \leq B_{G,\ L+m}. \tag{26}$$

We wish to show that $B_{H,\ L+m+1} \leq B_{G,\ L+m+1}$. Note that

$$B_{H,\ L+m+1} - B_{H,\ L+m} = 2H_{L+m} - H_{L+m+1}, \tag{27}$$

and similarly,

$$B_{G,\ L+m+1} - B_{G,\ L+m} = 2G_{L+m} - G_{L+m+1}. \tag{28}$$

We use Lemma 7.3, which yields for all $k \geq 0$, $H_{L+k+1} - G_{L+k+1} \geq 2\left(H_{L+k} - G_{L+k}\right)$. Applying it to (27) and (28), we see $B_{H,\ L+m+1} - B_{H,\ L+m} \leq B_{G,\ L+m+1} - B_{G,\ L+m}$. Summing this inequality to both sides of inequality (26), we conclude that $B_{H,L+m+1} \leq B_{G,L+m+1}$, as desired.

How many times can Lemma 2.5 be applied? The answer is all the way up to $[c_1, \ldots, c_{L-1} + c_L - 1, 1]$, as the last coefficient must remain positive to stay a PLRS.

Lemma 2.6 *Let $\{G_n\}$ be the sequence defined by $[c_1, \ldots, c_{L-1}, 1]$, and let $\{H_n\}$ be the sequence defined by $[c_1, \ldots, c_{L-1} + 1]$. If $\{G_n\}$ is incomplete, then $\{H_n\}$ must be incomplete as well.*

Remark 2.7 Despite the similarities, Lemma 2.6 is not directly implied by Lemma 2.5; both are necessary for the proof Theorem 1.14. Applying Lemma 2.5 $(c_L - 1)$ times proves that if $[c_1, \ldots, c_{L-1}, c_L]$ is incomplete, then $[c_1, \ldots, c_{L-1} + c_L - 1, 1]$ is incomplete; at this point, we cannot apply the lemma further while maintaining a positive final coefficient to meet the definition of a PLRS. Hence, the case of Lemma 2.6 must be dealt with separately, in order to arrive at the full result of Theorem 1.14.

Proof The proof is similar to that of Lemma 2.5. We aim to show that $B_{H,m} \leq B_{G,m}$ for all m. Clearly $B_{H,k} = B_{G,k}$ for $1 \leq k \leq L$. Further, for $k = L + 1$, we see

$$B_{G,L+1} - B_{H,L+1} = \sum_{i=1}^{L} G_i - G_{L+1} - \left(1 + \sum_{i=1}^{L-1} H_L - H_{L+1}\right) = H_{L+1} - G_{L+1} = c_1 > 0. \tag{29}$$

Now, let $m \geq 0$ be arbitrary, and suppose

$$B_{H,L+m} \leq B_{G,L+m}. \tag{30}$$

We wish to show that $B_{H,L+m+1} \leq B_{G,L+m+1}$. Note that

$$B_{H,L+m+1} - B_{H,L+m} = 2H_{L+m} - H_{L+m+1}, \tag{31}$$

and similarly

$$B_{G,L+m+1} - B_{G,L+m} = 2G_{L+m} - G_{L+m+1}. \tag{32}$$

We use Lemma 7.4, which yields for all $k \geq 0$, $H_{L+k+1} - G_{L+k+1} \geq 2(H_{L+k} - G_{L+k})$. Applying it to Eqs. (31) and (32), we see $B_{H,L+m+1} - B_{H,L+m} \leq B_{G,L+m+1} - B_{G,L+m}$. Summing this inequality to both sides of Inequality (30), we conclude that $B_{H,L+m+1} \leq B_{G,L+m+1}$, as desired.

Using these lemmas, we can now prove Theorem 1.14.

Proof (*Proof of Theorem* 1.14.) We apply Lemma 2.5 $c_L - 1$ times, to conclude that if $[c_1, \ldots, c_{L-1}, c_L]$ is incomplete, then $[c_1, \ldots, c_{L-1} + c_L - 1, 1]$ is incomplete. Finally, applying Lemma 2.6, we achieve the desired result.

3 Families of Sequences

If we recall Theorem 1.13, it says that given a complete PLRS, decreasing the last coefficient preserves its completeness. This raises a natural question: Given the first $L - 1$ coefficients $c_1, c_2, \ldots, c_{L-1}$, what is the maximal N such that $[c_1, c_2, \ldots, c_{L-1}, N]$ is complete? In this section we explore this question.

Using 1s and 0s as Initial Coefficients

Proof (*Proof of Theorem* 1.15.) First assume that $\{H_n\}$ is complete. By the definition of a PLRS, we can easily generate the first $k + 2$ terms of the sequence: $H_i = i$ for all $1 \leq i \leq k + 2$. We then have for all $n > k + 1$,

$$H_{n+1} = H_n + NH_{n-k-1}, \tag{1}$$

which implies that

$$H_{k+4} = H_{k+3} + NH_2 = H_{k+3} + 2N. \tag{2}$$

By Brown's criterion,

$$H_{k+4} \leq H_{k+3} + H_{k+2} + \cdots + H_1 + 1.$$

By (2),

$$H_{k+3} + 2N \leq H_{k+3} + H_{k+2} + \cdots + H_1 + 1,$$

and we obtain

$$\begin{aligned} 2N &\leq H_{k+2} + H_{k+1} + \cdots + H_1 + 1 \\ &= (k+2) + (k+1) + \cdots + 1 + 1 \\ &= \frac{(k+2)(k+3)}{2} + 1, \end{aligned}$$

and thus we find

$$N \leq \frac{(k+2)(k+3)}{4} + \frac{1}{2}. \tag{3}$$

Since N is an integer and $\lfloor (k+2)(k+3)/4 + 1/2 \rfloor = \lceil (k+2)(k+3)/4 \rceil$, we may take the floor of the right-hand side of Eq. (3), and then $N \leq \lceil (k+2)(k+3)/4 \rceil$.

We now prove that if $N \leq \lceil (k+2)(k+3)/4 \rceil$, then $\{H_n\}$ is complete. We first show that if $N = \lceil (k+2)(k+3)/4 \rceil$, then $\{H_n\}$ is complete. Taking the recurrence relation $H_{n+1} = H_n + NH_{n-k-1}$, and applying Brown's criterion gives

$$\begin{aligned} H_{n+1} &= H_n + NH_{n-k-1} \\ &\leq H_n + (N-2)H_{n-k-1} + H_{n-k-1} + H_{n-k-2} + \cdots + H_1 + 1. \end{aligned}$$

By Lemma 8.1, we can expand $(N-2)H_{n-k-1}$ and find that

$$H_{n+1} \leq H_n + H_{n-1} + \cdots + H_{n-k} + H_{n-k-1} + H_{n-k-2} + \cdots + H_1 + 1. \tag{4}$$

Hence, by Brown's criterion, the sequence $\{H_n\}$ is complete. Lastly, by Theorem 1.13, for all positive $N < \lceil (k+2)(k+3)/4 \rceil$, the sequence is also complete.

For coefficients as defined in Theorem 1.15, for sufficiently large L, if we switch any one of the coefficients from 0 to 1 except for the final zero, then the bound on N is at least as large, which we prove in the following corollary.

Corollary 3.1 *For* $L \geq 6$, *given that* $[1, 0, \ldots, 0, N]$ *is complete, with* $N = \lceil L(L+1)/4 \rceil$, *then* $[1, c_2, \ldots, c_{L-2}, 0, N]$ *is complete where* $c_i = 1$ *for one* $i \in \{2, \ldots, L-2\}$, *and the rest are* 0.

Proof We have the recurrence relation for fixed $i \in \{2, \ldots, L-2\}$:

$$H_{n+1} = H_n + H_{n-i+1} + NH_{n-L+1}. \tag{5}$$

Applying Brown's criterion yields

$$H_{n+1} \leq H_n + H_{n-i+1} + (N-2)H_{n-L+1} + H_{n-L+1} + H_{n-L} + \cdots + H_1 + 1. \tag{6}$$

We apply the result of Lemma 8.3, and see that

$$\leq H_n + H_{n-1} + \cdots + H_{n-L+2} + H_{n-L+1} + H_{n-L} + \cdots + H_{\_} + 1. \tag{7}$$

Hence, by Brown's criterion, the sequence is complete for all $L \geq 6$.

Proof (*Proof of Theorem* 1.16.) Suppose that $\{H_n\}$ is complete. Using the definition of a PLRS, the first $k+3$ terms of the sequence can be generated in the same way: $H_i = F_{i+1} - 1$ for all $1 \leq i \leq k+3$, where F_n is the Fibonacci sequence. Proceeding in a manner similar to the proof of Theorem 1.15, we see that

$$\begin{aligned} H_{k+4} &= H_{k+3} + H_{k+2} + NH_1 = F_{k+5} + N - 2, \\ H_{k+5} &= H_{k+4} + H_{k+3} + NH_2 = F_{k+6} + 3N - 3, \\ H_{k+6} &= H_{k+5} + H_{k+4} + NH_3 = F_{k+7} + 8N - 5. \end{aligned} \tag{8}$$

By applying Brown's criterion,

$$\begin{aligned} H_{k+6} &\leq H_{k+5} + H_{k+4} + \cdots + H_1 + 1 \\ &= F_{k+6} + 3N - 3 + F_{k+5} + N - 2 + \sum_{i=1}^{k+3} H_i + 1 \\ &= F_{k+7} + 4N - 5 + \sum_{i=1}^{k+3} (F_{i+1} - 1) + 1. \end{aligned} \tag{9}$$

Next,

$$F_{k+7} + 8N - 5 \leq F_{k+7} + 4N - 5 + \sum_{i=1}^{k+3} (F_{i+1} - 1) + F_1,$$

which implies

$$4N \leq \sum_{i=1}^{k+3} (F_{i+1} - 1) + F_1 = \sum_{i=1}^{k+4} F_i + (k+3) = F_{k+6} + (k+5). \tag{10}$$

Thus

$$N \leq \frac{F_{k+6} - k - 5}{4}, \tag{11}$$

and since N is an integer,

$$N \leq \left\lfloor \frac{F_{k+6} - k - 5}{4} \right\rfloor. \tag{12}$$

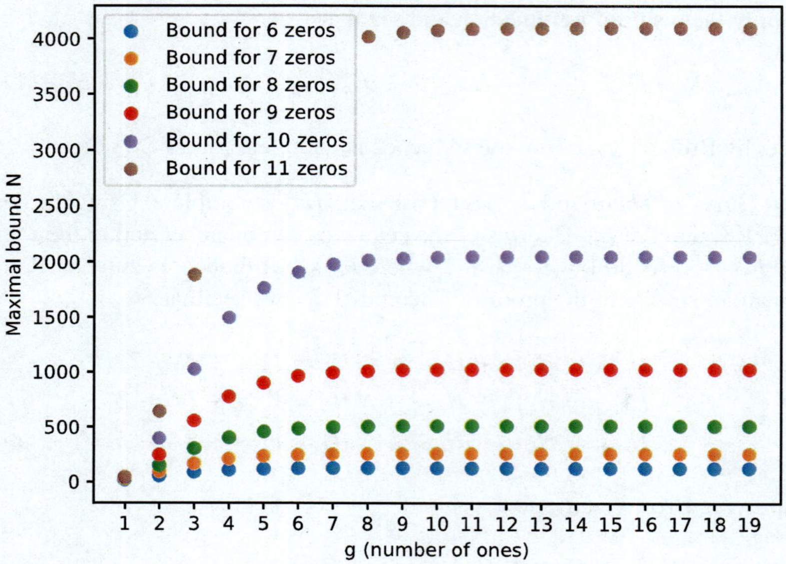

Fig. 1 $[\underbrace{1, \ldots, 1}_{g}, \underbrace{0, \ldots, 0}_{k}, N]$ with k and g varying, where each color represents a fixed k

Next, we show that if $N = \lfloor (F_{k+6} - k - 5)/4 \rfloor$, then $\{H_n\}$ is complete. The initial conditions can be found easily, and for the later terms we have

$$\begin{aligned} H_{n+1} &= H_n + H_{n-1} + NH_{n-k-2} \\ &\leq H_n + (N-2)H_{n-k-2} + H_{n-k-2} + H_{n-k-3} + \cdots + H_1 + 1. \end{aligned}$$

Using Lemma 8.4, we expand $(N-2)H_{n-k-2}$ and obtain

$$\leq H_n + H_{n-1} + H_{n-2} + \cdots + H_{n-k-1} + H_{n-k-2} + H_{n-k-3} + \cdots + H_1 + 1. \tag{13}$$

Hence, by Brown's criterion, this sequence is complete. Lastly, by Theorem 1.13, for all positive $N < \lfloor (F_{k+6} - k - 5)/4 \rfloor$, the sequence is also complete.

We want to find a more general result for $[\underbrace{1, \ldots, 1}_{g}, \underbrace{0, \ldots, 0}_{k}, N]$, as seen in Fig. 1.

Interestingly, we see that as we keep k fixed and increase g, the bound increases, and then stays constant from some value of g onward. This motivates the following conjecture.

Conjecture 3.2 If $[\underbrace{1, \ldots, 1}_{g}, \underbrace{0, \ldots, 0}_{k}, N]$ is complete, then so is $[\underbrace{1, \ldots, 1}_{g+1}, \underbrace{0, \ldots, 0}_{k}, N]$.

We have made some progress toward this conjecture; in fact, we show the precise bound for N for the case where $g \geq k$ in Theorem 1.17.

Theorem 3.3 *The PLRS $\{H_n\}$ generated by $[c_1, c_2, \ldots, c_L]$ is complete if*

$$\begin{cases} B_{H,n} \geq 0 & \text{for } n < L \\ B_{H,n} > 0 & \text{for } L \leq n \leq 2L-1. \end{cases} \tag{14}$$

Proof Consider $L \geq 2$; we see that if $c_1 \geq 2$, then the sequence is automatically incomplete, so we need only to consider $c_1 = 1$. For $B_n := B_{H,n}$, and we show by induction on n that $B_n > 0$ when $n \geq L$. Suppose $B_n > 0$ for $L \leq n \leq m$ (with $m \geq 2L - 1$). Then

$$\begin{aligned}
B_{m+1} &= 1 + \sum_{i=1}^{m} H_i - H_{m+1} \\
&= 1 + \sum_{i=1}^{L} H_i + \sum_{i=L+1}^{m} H_i - \left(H_m + \sum_{j=2}^{L} c_j H_{m+1-j} \right) \\
&= 1 + \sum_{i=1}^{L} H_i + \sum_{i=L+1}^{m} \left(H_{i-1} + \sum_{j=2}^{L} c_j H_{i-j} \right) - \left(H_m + \sum_{j=2}^{L} c_j H_{m+1-j} \right) \\
&= \left(1 + \sum_{i=1}^{m-1} H_i - H_m + H_L \right) + \sum_{j=2}^{L} c_j \left(\sum_{i=L+1}^{m} H_{i-j} - H_{m+1-j} \right) \\
&= (B_m + H_L) + \sum_{j=2}^{L} c_j \left(1 + \sum_{i=j+1}^{m} H_{i-j} - H_{m+1-j} - 1 - \sum_{i=j+1}^{L} H_{i-j} \right) \\
&= (B_m + H_L) + \sum_{j=2}^{L} c_j \left(B_{m+1-j} - 1 - \sum_{i=j+1}^{L} H_{i-j} \right) \\
&= B_m + \sum_{j=2}^{L} c_j (B_{m+1-j} - 1) + H_L - \sum_{i=3}^{L} \sum_{j=2}^{i-1} c_j H_{i-j} \\
&= B_m + \sum_{j=2}^{L} c_j (B_{m+1-j} - 1) + H_L - \sum_{i=3}^{L} (H_i - H_{i-1} - 1) \\
&= B_m + \sum_{j=2}^{L} c_j (B_{m+1-j} - 1) + (L-2) + H_L - \sum_{i=3}^{L} (H_i - H_{i-1}) \\
&= B_m + \sum_{j=2}^{L} c_j (B_{m+1-j} - 1) + L.
\end{aligned} \tag{15}$$

The last line is positive since $B_{m+1-j} - 1 \geq 0$ and $B_m, L > 0$. Our proof by induction is complete.

Lemma 3.4 *The PLRS* $\{H_i\}$ *generated by* $[\underbrace{1, \ldots, 1}_{g}, \underbrace{0, \ldots, 0}_{k}, 2^{k+1}]$ *is incomplete for* $g \geq k \geq 1$.

Proof Suppose this sequence is complete. Note that

$$H_{2g+2} = H_{2g+1} + \cdots + H_{g+2} + 2^{k+1} H_{g+1-k}. \tag{16}$$

By applying Brown's criterion to H_{2g+2}, we see that

$$2^{k+1} H_{g+1-k} \leq \sum_{i=1}^{g+1} H_i + 1. \tag{17}$$

Now, note k is positive, so that $g + 1 - k \leq g + 1$. Also, by the structure of the sequence, $H_i = 2^{i-1}$ for $i \leq g + 1$. Hence

$$2^{g+1} = 2^{k+1} H_{g+1-k} = 2^{k+1} 2^{g-k} \leq \sum_{i=1}^{g+1} 2^{i-1} + 1 = 2^{g+1}. \tag{18}$$

Therefore, one may substitute previous inequalities with equalities and obtain

$$H_{2g+2} = \sum_{i=1}^{2g+1} H_i + 1. \tag{19}$$

It follows immediately from (19) that

$$\sum_{i=1}^{2g+2} H_i + 1 = 2H_{2g+2}. \tag{20}$$

Now, consider

$$H_{2g+3} = H_{2g+2} + H_{2g+1} + \cdots + H_{g+3} + 2^{k+1} H_{g+2-k}. \tag{21}$$

Since $g + 2 - k \leq g + 1$ as $k \geq 1$, one gets

$$H_{g+2-k} M = 2^{g+1-k} 2^{k+1} = 2^{g+2} = 2(2^{g+1}) = 2H_{g+2}. \tag{22}$$

Hence

$$\begin{aligned}
H_{2g+3} &= H_{2g+2} + H_{2g+1} + \cdots + H_{g+3} + 2H_{g+2} \\
&= H_{2g+2} + \left(H_{2g+1} + \cdots + H_{g+3} + H_{g+2} + H_{g+2}\right) \\
&> H_{2g+2} + \left(H_{2g+1} + \cdots + H_{g+3} + H_{g+2} + H_{g+1-k}\right) \\
&= 2H_{2g+2} = \sum_{i=1}^{2g+2} H_i + 1.
\end{aligned} \tag{23}$$

So H_{2g+3} causes Brown's criterion to fail, rendering whole sequence incomplete.

We now show the stabilizing behavior of the bound mentioned above.

Lemma 3.5 *If* $g \geq k + \lceil \log_2 k \rceil$, *then* $[\underbrace{1, \ldots, 1}_{g}, \underbrace{0, \ldots, 0}_{k}, 2^{k+1} - 1]$ *is complete.*

Proof Define $\{F_n\} = [\underbrace{1, \ldots, 1}_{g}]$, and $\{H_n\} = [\underbrace{1, \ldots, 1}_{g}, \underbrace{0, \ldots, 0}_{k}, 2^{k+1} - 1]$. We can calculate the terms of $\{F_n\}$ and $\{H_n\}$ up to $2g + 1$. Namely,

$$\begin{aligned}
H_n &= F_n = 2^{n-1} && \text{when } 1 \leq n \leq g \\
H_{g+n} &= F_{g+n} + 2^{n-1} && \text{when } 1 \leq n \leq k+1 \\
H_{g+k+1+n} &= F_{g+k+1+n} + \left(2^{k+1} - 1\right)\left(2^n + 2^{n-2}(n-1)\right) && \text{when } 1 \leq n \leq g-k \\
F_{g+n} &= 2^{g+n-1} - 2^{n-2}(n+1) && \text{when } 1 \leq n \leq g.
\end{aligned} \tag{24}$$

The third and fourth lines are verified in Lemmas 8.6 and 8.7, respectively. We show that the conditions in Theorem 3.3 hold for $\{H_n\}$. We can verify directly that Brown's criterion holds for the first $(2g + 1)$ terms of $\{H_n\}$; in fact, for $B_n := B_{H,n}$, we get

$$\begin{cases} B_n \geq 0 & 1 \leq n \leq g + k \\ B_n > 0 & g + k + 1 \leq n \leq 2g + 1. \end{cases} \tag{25}$$

Thus, it remains to show that $B_n > 0$ for $2g + 2 \leq n \leq 2(g + k) - 1$.

Case 1: $2g + 2 \leq n \leq 2g + k + 1$
Define $b(n) := H_n - F_n$. Note that $b(n) \geq 0$, and by induction, $b(n) > 0$ for all $n \geq g + 1$. For $n \geq g + k + 2$,

$$\begin{aligned}
F_n + b(n) = H_n & \\
&= H_{n-1} + H_{n-2} + \cdots + H_{n-g} + \left(2^{k+1} - 1\right) H_{n-(g+k+1)} \\
&= \sum_{i=1}^{g} F_{n-i} + \sum_{i=1}^{g} b(n-i) + \left(2^{k+1} - 1\right) H_{n-(g+k+1)}.
\end{aligned}$$

Since $F_n = \sum_{i=1}^{g} F_{n-i}$,

$$b(n) = \sum_{i=1}^{g} b\,(n-i) + \left(2^{k+1}-1\right) H_{n-(g+k+1)}. \tag{26}$$

Thus, for any $n \geq 2g+2$,

$$\begin{aligned}
B_n &= 1 + \sum_{i=1}^{n-1} H_i - H_n \\
&= 1 + \sum_{i=1}^{n-1} (F_i + b(i)) - (F_n + b(n)) \\
&= \left(1 + \sum_{i=1}^{n-1} F_i - F_n\right) - \left(2^{k+1}-1\right) H_{n-(g+k+1)} + \sum_{i=g+1}^{n-(g+1)} b(i) \\
&> \left(1 + \sum_{i=1}^{n-1} F_i - F_n\right) - \left(2^{k+1}-1\right) H_{n-(g+k+1)}.
\end{aligned} \tag{27}$$

We are to show that the last term is nonnegative. As $n-(g+k+1) \leq g$,

$$\begin{aligned}
&1 + \sum_{i=1}^{n-1} F_i - F_n - \left(2^{k+1}-1\right) H_{n-(g+k+1)} \\
&= 1 + \sum_{i=1}^{n-(g+1)} F_i - \left(2^{k+1}-1\right) H_{n-(g+k+1)} \\
&= 1 + \sum_{i=1}^{g} F_i + \sum_{i=1}^{n-(2g+1)} F_{g+i} - \left(2^{k+1}-1\right) \cdot 2^{n-(g+k+1)-1} \\
&= 1 + \sum_{i=1}^{g} 2^{i-1} + \sum_{i=1}^{n-(2g+1)} \left(2^{g+i-1} - 2^{i-2}\,(i+1)\right) - 2^{n-g-1} + 2^{n-(g+k+1)-1} \\
&= 2^{n-(g+k+1)-1} - \sum_{i=1}^{n-(2g+1)} 2^{i-2}\,(i-1) - \sum_{i=1}^{n-(2g+1)} 2^{i-1} \\
&= 2^{n-(g+k+1)-1} - \left(2^{n-(2g+2)}\,(n-(2g+3)) + 1\right) - \left(2^{n-(2g+2)} - 1\right) \\
&= 2^{n-(g+k+1)-1} - 2^{n-(2g+2)}\,(n-(2g+2)) \\
&= 2^{n-(2g+2)}\left(2^{g-k} - (n-(2g+2))\right) \\
&\geq 2^{n-(2g+2)}\left(2^{g-k} - (k-1)\right) \\
&> 0.
\end{aligned} \tag{28}$$

Note that the last line comes from $g \geq k + \log_2 k$, which implies $2^{g-k} \geq k > k - 1$.

Case 2: $2g + k + 2 \leq n \leq 2g + 2k + 1$
We show that $B_{n+1} \geq B_n$ for $2g + k + 2 \leq n < 2g + 2k + 1$, and that $B_{2g+k+2} > 0$.

$$\begin{aligned}
B_{n+1} - B_n &= 2H_n - H_{n+1} \\
&= 2H_n - \left(\sum_{i=n-g+1}^{n} H_i + (2^{k+1} - 1)H_{n-(g+k)} \right) \\
&= \left(H_n - \sum_{i=n-g+1}^{n} H_i \right) + (2^{k+1} - 1)H_{n-(g+k)} \\
&= H_{n-g} - (2^{k+1} - 1)(H_{n-(g+k)} - H_{n-(g+k+1)}).
\end{aligned}$$

Replace n by $2g + k + 1 + m$, with $1 \leq m \leq k$

$$\begin{aligned}
&= H_{(g+k+1)+m} - (2^{k+1} - 1)(H_{g+m+1} - H_{g+m}) \\
&= H_{(g+k+1)+m} - (2^{k+1} - 1)(2^{g+m-1} - 2^{m-2}(m + 1)).
\end{aligned} \tag{29}$$

For $1 \leq m \leq g - k$, we have an explicit formula for $H_{(g+k+1)+m}$, so we can substitute directly to show that (29) is nonnegative. Thus, if $g - k \geq k$ (i.e., $g \geq 2k$), then this holds for all $1 \leq m \leq k$. If $g - k < k$ (i.e., $g < 2k$), then from Lemma 8.9, (29) is nonnegative. Thus, $B_{n+1} \geq B_n$ for all $2g + k + 2 \leq n \leq 2g + k + 1$. It remains to show that $B_{2g+k-2} > 0$, which we can do by directly substituting the explicit formulas. □

Combining these lemmas, we can prove the first part of Theorem 1.17.

Proof (*Proof of Theorem* 1.17.1.) From Lemmas 3.4 and 3.5, the bound for N is precisely $2^{k+1} - 1$ when $g \geq k + \lceil \log_2 k \rceil$.

Next, we consider when $k \leq g \leq k + \lceil \log_2(k) \rceil$, and prove the second part of Theorem 1.17 using similar methods.

Proof (*Proof of Theorem* 1.17.2.) First, we show that for $N > 2^{k+1} - \lceil k/2^{g-k} \rceil$, $\{H_i\}$ is incomplete, and suppose $k \geq 2$. Let us calculate the initial $L = g + k + 1$ terms of the sequence. Note

$$\begin{aligned}
H_n &= 2^{n-1} && \text{for all } 1 \leq n \leq g + 1 \\
H_{g+n} &= 2^{g+n-1} - 2^{n-2}(n - 1) && \text{for all } 1 \leq n \leq k + 1.
\end{aligned} \tag{30}$$

Let $B_i := B_{H,i}$. Then, we consider Brown's gap B_{2g+k+2},

$$\begin{aligned}
B_{2g+k+2} &= \left(1 + \sum_{i=1}^{2g+k+1} H_i\right) - H_{2g+k+2} \\
&= \left(1 + \sum_{i=1}^{2g+k+1} H_i\right) - \left(\sum_{i=g+k+2}^{2g+k+1} H_i + NH_{g+1}\right) \\
&= \left(1 + \sum_{i=1}^{g+k+1} H_i\right) - NH_{g+1} \\
&= 1 + \sum_{i=1}^{g} H_i + \sum_{i=g+1}^{g+k+1} H_i - NH_{g+1} \\
&= 1 + \sum_{i=1}^{g} 2^{i-1} + \sum_{i=1}^{k+1} \left(2^{g+i-1} - 2^{i-2}(i-1)\right) - 2^g N \\
&= 2^{g+k+1} - \sum_{i=1}^{k} 2^{i-1} i - 2^g N \\
&= 2^{g+k+1} - 2^k(k-1) - 1 - 2^g N.
\end{aligned}$$

Now, $N > 2^{k+1} - \lceil k/2^{g-k} \rceil$ by assumption so it follows that $N \geq 2^{k+1} - k/2^{g-k} + 1$, hence

$$\begin{aligned}
&\leq 2^{g+k+1} - 2^k(k-1) - 1 - 2^g \left(2^{k+1} - \frac{k}{2^{g-k}} + 1\right) \\
&= 2^k - 2^g - 1, \qquad (31)
\end{aligned}$$

which must be negative as $g \geq k$. So $\{H_n\}$ fails Brown's criterion at the $(2g+k+1)$st term, rendering the sequence incomplete.

Now we can show that for $N = 2^{k+1} - \lceil k/2^{g-k} \rceil$, $\{H_i\}$ is complete by Theorem 3.3. We can easily verify that $B_n \geq 0$ for all $1 \leq n \leq g+k+1$ and $B_{g+k+1} > 0$; it remains to show that $B_n > 0$ for $g+k+2 \leq n \leq 2g+2k+1$. We consider two cases.

Case 1: $2 \leq n - (g+k) \leq g+1$.

We want to show that $B_{n+1} \geq B_n$ for all $2 \leq n - (g+k) \leq g+1$ and that $B_{g+k+2} > 0$. Now,

$$\begin{aligned} B_n &= 1 + \sum_{i=1}^{n-1} H_i - H_n \\ &= 1 + \sum_{i=1}^{n-1} H_i - \left(\sum_{i=n-g}^{n-1} H_i + N H_{n-(g+k+1)} \right) \\ &= 1 + \sum_{i=1}^{n-g-1} H_i - N H_{n-(g+k+1)}. \end{aligned} \tag{32}$$

Then, note that

$$\begin{aligned} B_{n+1} - B_n &= H_{n-g} - N\left(H_{n-(g+k)} - H_{n-(g+k+1)}\right) \\ &= H_{n-g} - N\left(2^{n-(g+k+1)} - 2^{n-(g+k+2)}\right), \end{aligned}$$

and by assumption,

$$\begin{aligned} &= H_{n-g} - \left(2^{k+1} - \left\lceil \frac{k}{2^{g-k}} \right\rceil\right) 2^{n-(g+k+2)} \\ &= 2^{n-(g+k+2)} \left\lceil \frac{k}{2^{g-k}} \right\rceil - (2^{n-g-1} - H_{n-g}). \end{aligned} \tag{33}$$

If $n - g \leq g + 1$, then $2^{n-g-1} - H_{n-g} = 0$, so $B_{n+1} - B_n > 0$. If $g + 2 \leq n - g \leq g + k + 1$, then

$$2^{n-g-1} - H_{n-g} = 2^{n-2g-2}(n - 2g - 1) \leq 2^{n-(g+k+2)} \frac{k}{2^{g-k}} \leq 2^{n-(g+k+2)} \left\lceil \frac{k}{2^{g-k}} \right\rceil, \tag{34}$$

so that $B_{n+1} - B_n \geq 0$. In any case, $B_{n+1} \geq B_n$. We can verify directly that $B_{g+k+2} > 0$, completing this case.

Case 2: $g \leq n - (g + k) \leq g + k + 1$.
From the previous case, $B_{2g+k+2} \geq B_{2g+k+1} > 0$. Now,

$$\begin{aligned} B_n &= 1 + \sum_{i=1}^{n-g-1} H_i - N H_{n-(g+k+1)} \\ &= 1 + \sum_{i=1}^{n-2g-1} H_i + \sum_{i=n-2g}^{n-g-1} H_i - N H_{n-(g+k+1)} \\ &= 1 + \sum_{i=1}^{n-2g-1} H_i + H_{n-g} - N H_{n-(2g+k+1)} - N H_{n-(g+k+1)}. \end{aligned}$$

Substituting $n = 2g + k + 1 + m$ for $1 \le m \le k$,

$$
\begin{aligned}
&= 1 + \sum_{i=1}^{k+m} H_i + H_{g+k+1+m} - N(H_m + H_{g+m}) \\
&\ge H_{k+m+1} + H_{g+k+1+m} - N(2^{m-1} + 2^{g+m-1} - 2^{m-2}(m-1)). \qquad (35)
\end{aligned}
$$

Let $C_m := H_{k+m+1} + H_{g+k+1+m} - N(2^{m-1} + 2^{g+m-1} - 2^{m-2}(m-1))$, from Eq. (35). We show by strong induction that $C_m > 0$. By direct computation, $C_1 > 0$. Suppose it holds for all values from 1 to $m-1$ for $m \ge 2$. Then by the induction hypothesis,

$$
\begin{aligned}
H_{g+k+1+m} &= (H_{g+k+m} + \cdots + H_{g+k+2}) + (H_{g+k+1} + \cdots + H_{m+k+1}) + NH_m \\
&> \sum_{i=1}^{m-1} (N(2^{i-1} + 2^{g+1-i} - 2^{i-2}(i-1)) - H_{k+i+1}) + \\
&\qquad + \left(2^{g+k} + \cdots + 2^{m+k} - \sum_{i=1}^{k+1} 2^{i-2}(i-1)\right) + 2^{m-1}N \\
&= N(2^m - 1 + 2^{g+m+1} - 2^g - 2^{m-2}(m-3) - 1) - \\
&\qquad - \sum_{i=k+2}^{k+m} H_i + \left(2^{g+k+1} - 2^{m+k} - 2^k(k-1) - 1\right) \\
&\ge N(2^{m-1} + 2^{g+m-1} - 2^{m-2}(m-1)) - (2^g + 2 - 2^m)N - \\
&\qquad - \sum_{i=k+m-g}^{k+m} H_i + \left(2^{g+k+1} - 2^{m+k} - 2^k(k-1) - 1\right),
\end{aligned}
\qquad (36)
$$

where $H_i = 0$ for nonpositive i. Hence,

$$
\begin{aligned}
C_m &= H_{g+k+1+m} - N(2^{m-1} + 2^{g+m-1} - 2^{m-2}(m-1)) + H_{k+m+1} \\
&> \left(H_{k+m+1} - \sum_{i=k+m-g}^{k+m} H_i \right) + \left(2^{g+k+1} - 2^{m+k} - 2^k(k-1) - 1\right) - \\
&\qquad - (2^g + 2 - 2^m)N \\
&= 1 + \left(2^{g+k+1} - 2^{m+k} - 2^k(k-1) - 1\right) - (2^g + 2 - 2^m)\left(2^{k+1} - \left\lceil \frac{k}{2^{g-k}} \right\rceil\right) \\
&= 2^{m+k} - 2^k(k+3) + (2^g + 2 - 2^m)\left\lceil \frac{k}{2^{g-k}} \right\rceil \\
&\ge 2^{m+k} - 2^k(k+3) + \left(2^g + 2 - 2^m\right) \frac{k}{2^{g-k}}
\end{aligned}
$$

$$= 2^{m+k} - 3 \cdot 2^k - (2^m - 2)\frac{k}{2^{g-k}}$$
$$= (2^m - 3)\left(2^k - \frac{k}{2^{g-k}}\right) - \frac{k}{2^{g-k}}$$
$$\geq 2^k - \frac{2k}{2^{g-k}} \geq 2^k - 2k \geq 0. \quad (37)$$

This completes the induction, so $B_n \geq C_m > 0$.

Since both cases are satisfied, $\{H_i\}$ is complete.

Remark 3.6 The case $k = 1$ is characterized in Lemma 3.8.

The "2L—1 Conjecture"

We conjecture a strengthened version of Theorem 3.3 as follows.

Conjecture 3.7 The PLRS $\{H_n\}$ defined by $[c_1, \ldots, c_L]$ is complete if $B_{H,n} \geq 0$ for all $n \leq 2L - 1$, i.e., Brown's criterion holds for the first $2L - 1$ terms.

When using Brown's criterion, it would be very helpful to be able to know how many terms must be checked to be sure that a PLRS is complete. This conjecture, if true, would be a powerful tool to do so. We do not know yet if such a threshold exists for each L; however, if it does, then it is at least $2L - 1$, as shown by the following example, where $k + 2 = L$.

Lemma 3.8 $[1, \ldots, 1, 0, 4]$, *with* $k \geq 1$ *ones, is always incomplete. Moreover, it first fails Brown's criterion on the* $(2k + 3)$*rd term.*

Proof We have the recurrence relation $H_{n+1} = H_n + \cdots + H_{n-k+1} + 4H_{n-k-1}$. We show that the term in the $(2k + 3)$rd position in the sequence fails Brown's criterion. First,

$$H_{2k+3} = H_{2k+2} + \cdots + H_{k+3} + 4H_{k+1}. \quad (38)$$

Next, we observe that for $1 \leq j \leq k + 1$, then $H_j = 2^{j-1}$. Additionally, $H_{k+2} = 2^{k+1} - 1$. Thus,

$$2H_{k+1} = 2^{k+1} > 2^{k+1} - 1 = H_{k+2}. \quad (39)$$

We also note that $H_{k+1} = H_k + \cdots + H_1 + 1$. Putting everything together,

$$\begin{aligned} H_{2k+3} &= H_{2k+2} + \cdots + H_{k+3} + 4H_{k+1} \\ &= H_{2k+2} + \cdots + H_{k+3} + 3H_{k+1} + H_k + \cdots + H_1 + 1 \\ &> H_{2k+2} + \cdots + H_{k+3} + H_{k+2} + H_{k+1} + H_k + \cdots + H_1 + 1. \end{aligned} \quad (40)$$

Hence, we have shown that $[1, \ldots, 1, 0, 4]$, with $k \geq 1$ ones, is incomplete, as it fails Brown's criterion on the $(2k+3)$rd term.

We now show that Brown's criterion holds for the first $(2k+2)$ terms. For $1 \leq j \leq k+1$, we have $H_j = 2^{j-1}$, which satisfies the equality $H_{j+1} = H_j + \cdots + H_1 + 1$.

We consider when $k+2 \leq j \leq 2k+2$, that

$$H_{j+1} = H_j + \cdots + H_{j-k+1} + 4H_{j-k-1}.$$

Note that $H_{j-k-1} = H_{j-k+2} + \cdots + H_1 + 1$ as $1 \leq j-k-1 \leq k+1$,

$$H_{j+1} = H_j + \cdots + H_{j-k+1} + 2H_{j-k-1} + H_{j-k-1} + H_{j-k-2} + \cdots + H_1 + 1.$$

And as $2H_{j-k-1} = 2^{j-k-1} = H_{j-k}$, we see

$$H_{j+1} = H_j + \cdots + H_{j-k+1} + H_{j-k} + H_{j-k-1} + H_{j-k-1} + H_{j-k-2} + \cdots + H_1 + 1. \quad (41)$$

Hence, this equality satisfies Brown's criterion for terms $k+2 \leq j \leq 2k+2$.

Assuming this conjecture, we can further explore sequences of the form $[1, 0, \ldots, 0, 1, \ldots, 1, N]$. In Theorems 3.10 and 3.11, we show that the bound on N for $[1, \underbrace{0, \ldots, 0}_{L-m-2}, \underbrace{1, \ldots, 1}_{m}, N]$ strictly increases if we keep L fixed and increase m from 0 to $L-3$, i.e., switching the coefficients from 0 to 1 gradually from the end so that at least one 0 remains. We first state a following powerful lemma that is contingent on this conjecture.

Lemma 3.9 (Conditional) *Let $\{H_n\}$ defined by $[1, 0, \ldots, 0, 1, \ldots, 1, N]$ be a sequence of length L with m ones. Then, if the sequence is incomplete, it must fail Brown's criterion at the $(L+1)$st or $(L+2)$nd term. In other words, if $H_{L+1} \leq 1 + \sum_{i=1}^{L} H_i$ and $H_{L+2} \leq 1 + \sum_{i=1}^{L+1} H_i$, then $\{H_n\}$ is complete.*

The proof of this lemma is deferred to 8.11 of Appendix C.

Theorem 3.10 *Let $\{H_n\}$ be a PLRS with L coefficients defined by $[1, 0, \ldots, 0, \underbrace{1, \ldots, 1}_{m}, N]$, where $L \geq 2m+2$. Then $\{H_n\}$ is complete if and only if*

$$N \leq \left\lfloor \frac{(L-m)(L+m+1)}{4} + \frac{1}{48}m(m+1)(m+2)(m+3) + \frac{1-2m}{2} \right\rfloor. \quad (42)$$

Proof First, note for all $1 \leq n \leq L-m$, that $H_n = n$.

Now, we claim that for all $1 \leq k \leq m$,

$$H_{L-m+k} = L - m + \frac{1}{6}k(k+1)(k+2) + k. \quad (43)$$

We use induction, appealing to the identity that $\sum_{a=1}^{n} a(a+1)/2 = n(n+1)(n+2)/6$. We first see that

$$H_{L-m+1} = H_{L-m} + H_1 + 1 = L - m + 2 = L - m + \sum_{a=1}^{1} \frac{a(a+1)}{2} + 1. \quad (44)$$

Additionally,

$$\begin{aligned} H_{L-m+2} &= H_{L-m+1} + H_2 + H_1 + 1 \ = \ (L - m + 2) + 2 + 1 + 1 \\ &= L - m + \sum_{a=1}^{2} \frac{a(a+1)}{2} + 2. \end{aligned} \quad (45)$$

Now, suppose $H_{L-m+k} = L - m + \sum_{a=1}^{k} a(a+1)/2 + k$ for some $k < m$. Note that

$$H_{L-m+k+1} = H_{L-m+k} + H_{k+1} + \cdots + H_1 + 1. \quad (46)$$

Since we supposed $L \geq 2m + 2$, we see $k + 1 \leq m + 1 \leq L - m$, and thus for all $1 \leq i \leq k,\ H_i = i$. Thus,

$$\begin{aligned} H_{L-m+k+1} &= \left(L - m + \sum_{a=1}^{k} \frac{a(a+1)}{2} + k\right) + \frac{(k+1)(k+2)}{2} + 1 \\ &= L - m + \sum_{a=1}^{k+1} \frac{a(a+1)}{2} + k + 1. \end{aligned} \quad (47)$$

Thus, we have an explicit formula for H_i, for $1 \leq i \leq L$.

Note that $\{H_n\}$ is complete if and only if it fulfills Brown's criterion for the $(L+1)$st and $(L+2)$nd term. We show that $\{H_n\}$ fulfills the criterion for $L+2$ if and only if the bound above holds; it is not difficult to show that the bound for $L+1$ is less strict.

Indeed, we wish to reduce the inequality

$$H_{L+2} = H_{L+1} + H_{m+2} + \cdots + H_3 + 2N \leq 1 + \sum_{i=1}^{L+1} H_i \quad (48)$$

$$\iff H_{m+2} + \cdots + H_3 + 2N \leq 1 + \sum_{i=1}^{L} H_i. \quad (49)$$

Simplifying the left-hand side of inequality (49),

$$\begin{aligned} H_{m+2} + \cdots + H_3 + 2N &= H_{m+2} + \cdots + H_3 + (H_2 + H_1 - H_2 - H_1) + 2N \\ &= \frac{(m+2)(m+3)}{2} - 3 + 2N. \end{aligned} \quad (50)$$

Additionally,

$$\begin{aligned}1+\sum_{n=1}^{L} H_n &= 1+\sum_{n=1}^{L-m} H_n+\sum_{n=L-m+1}^{L} H_n\\ &= 1+\frac{(L-m)\,(L-m+1)}{2}+\sum_{n=1}^{m}\left(\frac{1}{6}n(n+1)(n+2)+n+L-m\right).\end{aligned} \tag{51}$$

We use the fact that $\sum_{n=1}^{m} n(n+1)(n+2)=m(m+1)(m+2)(m+3)/4$ to simplify Eq. 51 as follows:

$$\begin{aligned}&1+\frac{(L-m)\,(L-m+1)}{2}+\frac{m(m+1)}{2}+mL-m^2+\frac{1}{6}\sum_{n=1}^{m} n(n+1)(n+2)\\ =&1+\frac{(L-m)\,(L-m+1)}{2}+\frac{m(m+1)}{2}+mL-m^2+\frac{1}{24}m(m+1)(m+2)(m+3).\end{aligned} \tag{52}$$

Hence, the inequality is equivalent to

$$\begin{aligned}\frac{(m+2)\,(m+3)}{2}-3+2N \le 1+&\frac{(L-m)\,(L-m+1)}{2}+\frac{m(m+1)}{2}\\ &+mL-m^2+\frac{1}{24}m(m+1)(m+2)(m+3).\end{aligned} \tag{53}$$

Simplifying, this gives us

$$N \le \left\lfloor \frac{(L-m)\,(L+m+1)}{4}+\frac{1}{48}m(m+1)(m+2)(m+3)+\frac{1-2m}{2}\right\rfloor. \tag{54}$$

Theorem 3.11 *Let $\{G_n\}$ and $\{H_n\}$ be PLRS's with L coefficients defined by*

$$[1,0,\ldots,0,\underbrace{1,\ldots,1}_{m},N] \text{ and } [1,0,\ldots,0,\underbrace{1,\ldots,1}_{m+1},N+1],$$

respectively. Suppose $L-m\ge 4$ (so that at least one zero is present in $\{H_n\}$), $m\ge (L-1)/2$, and $\{G_n\}$ is complete. Then $\{H_n\}$ is also complete.

Proof As $\{G_n\}$ is complete, from Brown's criterion, we obtain

$$G_{L+2}=G_{L+1}+\sum_{i=3}^{m+2} G_i+NG_2 \le 1+\sum_{i=1}^{L+1} G_i, \tag{55}$$

which is equivalent to

$$2N \leq \sum_{i=m+3}^{L} G_i + 4. \tag{56}$$

From Lemma 3.9, it suffices to show that

$$\begin{cases} H_{L+1} \leq 1 + \sum_{i=1}^{L} H_i \\ H_{L+2} \leq 1 + \sum_{i=1}^{L+1} H_i, \end{cases} \tag{57}$$

or equivalently,

$$N \leq \sum_{i=m+3}^{L-1} H_i \tag{58}$$

and

$$2N \leq \sum_{i=m+4}^{L} H_i + 2. \tag{59}$$

We first show Eq. (58). Combining with Eq. (56), it suffices to show that

$$\sum_{i=m+3}^{L} G_i + 4 \leq 2 \sum_{i=m+4}^{L} H_i. \tag{60}$$

From Lemma 8.12,

$$\begin{cases} G_i \leq H_i & m+3 \leq i \leq L \\ G_i \leq H_{i-1} - 1 & 2(L-m) < i \leq L. \end{cases} \tag{61}$$

Thus,

$$\begin{aligned} \sum_{i=m+3}^{L} G_i + 4 &= \sum_{i=m+3}^{2(L-m)} G_i + \sum_{i=2(L-m)+1}^{L} G_i + 4 \\ &\leq \sum_{i=m+3}^{2(L-m)} H_i + \sum_{i=2(L-m)}^{L-1} H_i + (2m - L + 4) \\ &\leq 2 \sum_{i=m+4}^{L} H_i, \end{aligned} \tag{62}$$

the last inequality can be taken crudely. We then show (59). Similarly, combining with (56), it suffices to show that

$$\sum_{i=m+3}^{L} G_i + 2 \leq \sum_{i=m+4}^{L} H_i. \tag{63}$$

If $m + 3 \geq 2(L - m)$, then

$$\begin{aligned}
\sum_{i=m+4}^{L} H_i &= \sum_{i=m+4}^{L} \left(H_{i-1} + \sum_{j=1}^{i-L+m+1} H_j + 1 \right) \\
&\geq \sum_{i=m+4}^{L} (H_{i-1} + H_{i-L+m+2}) \text{ (Brown's criterion for the first terms)} \\
&= \sum_{i=m+3}^{L-1} H_i + \sum_{i=2m+6-L}^{m+2} H_i \geq \sum_{i=m+3}^{L-1} (G_{i+1} + 1) + H_{m+2} \\
&\geq \sum_{i=m+3}^{L} G_i + 2.
\end{aligned} \tag{64}$$

If $m + 3 < 2(L - m)$, then

$$\begin{aligned}
\sum_{i=m+3}^{L} G_i &= \sum_{i=m+3}^{2(L-m)-1} G_i + G_{2(L-m)} + \sum_{i=2(L-m)+1}^{L} G_i \\
&= \sum_{i=m+3}^{2(L-m)-1} (H_{i-1} + 1) + H_{2(L-m)-1} + \sum_{i=2(L-m)+1}^{L} G_i \\
&= \sum_{i=m+2}^{2(L-m)-1} H_i + (2L - 3(m + 1)) + \sum_{i=2(L-m)+1}^{L} G_i.
\end{aligned} \tag{65}$$

Thus, our original inequality, Eq. (59), holds if we can show that

$$H_{m+2} + H_{m+3} + (2L - 3(m + 1)) + \sum_{i=2(L-m)+1}^{L} G_i \leq \sum_{i=2(L-m)}^{L} H_i. \tag{66}$$

Similar to the previous case,

$$\begin{aligned}\sum_{i=2(L-m)}^{L} H_i \geq \sum_{i=2(L-m)}^{L} (H_{i-1} + H_{i-L+m+2}) \\ = \sum_{i=2(L-m)-1}^{L-1} H_i + \sum_{i=L-m+2}^{m+2} H_i \\ = \sum_{i=2(L-m)}^{L-1} H_i + H_{2(L-m)-1} + H_{m+2} + \sum_{i=L-m+2}^{m+1} H_i. \end{aligned} \quad (67)$$

As $2(L-m)-1 \geq m+3$ and $H_i \geq i$

$$\begin{aligned}\sum_{i=2(L-m)}^{L} H_i \geq \sum_{i=2(L-m)}^{L-1} (G_{i+1} + 1) + H_{m+3} + H_{m+2} + \sum_{L-m+2}^{m+1} i \\ = \sum_{i=2(L-m)+1}^{L} G_i + H_{m+3} + H_{m+2} + (2m - L + \sum_{i=L-m+2}^{m+1} i). \end{aligned} \quad (68)$$

From Lemma 8.13,

$$\sum_{i=2(L-m)}^{L} H_i \geq \sum_{i=2(L-m)+1}^{L} G_i + H_{m+3} + H_{m+2} + (2L - 3(m+1)). \quad (69)$$

4 An Analytical Approach

An Introduction to Principal Roots

We begin by restating some results from [8].

Lemma 4.1 *Let $P(x)$ be the characteristic polynomial of a recurrence relation with nonnegative coefficients and at least one positive coefficient, and let $S = \{m \mid c_m \neq 0\}$. Then*

1. *there exists exactly one positive root r, and this root has multiplicity* 1*;*
2. *every root $z \in \mathbb{C}$ satisfies $|z| \leq r$; and*
3. *if $\gcd(S) = 1$, then r is the unique root of greatest magnitude.*

Proof This is Lemma 2.1 from [8].

Remark 4.2 We refer to the unique positive root from Lemma 4.1 as the *principal root* of the recurrence sequence and corresponding characteristic polynomial.

Lemma 4.3 *Let $P(x)$ be the characteristic PLRS $\{H_n\}$ and let r_1 be its principal root. Then*

$$\lim_{n\to\infty} \frac{H_n}{r_1^n} = C \tag{1}$$

for some constant $C > 0$.

Proof Corollary 2.3 from [8] proves a stronger result than this, which immediately implies this lemma.

Lemma 4.4 *Let $P(x)$ be the characteristic polynomial of a PLRS $\{H_n\}$ with roots r_i, each of multiplicity m_i, where r_1 is the principal root. If*

$$H_n = a_1 r_1^n + \sum_{i=2}^{k} q_i(n) r_i^n, \tag{2}$$

where $q_i(x)$ is a polynomial of degree at most $m_i - 1$, then $a_1 > 0$.

Proof First, note that the set S of Lemma 4.1 contains 1 because $c_1 > 0$ in a PLRS. Therefore $\gcd(S) = 1$, and r_1 is the unique root of greatest magnitude. If $a_1 < 0$, then this implies that $H_n < 0$ for some n because the behavior of $a_1 r_1^n$ eventually dominates the expression for H_n in (2). If $a_1 = 0$, then

$$\lim_{n\to\infty} \frac{H_n}{r_1^n} = 0 \tag{3}$$

because r_1 is the unique root of greatest magnitude, so if $a_1 = 0$ then the behavior of H_n is bounded by geometric growth of the root of next greatest magnitude, which is necessarily smaller than r_1^n. Thus, $a_1 > 0$.

Applications to Completeness

Given these results, we see that the principal root of a PLRS serves as a measure for the rate of that sequence's growth. Guided by the simple heuristic that, generally, a sequence which grows slowly is more likely to be complete than a sequence which grows rapidly, we find bounds for the potential roots of a complete or incomplete PLRS. We aim to answer these questions: For any given L, what is the fastest growing complete PLRS with L coefficients? What is the slowest growing incomplete PLRS with L coefficients?[1]

[1] While the principal root of a PLRS has not been related to completeness before, there is previous work on bounding the principal root of other linear recurrence sequences in [6].

Lemma 4.5 *If $\{H_n\}$ is a complete PLRS and r_1 is its principal root, then $|r_1| \leq 2$.*

Proof Suppose that $|r_1| > 2$. Set

$$H_n = a_1 r_1^n + q_2(n) r_2^n + \cdots + q_r(n) r_k^n. \tag{4}$$

Since r_1 is the unique root of largest magnitude by Lemma 4.1, the behavior of $a_1 r_1^n$ dominates in the limit. By Lemma 4.4, $a_1 > 0$, so if $|r_1| > 2$, then eventually $|a_1 r_1^n| > 2^{n-1}$, and so there exists a large n for which $H_n > 2^{n-1}$. As the sequence $\{2^{n-1}\}$ is the complete PLRS with maximal terms by Theorem 2.1, we see $\{H_n\}$ must be incomplete.

Remark 4.6 The converse to this lemma does not hold. A counterexample is [1, 1, 1, 0, 4], which has principal root 2 but is not complete.

While the proof is simple, this lemma gives us an effective upper bound for the roots of a complete PLRS, regardless of length. Recall from Theorem 1.11 that for any L, the PLRS $\{H_n\}$ generated by the coefficients $[\underbrace{1, \ldots, 1}_{L-1}, 2]$ satisfies $H_n = 2^{n-1}$. This sequence naturally has a principal root of 2, and is complete. Similarly, for any $L \geq 1$, the sequence $[\underbrace{1, \ldots, 1}_{L}]$ is complete, and its principal root asymptotically approaches 2 as L grows.

We now focus on finding a lower bound for the roots of an incomplete sequence, which proves to be a more difficult problem.

Lemma 4.7 *For any $L \in \mathbb{Z}^+$, there exists a constant B_L, with $1 < B_L < 2$ such that if $\{H_n\}$ is a PLRS with principal root r_1 and $r_1 < B_L$, then $\{H_n\}$ is complete.*

Remark 4.8 This means that for any L, there exists a lower bound B_L on possible values of the principal root of an incomplete PLRS generated by $[c_1, \ldots, c_L]$.

Proof In order to show that such a B_L exists, it suffices to show that for any given L, there exists only finitely many incomplete positive linear recurrence sequences generated by $[c_1, \ldots, c_L]$ with principal root $r_1 < 2$.

Recall that the principal root r_1 of a PLRS is the single positive root of the characteristic polynomial $p(x) = x^L - \sum_{i=1}^{L} c_i x^{L-i}$.

As $\lim_{x \to \infty} p(x) = +\infty$, the fact that r_1 is the unique positive root of $p(x)$ implies that $r_1 < 2 \iff p(2) > 0$, by intermediate value theorem. Note that

$$p(2) = 2^L - \sum_{i=1}^{L} c_i 2^{L-i} > 0 \iff \sum_{i=1}^{L} c_i 2^{L-i} < 2^L. \tag{5}$$

As for all i, $c_i \geq 0$, so the inequality above cannot hold if there exists i such that $c_i \geq 2^i$. As the set $\{[c_1, \ldots, c_L] : 0 \leq c_i \leq 2^i \text{ for all } i\}$ of such sequences is finite, we are done.

The remainder of this section is a series of lemmas which build toward the following conjecture:

Conjecture 4.9 Let $N_L = \lceil L(L+1)/4 \rceil$, and let λ_L be the principal root of the sequence generated by $[1, \underbrace{0, \ldots, 0}_{L-2}, N_L + 1]$, i.e., the sole principal root of

$$p_L(x) = x^L - x^{L-1} - \left\lceil \frac{L(L+1)}{4} \right\rceil - 1. \tag{6}$$

If $[c_1, \ldots, c_L]$ generates an incomplete sequence, then its principal root is at least λ_L.

Remark 4.10 This conjecture is equivalent to stating $B_L = \lambda_L$ for all $L \geq 2$, where B_L is the bound proposed in Lemma 4.7.

Remark 4.11 Using Theorem 1.15, it is easy to see that the sequence generated by $[1, 0, \ldots, 0, N_L + 1]$ is incomplete; in fact, the value $N_L + 1$ is the minimal positive integer such that a sequence of this form is incomplete.

As a first step toward a proof of Conjecture 4.9, we prove Lemma 4.15, which addresses the case of sequences with a large sum in coefficients.

Definition 4.12 For positive integers S, L, we define the set of positive linear recurrence sequences

$$P_{L,S} := \left\{ \{H_n\} \text{ generated by } [c_1, \ldots, c_L] \,\middle|\, \sum_{i=1}^{L} c_i = S + 1 \right\}. \tag{7}$$

Lemma 4.13 *The sequence in* $P_{L,S}$ *with the minimal principal root is* $[1, 0, \ldots, 0, S]$.

Proof Consider a sequence generated by $s = [c_1, \ldots, c_L] \in P_{L,S}$, and let $r_1, \ldots, r_L$ be its roots, with $r_1 > 0$ the principal root. Since $|c_L| = \left|\prod_{i=1}^{L} r_L\right|$ is a positive integer, we know $r_1 > 1$.

Now, for any $1 \leq m \leq L$ consider a sequence generated by $s_m \in P_{L,S}$ of the form

$$[c_1, \ldots, c_{m-1}, c_m - 1, c_{m+1}, \ldots, c_L + 1]. \tag{8}$$

We claim that the principal root q_1 of s_m fulfills $q_1 < r_1$. Define the characteristic polynomials $f(x)$ and $g(x)$ for s and s_m, respectively, so that

$$f(x) = x^L - \sum_{i=1}^{L} c_i x^{L-i}, \tag{9}$$

and

$$
\begin{aligned}
g(x) &= x^L - \sum_{i=1}^{m-1} c_i x^{L-i} - (c_m - 1)\, x^m - \sum_{i=m+1}^{L-1} c_i c_i x^{L-i} - (c_L + 1) \\
&= x^L - \sum_{i=1}^{L} c_i x^{L-i} + x^m - 1. \qquad (10)
\end{aligned}
$$

As q_1 is the sole positive root of $g(x)$, and $g(x)$ is eventually positive, we notice that $q_1 < r_1$ if and only if $g(r_1) > 0$, which is equivalent to $g\,(r_1) > f\,(r_1)$.

Now,

$$
\begin{aligned}
g\,(r_1) > f\,(r_1) &\iff r_1^L - \textstyle\sum_{i=1}^{L} c_i r_1^{L-i} + r_1^m - 1 > r_1^L - \sum_{i=1}^{L} c_i r_1^{L-1} \\
&\iff r_1^m - 1 > 0 \qquad (11) \\
&\iff r_1 > 1.
\end{aligned}
$$

As $r_1 > 1$, the principal root q_1 of $g(x)$ is strictly less than that of $f(x)$.

As s was chosen arbitrarily, we see that the principal root of any sequence $s \in P_{L,S}$ can be strictly decreased by using the transformation $s \to s_m$ for any $1 \le m \le L$. Applying this transformation iteratively for all values of m, we inevitably end up with the minimal possible values of $c_1, \ldots, c_{L-1}$, namely, $c_1 = 1,\ c_2 = c_3 = \cdots = c_{L-1} = 0$, and the maximal possible value of c_L, namely, $c_L = S$.

Thus, as the principal root under these iterated transformations is strictly decreasing, we conclude that $[1, 0, \ldots, 0, S]$ has the smallest principal root of any element of $P_{L,S}$.

Lemma 4.14 *For any $S > 0$, the principal root of* $[1, 0, \ldots, 0, S]$ *is strictly less than that of* $[1, 0, \ldots, 0, S + 1]$.

Proof Let S be an arbitrary positive integer, and denote by $f(x)$, $g(x)$ and r_1, q_1 the characteristic polynomials and principal roots of $[1, 0, \ldots, 0, S + 1]$ and $[1, 0, \ldots, 0, S]$, respectively.

As before, $q_1 < r_1$ if and only if $g(r_1) > 0 = f(r_1)$. Note that

$$
\begin{aligned}
g(r_1) > f(r_1) &\iff r_1^L - r_1^{L-1} - S > r_1^L - r_1^{L-1} - (S + 1) \\
&\iff S + 1 > S. \qquad (12)
\end{aligned}
$$

Thus, $q_1 < r_1$, for any value of S.

Lemma 4.15 *Any sequence fulfilling $\sum_{i=1}^{L} c_i \ge N_L + 2$ has a principal root greater than or equal to that of*

$$[1, 0, \ldots, 0, N_L + 1]. \qquad (13)$$

Proof Recall from Theorem 1.15 that the sequence $[1, 0, \ldots, 0, N]$ is complete if and only if $N \le N_L$, for $N_L = \lceil L(L + 1)/4 \rceil$. Thus, an immediate corollary to this theorem is that the incomplete sequence of the form $[1, 0, \ldots, 0, N]$ with the minimal possible principal root is $[1, 0, \ldots, 0, N_L + 1]$.

Furthermore, if we have a sequence generated by $[c_1, \ldots, c_L]$ which fulfills $\sum_{i=1}^{L} c_i \geq N_L + 2$, Lemmas 4.13 and 4.14 present a sequence of algorithms which allow us to transform this sequence into the sequence generated by $[1, 0, \ldots, 0, N_L + 1]$, in such a way that each transformation strictly lowers the magnitude of the principal root.

Thus, any sequence satisfying $\sum_{i=1}^{L} c_i \geq N_L + 2$ has a principal root strictly greater than the principal root of $[1, 0, \ldots, 0, N_L + 1]$.

The following lemmas are working toward proving Conjecture 4.21, which addresses the second case of Conjecture 4.9, which addresses the roots of sequences $[c_1, \ldots, c_L]$ which fulfill $\sum_{i=1}^{L} c_i \leq N_L + 2$.

Lemma 4.16 *Suppose the sequence generated by $[c_1, \ldots, c_L]$ has principal root r, then for any $c_{L+1} \in \mathbb{Z}^+$, the sequence generated by $[c_1, \ldots, c_L, c_{L+1}]$ (in which we add an additional positive coefficient) and principal root q fulfills $r < q$.*

Proof Let $f(x), g(x)$ be the characteristic polynomials of the two sequences, so that

$$f(x) = x^L - \sum_{i=1}^{L} c_i x^{L-i}, \text{ and } g(x) = x^{L+1} - \sum_{i=1}^{L} c_i x^{L+1-i} - c_{L+1}. \tag{14}$$

Similar to previous arguments, by the intermediate value theorem, $r < q$ if and only if $g(r) < f(r) = 0$. Note that

$$\begin{aligned} g(r) < f(r) &\iff r^{L+1} - \textstyle\sum_{i=1}^{L} c_i x^{L+1-i} - c_{L+1} < r^L - \sum_{i=1}^{L} c_i r^{L-i} \\ &\iff c_{L+1} > r^{L+1} - r^L + \textstyle\sum_{i=1}^{L} c_i r^{L-i} - \sum_{i=1}^{L} c_i r^{L+1-i} \\ &\iff c_{L+1} > r^L (r-1) + \textstyle\sum_{i=1}^{L} c_i r^{L-i} (1-r) \\ &\iff c_{L+1} > (1-r)\left(r^L - \textstyle\sum_{i=1}^{L} c_i r^{L-i}\right) = (1-r) \cdot f(r) \\ &\iff c_{L+1} > (1-r) f(r) = (1-r) \cdot 0 = 0. \end{aligned} \tag{15}$$

Since $c_{L+1} \in \mathbb{Z}^+$ the last line holds, and so $r < q$.

Lemma 4.17 *Let λ_L be the principal root of*

$$x^L - x^{L-1} - N_L - 1. \tag{16}$$

Then, for any $L \geq 2$, $\lambda_L > \lambda_{L+1}$.

Proof Define $f(x)$ and $g(x)$ to be the characteristic polynomials of $[1, 0, \ldots, 0, N_L + 1]$ and $[1, 0, \ldots, 0, N_{L+1} + 1]$, of length L and $L + 1$, respectively, so that

$$f(x) = x^L - x^{L-1} - N_L - 1, \quad g(x) = x^{L+1} - x^L - N_{L+1} - 1. \tag{17}$$

As in previous proofs, we see that $\lambda_L > \lambda_{L+1} \iff g(\lambda_L) > f(\lambda_L) = 0$.

$$\begin{aligned} g(\lambda) > f(\lambda) &\iff \lambda^{L+1} - \lambda^L - N_{L+1} - 1 > \lambda^L - \lambda^{L-1} - N_L - 1 \\ &\iff \lambda^{L+1} - 2\lambda^L + \lambda^{L-1} > N_{L+1} - N_L \\ &\iff \lambda^{L-1}(\lambda - 1)^2 > N_{L+1} - N_L. \end{aligned} \tag{18}$$

Note that as $f(\lambda) = 0$, we have $\lambda^{L-1}(\lambda - 1) = N_L + 1$. Note that $N_{L+1} - N_L \leq (L+2)/2$, which can be shown by using the definition of N_L and checking all cases modulo 4. Thus it suffices to show that

$$(N_L + 1)(\lambda_L - 1) \geq \frac{L+2}{2}. \tag{19}$$

Using the value of N_L, it suffices to show

$$(\lambda_L - 1) \geq \frac{L+2}{L^2 + L + 4}. \tag{20}$$

The proof of Eq. 20 is just algebra, and is left to Appendix D, as Lemma 9.1.

Lemma 4.18 *For any $L \in \mathbb{N}$, let λ_L be the sole positive root of the polynomial*

$$p_L(x) = x^L - x^{L-1} - \left\lceil \frac{L(L+1)}{4} \right\rceil - 1. \tag{21}$$

Then $\lim_{L\to\infty} \lambda_L = 1$.

Proof We show that for any $\varepsilon > 0$, there exists an M large enough so that for all $L > M$, $p_L(1+\varepsilon) > 0$. As $p_L(x)$ has only one positive root λ_L and $p(x)$ is positive as $x \to \infty$, we see $p_L(1+\varepsilon) > 0$ implies $\lambda_L < 1 + \varepsilon$. If this is possible for arbitrary ε, then $\lambda_L \to 1$, as desired.

Fix an $\varepsilon > 0$. For any L, we may write

$$\begin{aligned} p_L(1+\varepsilon) &= (1+\varepsilon)^L - (1+\varepsilon)^{L-1} - \left\lceil \frac{L(L+1)}{4} \right\rceil - 1 \\ &= \sum_{n=0}^{L} \varepsilon^n \left(\binom{L}{n} - \binom{L-1}{n} \right) - \left\lceil \frac{L(L+1)}{4} \right\rceil - 1, \end{aligned} \tag{22}$$

where $\binom{L-1}{L}$ is 0. Using Pascal's rule, we can reduce (22) to

$$p_L(1+\varepsilon) = \sum_{n=1}^{L} \varepsilon^n \binom{L-1}{n-1} - \lceil L(L+1)/4 \rceil - 1. \tag{23}$$

This quantity can easily be shown to be positive (and in fact tends toward infinity) for large enough L. For example, we can take the trivial bound

$$\sum_{n=1}^{L} \varepsilon^n \binom{L-1}{n-1} > \varepsilon^4 \binom{L-1}{3}, \tag{24}$$

as the full sum must be larger than only its fourth summand.

Since ε^4 is simply a positive constant and $L(L+1) \ll \binom{L-1}{3}$, then for large enough L,

$$p_L(1+\varepsilon) > \varepsilon^4 \binom{L-1}{3} - \lceil L(L+1)/4 \rceil - 1 > 0. \tag{25}$$

Remark 4.19 Even in the event that Conjecture 4.9 is false, this gives us conclusive proof that we may find incomplete sequences whose roots are arbitrarily close to 1; since 1 is the minimum possible size for the root of a PLRS, this may be interpreted as proof that we may find arbitrarily slow-growing incomplete sequences, with coefficients of any length L.

Lemma 4.20 *Consider the sequence generated by* $[c_1, \ldots, c_L]$. *For any value* $m \in \mathbb{Z}^+$, *the principal root of* $[c_1, \ldots, c_L + m]$ *is greater than that of* $[c_1, \ldots, c_L, m]$.

Proof Let $f(x)$, $g(x)$, and r, q be the principal roots of $[c_1, \ldots, c_L + m]$ and $[c_1, \ldots, c_L, m]$, respectively. Since f, g each have a unique positive root, we see that $r > q \iff g(r) > f(r) = 0$. Note that

$$\begin{aligned} g(r) > 0 &\iff 0 = rf(r) < g(r) \\ &\iff r\left(r^L - \textstyle\sum_{i=1}^{L} c_i r^{L-i} - m\right) < r^{L+1} - \textstyle\sum_{i=1}^{L} c_i r^{L+1-i} - m \\ &\iff m < rm \\ &\iff r > 1. \end{aligned} \tag{26}$$

Thus, the inequality always holds, and so $r > q$, as desired.

Conjecture 4.21 Let λ_L be the principal root of $x^L - x^{L-1} - N_L - 1$. If the sequence generated by $[c_1, \ldots, c_L]$ is incomplete with $\sum_{i=1}^{L} c_i \leq \lceil L(L+1)/4 \rceil + 2$, then its principal root is at least λ_L.

We present a partial proof, which addresses all cases except what is denoted as Subcase 2.

Proof (*Partial proof.*) We use induction.

For $L = 2$, $N_L = \lceil 2 \cdot 3/4 \rceil = 2$, and so the coefficients $[c_1, c_2]$ fulfilling the requirement are of the form $c_1 + c_2 \leq 4$. The incomplete sequences of this form have coefficients [2, 1], [2, 2], [1, 3], and [3, 1]. Checking each case directly, we see that their principal roots are approximately 2.414, 2.731, 2.303, and 3.303, respectively. Of these, the root of $[1, 3] = [1, N_2 + 1]$ is the minimum; thus, the lemma holds for the base case.

Now, suppose the lemma holds for some value of $L \geq 2$; we show that it holds for $L + 1$.

Let $[c_1, \ldots, c_L, c_{L+1}]$ be an incomplete sequence with

$$\sum_{i=1}^{L+1} c_i \leq \lceil (L+1)(L+2)/4 \rceil + 2.$$

<u>$Case 1 : \sum_{i=1}^{L} c_i < N_L + 2$</u>
If this is the case, the following two subcases arise:

- Subcase 1: $[c_1, \ldots, c_L]$ is incomplete.
  If this is the case, then by our inductive hypothesis, since $\sum_{i=1}^{L} c_i \leq N_L + 2$, we must have that the principal root r of $[c_1, \ldots, c_L]$ is greater than or equal to λ_L. Hence, by Lemma 4.16, since the principal root q of $[c_1, \ldots, c_{L+1}]$ satisfies $q > r$, we have that $[c_1, \ldots, c_{L+1}]$ has principal root $q > \lambda_L$. Finally, by Lemma 4.17, we know $\lambda_L > \lambda_{L+1}$; thus, we have $q > r \geq \lambda_L > \lambda_{L+1}$, and the statement holds in this case.
- Subcase 2: $[c_1, \ldots, c_L]$ is complete:
  The proof of this subcase has not been found yet, hence why the statement remains a conjecture.

<u>$Case 2 : \sum_{i=1}^{L} c_i \geq N_L + 2$</u>: If this inequality holds, then as we have shown using the transformations developed in Lemmas 4.13 and 4.14, this implies that $[c_1, \ldots, c_L]$ has principal root at least λ. Applying Lemma 4.16, we see the principal root of $[c_1, \ldots, c_{L+1}]$ is strictly greater, and thus the statement holds in this case.

The results in this section provide us with an efficient way to verify completeness for PLRSs. Namely, for a sequence $[c_1, \ldots, c_L]$, we may evaluate its characteristic polynomials at the points B_L and 2, which provides the following information:

- If $p(2) < 0$, the sequence is incomplete.
- If $p(B_L) > 0$, the sequence is complete.
- If $p(2) \geq 0$ and $p(B_L) \leq 0$, then the principal root of the sequence lies in the interval $[B_L, 2]$, and so further inquiry is necessary to determine whether it is complete.

Computationally, evaluating a polynomial of degree L is an $\mathcal{O}(L^2)$ problem; generating a minimum of $2L$ terms of the sequences and checking Brown's criterion for each, on the other hand, is a $\mathcal{O}(2^L)$ problem. Thus, this method—even if inconclusive—provides a fast and efficient method to categorize sequences, and narrows our search to the interesting interval $[B_L, 2]$, in which both complete and incomplete sequences arise.

Denseness of Incomplete Roots

Having narrowed our search for principal roots of complete and incomplete sequences to the interval $[B_L, 2]$, it is only natural to ask how the roots of these sequences are distributed throughout the interval.

Lemma 4.22 *For fixed $L > 2$ and $k > 0$, define the three polynomials $f(x) = x^L - x^{L-1} - k$, $g(x) = x^L - x^{L-1} - (k+1)$, and $h(x) = x^L - x^{L-1} - (k+2)$. Let q, r, s be the sole positive roots of f, g, h, respectively, so that $1 < q < r < s$. Then,*

$$r - q > s - r. \tag{27}$$

Proof From the definition, we see that

$$\begin{aligned} q^L - q^{L-1} &= k, \\ r^L - r^{L-1} &= k+1, \\ s^L - s^{L-1} &= k+2. \end{aligned} \tag{28}$$

Now, define the polynomial $p(x) = x^L - x^{L-1}$. Taking first and second derivatives of p, we see $p'(x) = Lx^{L-1} - (L-1)x^{L-2}$, and $p''(x) = L(L-1)x^{L-2} - (L-1)(L-2)x^{L-3}$. In particular, for all $x \geq 1$, $p(x) \geq 0$, $p'(x) > 0$, and $p''(x) > 0$.

Thus, $p(x)$ is increasing and convex on $(1, \infty)$. By Eq. (28), we have $p(r) - p(q) = p(s) - p(r)$. Thus, as $s > r > q > 1$, we conclude $r - q > s - r$, as desired.

Theorem 4.23 *For any $L \geq 2$, let R_L be the set of roots of all incomplete PLRSs generated by L coefficients. Then, for any $\varepsilon > 0$, there exists an M such that for all $L > M$ and for any ε-ball $B_\varepsilon \subset (1, 2)$, $B_\varepsilon \cap R_L \neq \varnothing$.*

Proof Let $\varepsilon > 0$ be arbitrary. By Lemma 4.18, we may fix an M such that for all $L > M$, $1 < \lambda_L < 1 + \varepsilon$.

From previous work, we know the sequence of length L with coefficients $[1, 0, \ldots, 0, \lceil L(L+1)/4 \rceil + 1]$ is incomplete, as is any sequence of the form $[1, 0, \ldots, 0, k]$, with $k \geq \lceil L(L+1)/4 \rceil + 1$.

Note that λ_L is the root of $[1, 0, \ldots, 0, \lceil L(L+1)/4 \rceil + 1]$. Since $\lambda_L < 1 + \varepsilon$, it is clear that the root α of $[1, 0, \ldots, 0, \lceil L(L+1)/4 \rceil]$ fulfills $1 < \alpha < \lambda_L$, and so $\lambda - \alpha < \varepsilon$.

Now, we know the sequence $[1, 0, \ldots, 0, 2^{L-1}]$ has a root of size exactly 2. Applying Lemma 4.22 iteratively, any two sequences $[1, 0, \ldots, 0, k]$, $[1, 0, \ldots, 0, k+1]$ with $k \geq \lceil L(L+1)/4 \rceil$ and roots q, r must fulfill $r - q < \lambda_L - \alpha < \varepsilon$. Thus, any two consecutive sequences $[1, 0, \ldots, 0, k]$, $[1, 0, \ldots, 0, k+1]$ with $k \geq \lceil L(L+1)/4 \rceil + 1$ have roots with separation less than ε, and so the set of roots of sequences of the form $[1, 0, \ldots, 0, k]$ with $\lceil L(L+1) 4 \rceil + 1 \leq k \leq 2^{L-1}$ intercepts any ε-ball of $(1, 2)$. As this is a subset of R_L, we are done.

Corollary 4.24 *The set of principal roots of incomplete sequences* $R = \bigcup_{L=2}^{\infty} R_L$ *is dense in* $(1, 2)$.

We conjecture that a similar result can be shown about complete roots; however, this proof has proven more difficult, as examples of families of complete sequences are more fragile.

5 Open Questions

Here are conjectures and several other questions that future researches could investigate.

- Our results often focus on the final coefficient, such as in Theorems 1.13 and 1.14. Do these results have any analogues for coefficients that are not the last?
- Can Theorem 1.17 be extended to address when $g < k$?
- Are there other interesting families of PLRSs that can be fully characterized that have entries other than 0 and 1 as coefficients that are not the final coefficient?
- Are Conjectures 3.2 and 3.7 true?
- Is the missing component of the proof of Conjecture 4.9, i.e., Conjecture 4.21 true?

6 Brown's Criterion and a Corollary

Here are several proofs of important results for our paper. All results will be restated for the reader's convenience.

Theorem 6.1 (Brown [3]) *If* a_n *is a nondecreasing sequence, then* a_n *is complete if and only if* $a_1 = 1$ *and for all* $n > 1$,

$$a_{n+1} \leq 1 + \sum_{i=1}^{n} a_i. \tag{1}$$

Proof Let $\{a_n\}_{n=1}^{\infty}$ be a sequence of positive integers, not necessarily distinct, such that $a_1 = 1$ and

$$a_{n+1} \leq 1 + \sum_{i=1}^{n} a_i \tag{2}$$

for $n \in \{1, 2, \ldots\}$. Then for $0 < n < 1 + \sum_{i=1}^{k} a_i$ there exists $\{b_i\}_{i=1}^{k}$, $b_i \in \{0, 1\}$ such that $n = \sum_{i=1}^{k} b_i a_i$. We proceed by induction on k. The claim obviously holds for $k = 1$, so one may assume that it holds for $k = N$. Hence, we must show that $0 < n < 1 + \sum_{i=1}^{N+1} a_i$ implies the existence of $\{\gamma_i\}_{i=1}^{N+1}$, $\gamma_i \in \{0, 1\}$ such that $n = \sum_{i=1}^{N+1} \gamma_i a_i$. Due to the inductive hypothesis, we only consider values satisfying

$$1 + \sum_{i=1}^{N} a_i \leq n < 1 + \sum_{i=1}^{N+1} a_i. \tag{3}$$

Note that by assumption,

$$n - a_{N+1} \geq 1 + \sum_{i=1}^{N} a_i - a_{N+1} \geq 0. \tag{4}$$

Now, if $n - a_{N+1} = 0$, the conclusion follows. Otherwise,

$$0 < n - a_{N+1} < 1 + \sum_{i=1}^{N} a_i \tag{5}$$

implies the existence of $\{b_i\}_{i=1}^{N}$ such that $n - a_{N+1} = \sum_{i=1}^{N} b_i a_i$. Then the result is immediate on transposing a_{N+1} and identifying $\gamma_i = b_i$ for $i \in \{1, \ldots, N\}$ and $\gamma_{N+1} = 1$. This completes the sufficiency part of the proof.

For the necessity, assume that there exists $n_0 \geq 1$ such that $a_{n_0+1} \geq 1 + \sum_{i=1}^{n_0} a_i$. Then, however,

$$a_{n_0+1} > a_{n_0+1} - 1 > \sum_{i=1}^{n_0} a_i, \tag{6}$$

which implies that the positive integer $a_{n_0+1} - 1$ cannot be represented in the form $\sum_{i=1}^{k} b_i a_i$. This leads to a contradiction and completes the proof.

Corollary 6.2 *If a_n is a nondecreasing sequence such that $a_1 = 1$ and $a_n \leq 2a_{n-1}$ for all $n \geq 2$, then a_n is complete.*

Proof We argue by induction on n that a_n satisfies Brown's criterion when $n \geq 2$. As $a_1 = 1$, for the base case we have

$$a_2 \leq 2a_1 = 2 = a_1 + 1. \tag{7}$$

Now assume for inductive hypothesis that for some $n \geq 2$,

$$a_n \leq a_{n-1} + \cdots + a_1 + 1. \tag{8}$$

Then

$$a_{n+1} \leq 2a_n = a_n + a_n \leq a_n + a_{n-1} + \cdots + a_1 + 1, \tag{9}$$

completing the induction.

Example 6.3 The converse does not hold. A sequence may be complete and have some terms that are larger than the double of the previous term. One such example is the sequence generated by [1, 0, 1, 4], whose terms are {1, 2, 3, 5, 11, ...}. Here, 11 is more than twice 5, yet the sequence is still complete.

7 Lemmas for Sect. 2

Lemma 7.1 *Let $\{G_n\}$, $\{H_n\}$ be sequences defined by $[c_1, \ldots, c_L]$, $[c_1, \ldots, c_L, c_{L+1}]$, respectively, where c_{L+1} is any positive integer. For all $k \geq 2$,*

$$H_{L+k} - G_{L+k} \geq 2\left(H_{L+k-1} - G_{L+k-1}\right). \tag{1}$$

Proof We use strong induction.

We begin with the base case. First, recall that for all n such that $1 \leq n \leq L$, we know $H_n = G_n$. Further, note that

$$H_{L+1} = c_1 H_L + \cdots + c_L H_1 + 1 = c_1 G_L + \cdots + c_L G_1 + 1 = G_{L+1} + 1. \tag{2}$$

Using this fact, we compute

$$\begin{aligned} H_{L+2} &= c_1 H_{L+1} + c_2 H_L + \cdots + c_L H_2 + c_{L+1} H_1 \\ &= c_1\left(G_{L+1} + 1\right) + c_2 G_L + \cdots + c_L G_2 + c_{L+1} \\ &= G_{L+1} + c_1 + c_{L+1}. \end{aligned} \tag{3}$$

Thus, we have that

$$H_{L+2} - G_{L+2} = c_1 + c_{L+1} \geq 2 = 2(1) = 2\left(H_{L+1} - G_{L+1}\right). \tag{4}$$

For the inductive step, suppose for some m, the lemma holds for all $2 \leq k \leq m - 1$. We wish to show it holds for m, i.e.,

$$H_{L+m} - G_{L+m} \geq 2\left(H_{L+m-1} - G_{L+m-1}\right). \tag{5}$$

Expanding the terms using the recurrence definition, we see

$$H_{L+m} - G_{L+m} \geq 2\left(H_{L+m-1} - G_{L+m-1}\right) \tag{6}$$

which holds if and only if

$$\sum_{i=1}^{L} c_i H_{L+m-i} - \sum_{i=1}^{L} c_i G_{L+m-i} \geq 2\left(\sum_{i=1}^{L} c_i H_{L+m-1-i} - \sum_{i=1}^{L} c_i G_{L+m-1-i}\right). \tag{7}$$

Note that for all $i \geq m$, $H_{L+m-i} - G_{L+m-i} = 0$. We cancel out any such terms on both sides of the inequality above, simplifying to

$$\sum_{i=1}^{\min(m-1,L)} c_i \left(H_{L+m-i} - G_{L+m-i}\right) \geq \sum_{i=1}^{\min(m-1,L)} 2c_i \left(H_{L+m-1-i} - G_{L+m-1-i}\right). \quad (8)$$

We encourage the reader to note that for $m - 1 \leq L$ we preserve in the right-hand sum the term $2c_{m-1}\left(H_L - G_L\right) = 0$, so that both sides of the inequality have the same number of summands.

By our inductive hypothesis, we see that for all i,

$$c_i (H_{L+m-i} - G_{L+m-i}) \geq 2c_i \left(H_{L+m-1-i} - G_{L+m-1-i}\right). \quad (9)$$

Thus, inequality 8 holds, which completes the proof.

Lemma 7.2 *Consider sequences* $\{G_n\} = [c_1, c_2, \ldots, c_L]$ *and* $\{H_n\} = [c_1, c_2, \ldots, k_L]$, *where* $1 \leq k_L \leq c_L$. *For all* $k \in \mathbb{N}$,

$$H_{L+k+1} - 2H_{L+k} \leq G_{L+k+1} - 2G_{L+k}. \quad (10)$$

Proof We proceed by strong induction on k. For $k = 1$, we have

$$\begin{aligned} H_{L+2} - 2H_{L+1} &= (c_1 H_{L+1} + c_2 H_L + \cdots + k_L H_2) - 2\,(c_1 H_L + c_2 H_{L-1} + \cdots + k_L H_1) \\ &= (c_1 H_{L+1} + c_2 G_L + \cdots + k_L G_2) - 2\,(c_1 G_L + c_2 G_{L-1} + \cdots + k_L G_1) \\ &= G_{L+2} - (G_{L+1} - H_{L+1}) - (2c_L - 2k_L) - 2G_{L+1} - 2\,(c_L - k_L) \\ &\leq G_{L+2} - 2G_{L+1}. \end{aligned} \quad (11)$$

Assume the statement holds true for a natural number k. Now, note

$$\begin{aligned} &H_{L+k+2} - 2H_{L+k+1} \\ &\quad = (c_1 H_{L+k+1} + c_2 H_{L+k} + \cdots + k_L H_{k+2}) \\ &\quad - 2\,(c_1 H_{L+k} + c_2 H_{L+k-1} + \cdots + k_L H_{k+1}) \\ &\quad = c_1\,(H_{L+k+1} - 2H_{L+k}) + c_2\,(H_{L+k} - 2H_{L+k-1}) + \cdots + k_L\,(H_{k+2} - 2h_{k+1}) \\ &\quad \leq c_1\,(H_{L+k+1} - 2H_{L+k}) + c_2\,(H_{L+k} - 2H_{L+k-1}) + \cdots + c_L\,(H_{k+2} - 2H_{k+1}). \end{aligned}$$

By the inductive hypothesis,

$$\begin{aligned} &\leq c_1\,(G_{L+k+1} - 2G_{L+k}) + c_2\,(G_{L+k} - 2G_{L+k-1}) + \cdots + c_L\,(G_{k+2} - 2G_{k+1}) \\ &= G_{L+k+2} - 2G_{L+k+1}. \end{aligned} \quad (12)$$

Therefore, the statement holds by induction.

Lemma 7.3 *Let* $\{G_n\}$ *be the sequence defined by* $[c_1, \ldots, c_L]$*, and let* $\{H_n\}$ *be the sequence defined by* $[c_1, \ldots, c_{L-1}+1,\ c_L - 1]$*. Then, for all* $k \geq 0$*,*

$$H_{L+k+1} - G_{L+k+1} \geq 2\,(H_{L+k} - G_{L+k})\,. \tag{13}$$

Proof We use strong induction. We begin with the base case. First, since the first $L-2$ coefficients of $\{G_n\}$, $\{H_n\}$ are equivalent, we have that for all $1 \leq n \leq L-1$, $G_n = H_n$. We also see that

$$\begin{aligned} H_L &= c_1 H_{L-1} + \cdots + (c_{L-1}+1)\,H_1 + 1 = c_1 G_{L-1} + \cdots + (c_{L-1}+1)\,G_1 + 1 \\ &= G_L + G_1 = G_L + 1. \end{aligned} \tag{14}$$

Moreover,

$$\begin{aligned} H_{L+1} &= c_1 H_L + \cdots + (c_{L-1}+1)\,H_2 + (c_L - 1)\,H_1 \\ &= c_1\,(G_L + 1) + \cdots + (c_{L-1}+1)\,G_2 + (c_L - 1)\,G_1 \\ &= c_1 + G_2 - G_1 + \textstyle\sum_{i=1}^{L} c_i G_{L+1-i} \\ &= c_1 + c_1 + G_{L+1} \\ &= 2c_1 + G_{L+1}. \end{aligned} \tag{15}$$

Thus, we see that

$$H_{L+1} - G_{L+1} = 2c_1 \geq 2 = 2(1) = 2\,(H_L - G_L)\,, \tag{16}$$

and so the base case holds.

For the induction step, suppose our lemma holds for all $0 \leq k \leq m$. We wish to show this holds for $m+1$, so that $H_{L+m+1} - G_{L+m+1} \geq 2\,(H_{L+m} - G_{L+m})$.

Since $\{G_n\}$ and $\{H_n\}$ are PLRS, we expand the terms in questions using their respective recurrence relations to see that $H_{L+m+1} - G_{L+m+1} \geq 2\,(H_{L+m} - G_{L+m})$ if and only if

$$\begin{aligned} &\sum_{i=1}^{L} c_i H_{L+m+1-i} + H_{m+2} - H_{m+1} - \sum_{i=1}^{L} c_i G_{L+m+1-i} \\ &\qquad \geq 2\left(\sum_{i=1}^{L} c_i H_{L+m-1} + H_{m+1} - H_m - \sum_{i=1}^{L} c_i G_{L+m-i}\right). \end{aligned} \tag{17}$$

We note that by the induction hypothesis, we have that for all i,

$$c_i\,(H_{L+m+1-i} - G_{L+m+1-i}) \geq 2c_i\,(H_{L+m-i} - G_{L+m-i})\,. \tag{18}$$

Moreover, we have that $H_{m+2} - H_{m+1} \geq H_{m+1} - H_m$, simply because we know that gaps in a PLRS grow. Combining these two statements, we have that inequality 17 holds, and so our inductive step is complete.

Lemma 7.4 *Let* $\{G_n\}$ *be the sequence defined by* $[c_1, \ldots, c_{L-1}, 1]$, *and let* $\{H_n\}$ *be the sequence defined by* $[c_1, \ldots, c_{L-1} + 1]$. *Then, for all* $k \geq 1$,

$$H_{L+k+1} - G_{L+k+1} \geq 2\,(H_{L+k} - G_{L+k})\,. \tag{19}$$

Proof The proof is similar to that of Lemma 7.3 and so we repeat our use of strong induction.

We begin with the base case. First, since first $L - 2$ coefficients of $\{G_n\}$, $\{H_n\}$ are equivalent, we have that for all $1 \leq n \leq L - 1$, $G_n = H_n$. In fact, even more can be said. $G_L = H_L$, as

$$\begin{aligned} H_L &= c_1 H_{L-1} + \cdots + (c_{L-1} + 1)\, H_1 \;=\; c_1 G_{L-1} + \cdots + (c_{L-1} + 1)\, G_1 \\ &= (G_L - 1) + G_1 \;=\; G_L. \end{aligned} \tag{20}$$

Hence,

$$\begin{aligned} H_{L+1} &= c_1 H_L + \cdots + (c_{L-1} + 1) H_2 = c_1 G_L + \cdots + (c_{L-1} + 1)\, G_2 \\ &= G_{L+1} - G_1 + G_2 = G_{L+1} - (1) + (c_1 + 1) = G_{L+1} + c_1. \end{aligned} \tag{21}$$

And so we see that

$$H_{L+1} - G_{L+1} = c_1 > 0 = 2\,(H_L - G_L)\,. \tag{22}$$

For the induction step, suppose for some m that our lemma holds for all $0 \leq k \leq m$. We wish to show this holds for $m + 1$, so that $H_{L+m+1} - G_{L+m+1} \geq 2\,(H_{L+m} - G_{L+m})$.

Since $\{G_n\}$ and $\{H_n\}$ are PLRS, we expand the terms in question using their respective recurrence relations. On this basis, we can claim that $H_{L+m+1} - G_{L+m+1} \geq 2\,(H_{L+m} - G_{L+m})$ if and only if

$$\begin{aligned} &\sum_{i=1}^{L-1} c_i H_{L+m+1-i} + H_{m+2} - \sum_{i=1}^{L-1} c_i G_{L+m+1-i} - G_{m+1} \\ &\qquad \geq 2\left(\sum_{i=1}^{L-1} c_i H_{L+m-i} + H_m - \sum_{i=1}^{L} c_i G_{L+m-i} - G_{m-1}\right). \end{aligned} \tag{23}$$

By the induction hypothesis, we have that for all i,

$$c_i H_{L+m+1-i} - c_i G_{L+m+1-i} \geq 2\left(c_i H_{L+m-i} - G_{L+m-i}\right). \tag{24}$$

However, we can also show that $H_{m+2} - G_{m+1} \geq 2\left(H_{m+1} - G_m\right)$. By rewriting this as $H_{m+2} - 2H_{m+1} \geq G_{m+1} - 2G_m$, we see that for $m \leq L-1$, both sides are equal. For $m \geq L+1$, it suffices to note that $\{H_n\}$ grows faster, and thus so must the gaps between consecutive terms. By combining these two observations, inequality (23) holds, which completes the proof.

8 Lemmas for Sect. 3

Lemma 8.1 *For the PLRS $H_{n+1} = H_n + NH_{n-k-1}$, with $N = \lceil (k+2)(k+3)/4 \rceil$, then*

$$(N-2)H_{n-k-1} \leq H_{n-1} + \cdots + H_{n-k}. \tag{1}$$

Proof By induction on n. Consider the base case, for $n = k+2$: $H_{n-k-1} = H_1 = 1$, $H_{n-k} = H_2 = 2, \ldots, H_{n-1} = H_{k+1} = k+1$.

$$\begin{aligned}
(N-2)\, H_{n-k-1} \leq H_{n-1} + \cdots + H_{n-k} &\iff (N-2) \leq 2 + 3 + \cdots + k + (k+1) \\
&\iff \left\lfloor \frac{(k+2)\,(k+3)}{4} + \frac{1}{2} \right\rfloor \leq \frac{(k+1)\,(k+2)}{2} + 1 \\
&\Longleftarrow \frac{(k+2)\,(k+3)+2}{4} \leq \frac{k^2+3k+2}{2} + 1 \\
&\iff k^2 + 5k + 8 \leq 2k^2 + 6k + 8 \\
&\iff 0 \leq k^2 - k.
\end{aligned} \tag{2}$$

Hence, the base case holds for $k \geq 0$.

For the induction hypothesis, assume the following holds for arbitrary, fixed n:

$$(N-2)H_{n-k-1} \leq H_{n-1} + \cdots + H_{n-k}. \tag{3}$$

For the induction step, we wish to show the following:

$$(N-2)H_{n-k} \leq H_n + \cdots + H_{n-k+1}. \tag{4}$$

We write, using the recurrence relation, that

$$(N-2)H_{n-k+1} = (N-2)H_{n-k} + N(N-2)H_{n-2k-2}.$$

And by the induction hypothesis,

$$
\begin{aligned}
(N-2)H_{n-k+1} &\leq H_{n-1} + \cdots + H_{n-k} + N(H_{n-k-2} + \cdots + H_{n-2k-1}) \\
&= \sum_{i=1}^{k} H_{n-i} + NH_{n-k-1-i} \\
&= \sum_{i=1}^{k} H_{n-i+1}. \qquad (5)
\end{aligned}
$$

Hence, the claim is true for all $n \geq k+1, k \geq 0$.

Lemma 8.2 *Let* $N = \lceil L(L+1)/4 \rceil$, *and consider the recurrence relation* $[1, c_2, \ldots, c_{L-2}, 0, N]$ *where* $c_i = 1$ *for one* $i \in \{2, \ldots, L-2\}$, *and the rest are* 0. *For fixed* $i \in \{2, \ldots, L-2\}$, *and* $L \geq 6$ *then* $H_{L-i+1} + (N-2)H_1 \leq H_{L-1} + \cdots + H_2$.

Proof This is equivalent to showing that

$$
H_{L-i+1} + \left\lceil \frac{L(L+1)}{4} \right\rceil \leq H_{L-1} + \cdots + H_2 + H_1 + 1. \qquad (6)
$$

Note that $i \in \{2, \ldots, L-2\}$ and recall that each term in the sequence must increase from the previous. So $H_{L-i+1} \in \{H_{L-1}, \ldots, H_3\}$, and the largest possibility is $H_{L-i+1} = H_{L-1}$ when $i = 2$. Thus,

$$
H_{L-i+1} + \left\lceil \frac{L(L+1)}{4} \right\rceil \leq H_{L-1} + \left\lceil \frac{L(L+1)}{4} \right\rceil. \qquad (7)
$$

So we need to only show that

$$
H_{L-1} + \left\lceil \frac{L(L+1)}{4} \right\rceil \leq H_{L-1} + \cdots + H_2 + H_1 + 1, \qquad (8)
$$

which is equivalent to

$$
\left\lceil \frac{L(L+1)}{4} \right\rceil \leq H_{L-2} + \cdots + H_2 + H_1 + 1. \qquad (9)
$$

Note that $H_i \geq i$, so for $L \geq 6$,

$$\begin{aligned} H_{L-2} + \cdots + H_1 + 1 &\geq L - 2 + \cdots + 1 + 1 \\ &= \frac{(L-2)(L-1)}{2} + 1 \\ &= \frac{(L-2)(L-1)+1}{2} + \frac{1}{2} \\ &\geq \frac{L(L+1)}{4} + \frac{1}{2} \\ &\geq \left\lceil \frac{L(L+1)}{4} \right\rceil. \end{aligned} \tag{10}$$

Thus, we have shown Eq. (9), and the proof is complete.

Lemma 8.3 *Let* $N = \lceil L(L+1)/4 \rceil$, *and consider the recurrence relation* $[1, c_2, \ldots, c_{L-2}, 0, N]$ *where* $c_i = 1$ *for one* $i \in \{2, \ldots, L-2\}$, *and the rest are* 0. *For fixed* $i \in \{2, \ldots, L-2\}$, *and* $L \geq 6$, *then for any* $n \geq L$,

$$H_{n-i+1} + (N-2)H_{n-L+1} \leq H_{n-1} + \cdots + H_{n-L+2}.$$

Proof By induction on n. The base case, $n = L$, was shown in Lemma 8.2. For the induction hypothesis, assume the claim in true for fixed n. We wish to show that the claim holds for $n+1$.

$$\begin{aligned} H_{n-i+2} + (N-2)H_{n-L+2} = H_{n-i+1} + H_{n-2i+2} + NH_{n-L-i+2} + \\ + (N-2)(H_{n-L+1} + H_{n-L-i+2} + NH_{n-2L+2}). \end{aligned}$$

By applying the induction hypothesis,

$$\begin{aligned} H_{n-i+2} + (N-2)H_{n-L+2} &\leq (H_{n-1} + \cdots + H_{n-L+2}) + (H_{n-i} + \cdots + H_{n-L-i+3}) \\ &\quad + N(H_{n-L} + \cdots + H_{n-2L+3}) \\ &= H_n + \cdots + H_{n-L+3}. \end{aligned} \tag{11}$$

Hence, the claim is true for all $n \geq L$.

Lemma 8.4 *For the PLRS* $\{H_n\}$ *generated by* $[1, 1, 0, \ldots, 0, N]$, *if* k *is the number of zeros and* $N = \lfloor (F_{k+6} - (k+5))/4 \rfloor$, *then*

$$(N-2)H_{n-k-2} \leq H_{n-2} + \cdots + H_{n-k-1}. \tag{12}$$

Proof In a similar manner to Lemma 8.1, the statement is proved by induction on n.

For the base case, $n = k+3$, so that $H_{n-k-2} = H_1$ and $\{H_{n-1}, \ldots, H_{n-k-1}\} = \{H_{k+2}, \ldots, H_2\}$. Hence

$$
\begin{aligned}
(N-2)H_1 \leq H_{k+2} + \cdots + H_2 &\iff (N-2) \leq (F_{k+3}-1) + \cdots + (F_3 - 1) \\
&\iff (N-2) \leq \sum_{i=1}^{k+3} F_i - (F_1 + F_2 + (k+1)) \\
&\iff \left\lfloor \frac{F_{k+6} - (k+5)}{4} \right\rfloor \leq \frac{F_{k+6} - (k+5)}{4} \\
&\qquad \leq F_{k+5} - (k+4) \\
&\iff F_{k+6} - (k+5) \leq 4(F_{k+5} - (k+4)) \\
&\iff F_{k+5} + F_{k+4} \leq 4F_{k+5} - 3k + 11 \\
&\iff 3k + 11 \leq 3F_{k+5} - F_{k+4} \\
&\iff 3k + 11 \leq 2F_{k+4} + 3F_{k+3}, \qquad (13)
\end{aligned}
$$

where the last line is true for all $k \geq 0$ by induction on k. Now, suppose the statement is true for some $n \geq k+3$, so that $(N-2)H_{n-k-2} \leq H_{n-2} + \cdots + H_{n-k-1}$. Thus, for the inductive step, note using the recursive definition that

$$
(N-2)H_{n-k-1} = (N-2)H_{n-k-2} + (N-2)H_{n-k-3} + N(N-2)H_{n-2k-4}. \quad (14)
$$

By the inductive hypothesis,

$$
\begin{aligned}
(N-2)H_{n-k-1} &\leq H_{n-2} + \cdots + H_{n-k-1} + (N-2)H_{n-k-3} + N(N-2)H_{n-2k-4} \\
&\leq \sum_{i=n-k}^{n-2} H_i + H_{n-k-1} + (N-2)H_{n-k-3} + N(N-2)H_{n-2k-4}.
\end{aligned} \quad (15)
$$

Note for all positive k we have that $n-k-1 \leq n-2$ as well as $n-k-3 < n-3$ and $n-2k-4 < n-k-3$. Since the sequence is nondecreasing and N is a positive integer,

$$
(N-2)H_{n-k-1} \leq \sum_{i=n-k+1}^{n-2} H_i + NH_{n-2} + NH_{n-3} + N(N-2)H_{n-k-3} = \sum_{i=n-k}^{n-1} H_i, \quad (16)
$$

which completes the induction on n.

Lemma 8.5 $\sum_{i=1}^{n} 2^{i-1} i = 2^n(n-1)+1$.

Proof By induction on n.

Lemma 8.6 *Define* $\{F_n\} = [\underbrace{1, \ldots, 1}_{g}]$ *and* $\{H_n\} = [\underbrace{1, \ldots, 1}_{g}, \underbrace{0, \ldots, 0}_{k}, 2^{k+1}-1]$. *Then* $H_{g+k+1+n} = F_{g+k+1+n} + (2^{k+1}-1)(2^n + 2^{n-2}(n-1))$ *when* $1 \leq n \leq g-k$.

Proof Define $a(n)$ so that $H_{g+k+1+n} = F_{g+k+1+n} + a(n)$ for $1 \leq n \leq g-k$. Now,

$$\begin{aligned} H_{g+k+1+n} &= H_{g+k+n} + \cdots + H_{k+1+n} + (2^{k+1}-1)H_n \\ &= \sum_{i=k+1+n}^{g+k+n} F_i + \sum_{i=1}^{n-1} a(i) + \sum_{i=1}^{k+1} 2^{i-1} + (2^{k+1}-1)2^{n-1} \end{aligned} \tag{17}$$

(since $k+1+n \leq g+1$, $(H_i - F_i)$ spans all the indices from $k+1+n \leq g+1$ to $g+k+n$)

$$= F_{g+k+n+1} + \sum_{i=1}^{n-1} a(i) + (2^{k+1}-1)(2^{n-1}+1). \tag{18}$$

Therefore,

$$a(n) = \sum_{i=1}^{n-1} a(i) + (2^{k+1}-1)(2^{n-1}+1) \tag{19}$$

and

$$a(n-1) = \sum_{i=1}^{n-2} a(i) + (2^{k+1}-1)(2^{n-2}+1). \tag{20}$$

Hence

$$a(n) = 2a(n-1) + (2^{k+1}-1)(2^{n-2}). \tag{21}$$

Since $a(1) = 2(2^{k+1}-1)$, by induction we have

$$a(n) = (2^{k+1}-1)(2^n + 2^{n-2}(n-1)). \tag{22}$$

Lemma 8.7 *Define* $\{F_n\} = [\underbrace{1, \ldots, 1}_{g}]$ *and* $\{H_n\} = [\underbrace{1, \ldots, 1}_{g}, \underbrace{0, \ldots, 0}_{k}, 2^{k+1}-1$. *Then* $F_{g+n} = 2^{g+n-1} - 2^{n-2}(n+1)$ *when* $1 \leq n \leq g$.

Proof Set $F_{g+n} = 2^{g+n-1} - a(n)$ for $1 \leq n \leq g$. Then

$$\begin{aligned} F_{g+n} &= F_{g+n-1} + \cdots + F_n \\ &= 2^{g+n-2} + 2^{g+n-3} + \cdots + 2^{n-1} - (a(n-1) + \cdots + a(1)) \\ &= 2^{g+n-1} - \left(2^{n-1} + \sum_{i=1}^{n-1} a(i)\right). \end{aligned} \tag{23}$$

Therefore,

$$a(n) = 2^{n-1} + \sum_{i=1}^{n-1} a(i) \tag{24}$$

and

$$a(n-1) = 2^{n-2} + \sum_{i=1}^{n-2} a(i). \tag{25}$$

Hence

$$a(n) = 2^{n-1} + 2a(n-1) - 2^{n-2} = 2a(n-1) + 2^{n-2}. \tag{26}$$

Since $a(1) = 1$, by induction, we have $a(n) = 2^{n-2}(n+1)$.

Lemma 8.8 *For* $k + \lceil \log_2 k \rceil \leq g < 2k$, $\{H_n\}$ *defined as* $[\underbrace{1, \ldots, 1}, \underbrace{0, \ldots, 0}_{g}, 2^{k+1} - 1]$, *we have*

$$2^{g+k+1} - \sum_{i=k+n+2}^{g+k+1} H_i \leq 2^g + 2^{k+n+2} - 2^{n+1} \tag{27}$$

for all $g - k \leq n \leq k$.

Proof By induction on n. Suppose it holds for some $n \geq g - k$. Then

$$2^{g+k+1} - \sum_{i=k+n+3}^{g+k+1} H_i = 2^{g+k+1} - \sum_{i=k+n+2}^{g+k+1} H_i + H_{k+n+2}.$$

By the induction hypothesis,

$$2^{g+k+1} - \sum_{i=k+n+3}^{g+k+1} H_i \leq 2^g + 2^{k+n+2} - 2^{n+1} + H_{k+n+2}.$$

As we can check explicitly that $H_{k+n+2} \leq 2^{k+n+1} \leq 2^{k+n+2} - 2^{n+1}$, we see

$$\begin{aligned} 2^{g+k+1} - \sum_{i=k+n+3}^{g+k+1} H_i &\leq 2^g + 2^{k+n+2} - 2^{n+1} + (2^{k+n+2} - 2^{n+1}) \\ &= 2^g + 2^{k+n+3} - 2^{n+2}. \end{aligned} \tag{28}$$

It remains to show for the base case $n = g - k$. This can be shown directly from the given formulas in Theorem 3.5.

Lemma 8.9 *For* $\{H_n\}$ *defined as in Lemma 8.8 and the same conditions on g and k, we have*

$$H_{(g+k+1)+n} \geq (2^{k+1} - 1)(2^{g+n-1} - 2^{n-2}(n+1)) \tag{29}$$

for all $1 \leq n \leq k$.

Proof This is equivalent to showing that

$$2^{g+k+n} - H_{(g+k+1)+n} \leq 2^{g+n-1} + 2^{k+n-1}(n+1) - 2^{n-2}(n+1) \quad \text{for all } 1 \leq n \leq k, \tag{30}$$

which we proceed by strong induction on n. The case $1 \leq n \leq g-k$ has been established in Theorem 3.5, so we suppose this holds for all $n \leq m$ for some $g-k \leq m < k$. Then

$$\begin{aligned}
&2^{g+k+(m+1)} - H_{(g+k+1)+(m+1)} \\
&= 2^{g+k+m+1} - \left(\sum_{i=1}^{m} H_{(g+k+1)+i} + \sum_{i=k+m+2}^{g+k+1} H_i + (2^{k+1}-1)H_{m+1} \right) \\
&= \sum_{i=1}^{m} \left(2^{g+k+i} - H_{(g+k+1)+i}\right) + \left(2^{g+k+1} - \sum_{i=k+m+2}^{g+k+1} H_i \right) - (2^{k-1}-1)2^m.
\end{aligned}$$

By the inductive hypothesis and Lemma 8.8,

$$\begin{aligned}
&\leq \sum_{i=1}^{m} \left(2^{g+i-1} + 2^{k+i-1}(i+1) - 2^{i-2}(i+1)\right) + \left(2^g + 2^{k+m+2} - 2^{m+1}\right) \\
&\quad - (2^{k+1}-1)2^m \\
&= 2^{g+(m+1)-1} + 2^{k+(m+1)-1}((m+1)+1) - 2^{(m+1)-2}((m+1)+1). \qquad (31)
\end{aligned}$$

Our proof by induction is complete.

Lemma 8.10 *The sequence generated by* $[1, \ldots, 1, 0, 3]$, *with* $k \geq 1$ *ones, is always complete.*

Proof By strong induction on n.

For the base case, with $n = 1$, we have $H_2 = H_1 + 1$.

For the induction hypothesis, assume that for some n, Brown's criterion holds for all $m < n$, i.e., assume $H_{m+1} \leq 1 + H_1 + \cdots + H_m$ for all $m < n$.

For the induction step, we start with the recurrence relation and apply the induction hypothesis:

$$\begin{aligned}
H_{n+1} &= H_n + \cdots + H_{n-k+1} + 3H_{n-k-1} \\
&\leq H_n + \cdots + H_{n-k+1} + H_{n-k} + 2H_{n-k-1} \\
&\leq H_n + \cdots + H_{n-k+1} + H_{n-k} + H_{n-k-1} + H_{n-k-2} + \cdots + H_1 + 1. \qquad (32)
\end{aligned}$$

Hence, by Brown's criterion, the sequence is complete. By strong induction, the lemma is proved.

Lemma 8.11 *Let* $\{H_n\}$ *defined by* $[1, 0, \ldots, 0, \underbrace{1, \ldots, 1}_{m}, N]$ *be a PLRS with* L *coefficients. Then, if the sequence is incomplete, it must fail Brown's criterion* *at the* $L+1$*-th or* $L+2$*-th term. In other words, if* $H_{L+1} \leq 1 + \sum_{i=1}^{L} H_i$ *and* $H_{L+2} \leq 1 + \sum_{i=1}^{L+1} H_i$*, then* $\{H_n\}$ *is complete.*

Proof Let $\{H_n\}$ be defined as above; it is clear that the first L terms pass Brown's criterion. Now suppose the sequence passes Brown's criterion at the $L+1$-ist and $L+2$-nd term, so that

$$H_{L+1} \leq \sum_{i=1}^{L} H_i + 1, \quad H_{L+2} \leq \sum_{i=1}^{L+1} H_i + 1. \tag{33}$$

We show that $\{H_n\}$ is complete.

We show by induction that if $\{H_n\}$ satisfies Brown's criterion at the $L+2$nd term, then it satisfies Brown's criterion at the $L+k$th term, for any $2 \leq k \leq L-1$. We assume our base case of $k=2$ by hypothesis, so only the induction step remains to be shown.

Suppose for some k that

$$H_{L+k} = H_{L+k-1} + H_{k+m} + \cdots + H_{k+1} + NH_k \leq \sum_{i=1}^{L+k-1} H_i + 1. \tag{34}$$

We wish to show that

$$H_{L+k+1} = H_{L+k} + H_{k+m+1} + \cdots + H_{k+2} + NH_{k+1} \leq \sum_{i=1}^{L+k} H_i + 1. \tag{35}$$

Looking at the difference between Eqs. (34) and (35), we see it suffices to show that

$$(H_{L+k} - H_{L+k-1}) + N\,(H_{k+1} - H_k) + H_{k+m+1} - H_{k+1} \leq H_{L+k}. \tag{36}$$

Or equivalently,

$$N\,(H_{k+1} - H_k) + H_{k+m+1} - H_{k+1} \leq H_{L+k-1}. \tag{37}$$

Expanding H_{L+k-1},

$$H_{L+k-1} = H_{L+k-2} + H_{k+m-1} + \cdots + H_k + NH_{k-1}. \tag{38}$$

We can repeatedly expand the largest term of expression (38), giving us longer partial sums. In particular, applying the process k times, we see

$$\begin{aligned}
H_{L+k-1} &= H_{L+k-2} + \sum_{i=k}^{k+m-1} H_i + NH_{k-1} \\
&= \left(H_{L+k-3} + \sum_{i=k-1}^{k+m-2} H_i + NH_{k-2} \right) + \sum_{i=k}^{k+m-1} H_i + NH_{k-1} \\
&= \left(H_{L+k-3} + \sum_{i=k-2}^{k+m-3} H_i + NH_{k-3} \right) + \sum_{i=k-1}^{k+m-2} H_i + \sum_{i=k}^{k+m-1} H_i + N(H_{k-1} + H_{k-2}) \\
&\vdots \\
&= H_{L-1} + \sum_{i=1}^{m} H_i + \cdots + \sum_{i=k}^{k+m-1} H_i + N\left(H_{k-1} + H_{k-2} + \cdots + H_1\right) \\
&= H_{L-1} + \sum_{a=1}^{k} \sum_{i=a}^{a+m-1} H_i + N\left(\sum_{i=1}^{k-1} H_i \right). \qquad (39)
\end{aligned}$$

Thus, inequality (37) becomes

$$N\left(H_{k+1} - H_k - H_{k-1} - \cdots - H_2 - H_1\right) + H_{k+m+1} \leq H_{L-1} + H_{k+1} + \sum_{a=1}^{k} \sum_{i=a}^{a+m-1} H_i. \qquad (40)$$

Assuming $k < L$, we can write $H_{k+1} = H_k + H_{k-L+m+1} + \cdots + H_{k-L+2}$, where for the sake of notation we define $H_0 = 1$, and $H_j = 0$ for all $j < 0$. Thus,

$$\begin{aligned}
N\left(H_{k+1} - H_k - H_{k-1} - \cdots - H_2 - H_1\right) &= -N\left(H_{k-1} + H_{k-2} + \cdots + H_{k-L+m+2}\right) \\
&< -NH_{k-1}, \qquad (41)
\end{aligned}$$

and so for inequality (40) it suffices to show

$$H_{k+m+1} \leq H_{L-1} + H_{k+1} + \sum_{a=1}^{k} \sum_{i=a}^{a+m-1} H_i + NH_{k-1}. \qquad (42)$$

Expanding the left-hand side, we wish to show

$$\begin{aligned}
H_{k+m+1} &= H_{k+m} + H_{k-L+2m+1} + \cdots + H_{k-L+1} + NH_{k-L+m} \\
&\leq H_{L-1} + H_{k+1} + \sum_{a=1}^{k} \sum_{i=a}^{a+m-1} H_i + NH_{k-1}. \qquad (43)
\end{aligned}$$

As $k - L + m \leq k - 3 < k - 1$, we see $NH_{k-L+m} < NH_{k-1}$, and so we need only to show

$$H_{k+m} + H_{k-L+2m+1} + \cdots + H_{k-L+1} \le H_{L-1} + H_{k+1} + \sum_{a=1}^{k} \sum_{i=a}^{a+m-1} H_i. \tag{44}$$

As $k \ge 2$, we note that in the double sum $\sum_{a=1}^{k} \sum_{i=a}^{a+m-1} H_i$, the summands H_1, H_{k+m-1} are present exactly once, and for any $1 < i < k+m-1$, the summand H_i is present at least twice. Thus, we can take the crude bound

$$\sum_{a=1}^{k} \sum_{i=a}^{a+m-1} H_i \ge \sum_{i=1}^{k+m-1} H_i + \sum_{i=2}^{k+m-2} H_i. \tag{45}$$

Applying this bound on the right-hand side of inequality (44), and taking the trivial bounds $H_{L-1} > H_1 + 1$, $H_{k+1} > 1$, we see

$$H_{L-1} + H_{k+1} + \sum_{a=1}^{k} \sum_{i=a}^{a+m-1} H_i \ge \left(\sum_{i=1}^{k+m-1} H_i + 1 \right) + \left(\sum_{i=1}^{k+m-2} H_i + 1 \right). \tag{46}$$

As we assumed $\{H_n\}$ fulfills Brown's criterion for terms below $L + k$, we know

$$H_{k+m} \le \sum_{i=1}^{k+m-1} H_i + 1. \tag{47}$$

Finally, it is clear that

$$H_{k-L+2m+1} + \cdots + H_{k-L+1} \le \sum_{i=1}^{k+m-2} H_i + 1, \tag{48}$$

as no indices on the left sum are repeated. Combining these two facts, we have

$$\begin{aligned} H_{k+m} + H_{k-L+2m+1} + \cdots + H_{k-L+1} &\le \left(\sum_{i=1}^{k+m-1} H_i + 1 \right) + \left(\sum_{i=1}^{k+m-2} H_i + 1 \right) \\ &\le H_{L-1} + H_{k+1} + \sum_{a=1}^{k} \sum_{i=a}^{a+m-1} H_i. \end{aligned} \tag{49}$$

Thus inequality (44) holds, and we are done.

Lemma 8.12 *Let* $\{G_n\}$ *and* $\{H_n\}$ *be PLRS with L coefficients defined by* $[1, 0, \dots, 0, \underbrace{1, \dots, 1}_{m}, M]$ *and* $[1, 0, \dots, 0, \underbrace{1, \dots, 1}_{m+1}, N]$*, respectively. For* $(L-1)/2 \le m \le L-4$,

$$\begin{cases} H_{i-1} = G_i - 1 & 2 \leq i < 2(L-m) \\ H_{i-1} = G_i & i = 2(L-m) \\ H_{i-1} > G_i & 2(L-m) < i \leq L. \end{cases} \tag{50}$$

Proof From the proof of Lemma 3.10, we get

$$\begin{cases} H_n = n,\ H_{(L-m-1)+n} = L - m - 1 + n + \frac{n(n+1)(n+2)}{6} & 1 \leq n \leq L - m - 1 \\ G_n = n,\ G_{L-m+n} = L - m + n + \frac{n(n+1)(n+2)}{6} & 1 \leq n \leq L - m. \end{cases} \tag{51}$$

From these explicit formulas, $H_{i-1} = G_i - 1$ for all $2 \leq i \leq 2(L-m) - 1$. Now,

$$\begin{aligned} H_{2(L-m)-1} &= H_{2(L-m-1)} + H_{L-m} + \sum_{i=1}^{L-m-1} H_i + 1 \\ &= 2(L-m-1) + \frac{(L-m-1)(L-m)(L-m+1)}{6} + 1 + \sum_{i=1}^{L-m-i} G_i + 1 \\ &= 2(L-m) - 1 + \frac{(L-m-1)(L-m)(L-m+1)}{6} + \sum_{i=1}^{L-m-i} G_i + 1 \\ &= G_{2(L-m)-1} + \sum_{i=1}^{L-m-i} G_i + 1 \\ &= G_{2(L-m)}. \end{aligned} \tag{52}$$

Similarly, by writing out explicit formulas, one can show that $H_{2(L-m)} > G_{2(L-m)+1}$. Also, it is clear that $H_i \geq G_i$ for all i. Therefore, for any $2(L-m) + 1 < k \leq L$,

$$\begin{aligned} H_{k-1} - G_k &= (H_{k-2} - G_{k-1}) + \sum_{i=1}^{k-(L-m)} (H_i - G_i) \\ &\geq H_{k-2} - G_{k-1}, \end{aligned} \tag{53}$$

and the last inequality follows by induction on k.

Lemma 8.13 *If* $m + 3 < 2(L-m)$ *and* $m \geq (L-1)/2$, *then*

$$2m - L + \sum_{i=L-m+2}^{m+1} i \geq 2L - 3(m+1) + 2. \tag{54}$$

Proof This is equivalent to

$$\frac{(m+1)(m+2)}{2} - \frac{(L-m+1)(L-m+2)}{2} \geq 3L - 5m - 1, \tag{55}$$

which simplifies to

$$L(2m - L) + 16m + 2 \geq 8L, \tag{56}$$

which is true since $L(2m - L) > 6$ and $2m + 1 \geq L$.

9 Lemmas for Sect. 4

Lemma 9.1 *For any $L \in \mathbb{Z}^+$, let λ_L be the principal root of $x^L - x^{L-1} - N_L - 1$. Then*

$$(\lambda_L - 1) \geq \frac{L+2}{L^2 + L + 4}. \tag{1}$$

Proof Since $1/(L^2 + L + 4) > 1/(L^2 + 4L + 4) = 1/(L+2)^2$, it suffices to show

$$(\lambda_L - 1) \geq \frac{L+2}{L^2 + 4L + 4} = \frac{1}{L+2}, \tag{2}$$

or equivalently, $\lambda_L \geq (L+3)/(L+2)$. This inequality holds if and only if $f((L+3)/(L+2)) \leq 0$, i.e.,

$$\left(\frac{L+3}{L+2}\right)^L - \left(\frac{L+3}{L+2}\right)^{L-1} - \left\lceil \frac{L\,(L+1)}{4} \right\rceil - 1 \leq 0, \tag{3}$$

or,

$$\frac{(L+3)^{L-1}}{(L+2)^L} \leq \left\lceil \frac{L\,(L+1)}{4} \right\rceil + 1. \tag{4}$$

It can be checked that for all $L \geq 1$, a stronger condition holds, that

$$\frac{(L+3)^{L-1}}{(L+2)^L} < 1, \tag{5}$$

completing the proof.

References

1. Olivia Beckwith, Amanda Bower, Louis Gaudet, Rachel Insoft, Shiyu Li, Steven J. Miller, and Philip Tosteson. *The Average Gap Distribution for Generalized Zeckendorf Decompositions*, Fibonacci Quarterly **51** (2013), 13–27.
2. Elżbieta Bołdyriew, John Haviland, Phúc Lâm, John Lentfer, Steven J. Miller, and Fernando Trejos Suárez. *Introduction to Completeness of Positive Linear Recurrence Sequences*, Fibonacci Quarterly **5** (2020), 77–90.

3. J. L. Brown. *Note on complete sequences of integers*, American Mathematical Monthly **68** (1961), no. 6, 557.
4. Aviezri S. Fraenkel, *Systems of numeration*, American Mathematical Monthly **92** (1985), no. 2, 105–114.
5. P. J. Grabner and R. F. Tichy, *Contributions to digit expansions with respect to linear recurrences*, J. Number Theory **36** (1990), no. 2, 160–169.
6. Daniele A. Gewurz and Francesca Merola, *Numeration and enumeration*, European Journal of Combinatorics **32** (2012), no. 7, 1547–1556.
7. V. E. Hoggatt and C. King, problem E 1424, American Mathematical Monthly **67** (1960), no. 6, 593.
8. Thomas C. Martinez, Steven J. Miller, Clay Mizgerd, Jack Murphy, and Chenyang Sun. *Generalizing Zeckendorf's Theorem to Homogeneous Linear Recurrences II*, Fibonacci Quarterly, to appear (2020).
9. S. J. Miller and Y. Wang, *From Fibonacci numbers to Central Limit Type Theorems*, Journal of Combinatorial Theory, Series A **119** (2012), no. 7, 1398–1413.
10. E. Zeckendorf, *Représentation des nombres naturels par une somme des nombres de Fibonacci ou de nombres de Lucas*, Bulletin de la Société Royale des Sciences de Liège **41** (1972), 179–182.

Length Density and Numerical Semigroups

Cole Brower, Scott Chapman, Travis Kulhanek, Joseph McDonough, Christopher O'Neill, Vody Pavlyuk, and Vadim Ponomarenko

1 Introduction

A numerical semigroup S is an additively closed subset of $\mathbb{Z}_{\geq 0}$ containing 0, usually specified using a generating set $n_1, \ldots, n_k$, i.e.,

$$S = \langle n_1, \ldots, n_k \rangle = \{z_1 n_1 + z_2 n_2 + \cdots + z_k n_k \mid z_1, \ldots, z_k \in \mathbb{Z}_{\geq 0}\}.$$

A *factorization* of an element of S is an expression of the form $z_1 n_1 + \cdots + z_k n_k$. Many classical problems surrounding factorizations in semigroup theory involve so-called factorization invariants, which are arithmetic quantities, often combinatorial in nature, that capture some precise aspect of non-uniqueness. For instance, the elasticity invariant equals the quotient of the maximum and minimum factorization lengths of an element $n \in S$, and the delta set invariant $\Delta_S(n)$ contains the "gaps" in the set of possible factorization lengths of n.

C. Brower · J. McDonough · C. O'Neill · V. Pavlyuk · V. Ponomarenko
Mathematics Department, San Diego State University, San Diego, CA 92182, USA
e-mail: joe.mcdonough@live.com

C. O'Neill
e-mail: cdoneill@sdsu.edu

V. Ponomarenko
e-mail: vponomarenko@sdsu.edu

S. Chapman (✉)
Department of Mathematics and Statistics, Sam Houston State University, Huntsville, TX 77341, USA
e-mail: scott.chapman@shsu.edu

T. Kulhanek
Mathematics Department, University of California Los Angeles, Los Angeles, CA 90095, USA

M. B. Nathanson (ed.), *Combinatorial and Additive Number Theory V*,
Springer Proceedings in Mathematics & Statistics 395,
https://doi.org/10.1007/978-3-031-10796-2_3

A recent paper [10] introduced a new invariant, known as *length density*, that measures the sparseness of the length set, defined in such a way that $\mathrm{LD}_S(n) = 1$ if and only if the length set of n is a full interval. The length density of any numerical semigroup S has the following immediate upper and lower bounds [10, Proposition 2.3]:

$$\frac{1}{\max \Delta(S)} \leq \mathrm{LD}(S) \leq \frac{1}{\min \Delta(S)}.$$

It is not hard to see that the second inequality above is strict precisely when $|\Delta(S)| > 1$. The strictness of the first inequality, on the other hand, turns out to be more nuanced. With this in mind, we introduce and study the following property: we say S is *tasty*[1] if $\mathrm{LD}(S) > 1/\max \Delta(S)$, and *bland* otherwise. In order for S to be bland, some $n \in S$ must have $\Delta_S(n) = \{\max \Delta(S)\}$, a phenomenon that has been studied in the context of Krull monoids [8]. As such, determining when S is tasty necessitates sufficient control over the delta sets of elements of S.

The aim of the present manuscript is to investigate how common the tasty and bland properties are among numerical semigroups. Our results are as follows. After reviewing the necessary background in Sect. 2, we provide in Sect. 3 an algorithm for computing $\mathrm{LD}(S)$ for any numerical semigroup S and characterize the asymptotic behavior of length density for large elements of S. In the remaining sections, we examine length density, and, in particular, the tastiness property, for several well-studied families of numerical semigroups. Several interesting consequences are also obtained as given below:

- Any numerical semigroup with at most three generators has its length density achieved at a Betti element (Theorem 4.5). Note that a four-generated numerical semigroup without this property is known (Example 4.6).
- If S has maximal embedding dimension and prime multiplicity, then the delta set of any Betti element containing $\max \Delta(S)$ must be a singleton. This resembles a property of certain Krull monoids [8] that generally does not hold for numerical semigroups.

2 Background

We open with a series of definitions.

Definition 2.1 A *numerical semigroup* is a subset of $\mathbb{Z}_{\geq 0}$ of the form

$$S = \langle n_1, \ldots, n_k \rangle = \{z_1 n_1 + \cdots + z_k n_k : z_1, \ldots, z_k \in \mathbb{Z}_{\geq 0}\}$$

for the semigroup generated by $n_1, \ldots, n_k$. A *factorization* of $n \in S$ is an expression

[1] This term was chosen since the McNugget semigroup is tasty; see Example 3.1.

$$n = z_1 n_1 + \cdots + z_k n_k$$

of n as a sum of generators of S, and the *length* of a factorization is the sum $z_1 + \cdots + z_k$. The *set of factorizations* of n is the set

$$\mathsf{Z}_S(n) = \{z \in \mathbb{Z}_{\geq 0}^k : n = z_1 n_1 + \cdots + z_k n_k\}$$

viewed as a subset of $\mathbb{Z}_{\geq 0}^k$, and the *length set* of n is the set

$$\mathsf{L}_S(n) = \{z_1 + \cdots + z_k : z \in \mathsf{Z}_S(n)\},$$

of all possible factorization lengths of n. Writing $\mathsf{L}_S(n) = \{\ell_1 < \cdots < \ell_r\}$, define

$$\Delta_S(n) = \{\ell_i - \ell_{i-1} : 2 \leq i \leq r\} \quad \text{and} \quad \Delta(S) = \bigcup_{n \in S} \Delta_S(n)$$

as the *delta sets* of n and S, respectively.

Definition 2.2 Fix a numerical semigroup $S = \langle n_1, \ldots, n_k \rangle$. The *factorization homomorphism* of S is the function $\varphi_S : \mathbb{Z}_{\geq 0} \to S$ given by

$$\varphi_S(z_1, \ldots, z_k) = z_1 n_1 + \cdots + z_k n_k$$

sending each k-tuple to the element of S it is a factorization of. The *kernel* of φ_S is the equivalence relation $\sim\, = \ker \varphi_S$ that sets $z \sim z'$ whenever $\varphi_S(z) = \varphi_S(z')$. The kernel $\sim$ is in fact a *congruence*, meaning that $z \sim z'$ implies $(z + z'') \sim (z' + z'')$ for all $z, z', z'' \in \mathbb{N}^k$. A subset $\rho \subset \ker \varphi_S$, viewed as a subset of $\mathbb{N}^k \times \mathbb{N}^k$, is a *presentation* of S if $\ker \varphi_S$ is the smallest congruence on $\mathbb{N}^k$ containing ρ. We say ρ is a *minimal presentation* if it is minimal with respect to containment among presentations for S.

Definition 2.3 The *factorization graph* of an element n of a numerical semigroup S is the graph ∇_n whose vertices are the elements of $\mathsf{Z}_S(n)$ in which two factorizations $z, z' \in \mathsf{Z}_S(n)$ are connected by an edge if $z_i > 0$ and $z'_i > 0$ for some i. If ∇_n is disconnected, we say n is a *Betti element* of S, and we write $\mathrm{Betti}(S)$ for the set of Betti elements of S.

The following theorem is a summary of multiple results which can be found in [1, Sect. 5.3].

Theorem 2.4 *A presentation ρ of a finitely generated semigroup $S = \langle n_0, \ldots, n_k \rangle$ is minimal if and only if for every $n \in S$, (i) the number of connected components in the graph ∇_n is one more than the number of relations in ρ containing factorizations of n and (ii) adding an edge to ∇_n corresponding to each such relation in ρ yields a connected graph.*

Our next definition yields our main property of interest.

Definition 2.5 Fix a numerical semigroup $S = \langle n_1, \ldots, n_k \rangle$. The *length density* of $n \in S$ with $|\mathsf{L}_S(n)| \geq 2$ is given by

$$\mathrm{LD}_S(n) = \frac{|\mathsf{L}(n)| - 1}{\max \mathsf{L}_S(n) - \min \mathsf{L}_S(n)}$$

and the *length density* of S is

$$\mathrm{LD}(S) = \inf\{\mathrm{LD}_S(n) : n \in S,\ |\mathsf{L}_S(n)| \geq 2\}.$$

It is not hard to prove (see [10]) that

$$\frac{1}{\max \Delta(S)} \leq \mathrm{LD}(S) \leq \frac{1}{\min \Delta(S)}.$$

We say S is *tasty* if

$$\mathrm{LD}(S) > \frac{1}{\max \Delta(S)}$$

and *bland* otherwise.

Before moving forward, we list a few basic results from [10] concerning the length density.

Theorem 2.6 ([10, Theorem 3.4 and Proposition 3.2]) *For any numerical semigroup S, some $n \in S$ satisfies $\mathrm{LD}_S(n) = \mathrm{LD}(S)$. Additionally, S is bland if and only if*

$$\mathrm{LD}_S(b) = \frac{1}{\max \Delta(S)}$$

for some $b \in \mathrm{Betti}(S)$, which in particular occurs if $\Delta_S(b) = \{\max \Delta(S)\}$.

3 Asymptotics and Computation

In this section, we characterize asymptotic behavior of $\mathrm{LD}_S(n)$ for a given numerical semigroup S. Like many other factorization invariants [17, 18], for large n, the function $\mathrm{LD}_S(n)$ coincides with a quotient of quasilinear functions of n (Theorem 3.4). The primary consequence is an explicit upper bound on the smallest element of S with length density $\mathrm{LD}(S)$ (Corollary 3.5), and thus an algorithm to compute $\mathrm{LD}(S)$ (Remark 3.6).

Example 3.1 Figure 1 depicts the function $\mathrm{LD}_S(n)$ for $S = \langle 6, 9, 20 \rangle$. Note in particular that S is tasty since $\mathrm{LD}(S) = \frac{4}{7} > 1/\max \Delta(S) = \frac{1}{4}$.

Throughout this section, suppose $S = \langle n_1 < \cdots < n_k \rangle$ is a numerical semigroup with $\gcd(n_1, \ldots, n_k) = 1$, and let

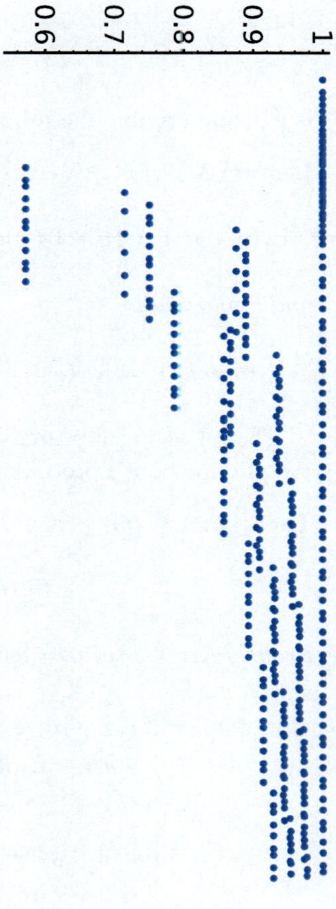

Fig. 1 A plot with a point at $(n, \mathrm{LD}_S(n))$ for each $n \in S = \langle 6, 9, 20 \rangle$

$$d = \min \Delta(S) \qquad \text{and} \qquad L = \frac{n_k - n_1}{d \gcd(n_1, n_k)}.$$

Note $L \in \mathbb{Z}$ since $d = \gcd(n_2 - n_1, \ldots, n_k - n_{k-1})$ by [5, Proposition 2.9].

We begin by recalling some pertinent results from [12], wherein the authors identify a constant N_S, given as a (large) formula in terms of $n_1, \ldots, n_k$, such that

$$\Delta_S(n + \mathrm{lcm}(n_1, n_k)) = \Delta_S(n)$$

for all $n \geq N_S$ [12, Corollary 14]. Letting

$$\begin{aligned}
\mathsf{L}_1(n) &= \{\ell \in \mathsf{L}_S(n) \mid \tfrac{1}{n_1}n + N_S(\tfrac{1}{n_2} - \tfrac{1}{n_1}) < \ell \leq \tfrac{1}{n_1}n\},\\
\mathsf{L}_2(n) &= \{\ell \in \mathsf{L}_S(n) \mid \tfrac{1}{n_k}n + N_S(\tfrac{1}{n_{k-1}} - \tfrac{1}{n_k}) \leq \ell \leq \tfrac{1}{n_1}n + N_S(\tfrac{1}{n_2} - \tfrac{1}{n_1})\},\\
\mathsf{L}_3(n) &= \{\ell \in \mathsf{L}_S(n) \mid \tfrac{1}{n_k}n \leq \ell < \tfrac{1}{n_k}n + N_S(\tfrac{1}{n_{k-1}} - \tfrac{1}{n_k})\}
\end{aligned}$$

so that

$$\mathsf{L}_S(n) = \mathsf{L}_1(n) \cup \mathsf{L}_2(n) \cup \mathsf{L}_3(n)$$

is a disjoint union, the following also follows directly from the work in [12].

Lemma 3.2 *If $n \geq N_S$, then*

$$\mathsf{L}_1(n+n_1) = \{\ell + 1 \mid \ell \in \mathsf{L}_1(n)\} \quad \textit{and} \quad \mathsf{L}_3(n+n_k) = \{\ell + 1 \mid \ell \in \mathsf{L}_3(n)\},$$

and in particular,

$$|\mathsf{L}_1(n+n_1)| = |\mathsf{L}_1(n)|, \qquad |\mathsf{L}_3(n+n_k)| = |\mathsf{L}_3(n)|, \qquad \textit{and} \qquad \Delta(\mathsf{L}_2(n)) = \{d\}.$$

The following appeared as [17, Theorem 4.5] but without an explicit lower bound. We provide here a proof that N_S is such a lower bound.

Theorem 3.3 *If $n \geq N_S$, then*

$$|\mathsf{L}(n + \mathrm{lcm}(n_1, n_k))| = |\mathsf{L}(n)| + L.$$

Proof First, we set $p = \mathrm{lcm}(n_1, n_r)$. Using Lemma 3.2, we have

$$\begin{aligned}
|\mathsf{L}(n+p)| &= |\mathsf{L}_1(n+p)| + |\mathsf{L}_2(n+p)| + |\mathsf{L}_3(n+p)| \\
&= |\mathsf{L}_1(n)| + |\mathsf{L}_2(n+p)| + |\mathsf{L}_3(n)| \\
&= |\mathsf{L}(n)| + |\mathsf{L}_2(n+p)| - |\mathsf{L}_2(n)| \\
&= |\mathsf{L}(n)| + \tfrac{1}{d}(\max \mathsf{L}_2(n+p) - \min \mathsf{L}_2(n+p)) - \tfrac{1}{d}(\max \mathsf{L}_2(n) - \min \mathsf{L}_2(n)) \\
&= |\mathsf{L}(n)| + \tfrac{1}{d}(\min \mathsf{L}_1(n+p) - \max \mathsf{L}_3(n+p)) - \tfrac{1}{d}(\min \mathsf{L}_1(n) - \max \mathsf{L}_3(n)) \\
&= |\mathsf{L}(n)| + \tfrac{1}{d}(\min \mathsf{L}_1(n) - \max \mathsf{L}_3(n) + dL) - \tfrac{1}{d}(\min \mathsf{L}_1(n) - \max \mathsf{L}_3(n)) \\
&= |\mathsf{L}(n)| + L
\end{aligned}$$

as desired. □

Theorem 3.4 *If $n \geq N_S$, then*

$$\mathrm{LD}(n + \mathrm{lcm}(n_1, n_k)) = \frac{|\mathsf{L}(n)| - 1 + L}{\max \mathsf{L}(n) - \min \mathsf{L}(n) + dL}.$$

In particular, $\mathrm{LD}(n)$ is a quotient of eventually quasilinear functions of n, each with period dividing $\mathrm{lcm}(n_1, n_k)$.

Proof First, let $p = \mathrm{lcm}(n_1, n_k)$. It follows from [4, Theorems 4.2 and 4.3] that

$$\max \mathsf{L}(n+n_1) = \max \mathsf{L}(n) + 1 \qquad \text{and} \qquad \min \mathsf{L}(n+n_k) = \min \mathsf{L}(n) + 1$$

hold for all $n \geq n_k^2$, and since $N_S > n_k^2$, one readily checks that

$$\begin{aligned}\mathsf{LD}(n+p) &= \frac{|\mathsf{L}(s+p)|-1}{\max(\mathsf{L}(s+p))-\min(\mathsf{L}(s+p))}\\ &= \frac{|\mathsf{L}(s)|-1+L}{(\max(\mathsf{L}(s))+\frac{1}{n_1}p)-(\min(\mathsf{L}(s))+\frac{1}{n_k}p)}\\ &= \frac{|\mathsf{L}(s)|-1+L}{\max(\mathsf{L}(s))-\min(\mathsf{L}(s))+dL}\end{aligned}$$

holds for all $n \geq N_S$. □

Our final result of the section is a corollary of Theorem 3.4 that is topological in nature, reminiscent of [9, Theorems 2.1–2.2 and Corollary 2.3] and [4, Corollary 4.5] concerning the set of elasticities of a numerical semigroup.

Corollary 3.5 *We have*

$$\mathsf{LD}(S) = \min\{\mathsf{LD}(n) \mid n \in S \text{ and } n < N_S + \operatorname{lcm}(n_1, n_k)\}.$$

Additionally, letting $R(S) = \{\mathsf{LD}(n) : n \in S\}$, *the set* $R(S) \cap [0, \alpha)$ *is finite for each* $\alpha \in [0, \frac{1}{d}]$, *and the only possible accumulation point of* $R(S)$ *is*

$$\sup R(S) = \lim_{n\to\infty} \mathsf{LD}(n) = \tfrac{1}{d}.$$

Proof All three claims follow from Theorem 3.4 and the observation that

$$\mathsf{LD}(n + \operatorname{lcm}(n_1, n_k)) \geq \mathsf{LD}(n)$$

for all $n \geq N_S$. □

Remark 3.6 Corollary 3.5 yields a method of computing $\mathsf{LD}(S)$ from the generators of S. Indeed, N_S can be immediately computed from the formula in [12, Sect. 3], and the length sets of all $n \leq N_S + \operatorname{lcm}(n_1, n_k)$ can be computed relatively quickly using the methods in [3, Sect. 3]. Additionally, Theorem 3.4 yields an algorithm to compute $\mathsf{LD}_S(n)$ for $n \geq N_S$ whose runtime does not depend on n, since one only needs to compute the length set of an appropriate element between N_S and $N_S + \operatorname{lcm}(n_1, n_k)$.

4 Families of Numerical Semigroups

In this section, we classify the length density and tastiness of numerical semigroups residing in one of several well-studied families.

Supersymmetric Numerical Semigroups

A numerical semigroup is called *supersymmetric* if it has exactly one Betti element. By [13, Theorem 12], this occurs if and only if S can be written in the form

$$S = \langle \tfrac{s}{t_1}, \tfrac{s}{t_2}, \ldots, \tfrac{s}{t_k} \rangle$$

for some pairwise coprime $t_1 > \cdots > t_k$ with $s = t_1 t_2 \cdots t_k$. Moreover, s is the unique Betti element in this case and has length set $\mathsf{L}(s) = \{t_k, \ldots, t_1\}$.

Lemma 4.1 *For any $m \geq 1$, $|\mathsf{L}(ms)| \geq mk - m + 1$.*

Proof We proceed by induction on m. For the base case, notice that

$$|\mathsf{L}(s)| = |\{t_k, \ldots, t_1\}| \geq (1)k - (1) + 1.$$

Next, assuming $|\mathsf{L}(ms)| \geq mk - m + 1$, we want to show

$$|\mathsf{L}((m+1)s)| \geq (m+1)n - (m+1) + 1.$$

Since $(m+1)s = ms + s$, it follows that

$$T := \mathsf{L}(ms) + t_k \subseteq \mathsf{L}((m+1)s)$$

has at least $mn - m + 1$ elements by our inductive hypothesis. Let ℓ denote the greatest element of $\mathsf{L}(ms)$, and fix a factorization $(a_1, \ldots, a_k)$ of ms of length ℓ. Then for each i, $(a_1, \ldots, a_i + t_i, \ldots, a_k)$ is a length $\ell + t_i$ factorization of $(m+1)s$. Since $\ell + t_k = \max T$,

$$\ell + t_1, \ldots, \ell + t_{k-1} \in \mathsf{L}((m+1)s) \setminus T,$$

from which we conclude

$$|\mathsf{L}((m+1)s)| \geq (mk - m + 1) + k - 1 = (m+1)k - (m+1) + 1,$$

as desired. □

Theorem 4.2 *If $n \in S$ has at least two factorizations, then* $\mathrm{LD}(n) \geq \mathrm{LD}(s)$. *Moreover,*

$$\mathrm{LD}(S) = \mathrm{LD}(s) = \frac{k-1}{t_1 - t_k},$$

and S is tasty if and only if $t_1, t_2, \ldots, t_k$ does not form an arithmetic progression.

Proof By [13, Theorem 12], we can write $n = ms + r$ with $m \geq 1$ such that $r \in S$ has unique factorization, say with length ℓ. In this case, $\mathsf{L}(n) = \mathsf{L}(ms) + \ell$. Since $\max \mathsf{L}(ms) = mt_1$ and $\min \mathsf{L}(ms) = mt_n$, Lemma 4.1 yields

$$\mathrm{LD}(n) = \mathrm{LD}(ms) = \frac{|\mathsf{L}(ms)| - 1}{mt_1 - mt_k} \geq \frac{(mk - m + 1) - 1}{mt_1 - mt_k} = \frac{k - 1}{t_1 - t_k} = \mathrm{LD}(s).$$

The final claim now follows from the fact that $\Delta(s)$ is a singleton if and only if

$$\mathsf{L}(s) = \{t_k, \ldots, t_1\}$$

forms an arithmetic progression. □

Embedding Dimension 3 Numerical Semigroups

Numerical semigroups with three generators have been well studied, and it is known that their factorization structure is largely determined by whether there are 1, 2, or 3 Betti elements. Throughout this subsection, we use the notation from [19, Chap. 9] and refer the reader there for a thorough overview of this dichotomy.

Proposition 4.3 *If* $S = \langle n_1, n_2, n_3 \rangle$ *has three Betti elements, then* S *is bland.*

Proof Each $b \in \mathrm{Betti}(S)$ has exactly two factorizations, which, after relabeling the generators accordingly, have the form

$$b = c_1 n_1 = r_2 n_2 + r_3 n_3.$$

As such, either $\Delta(b) = \emptyset$ or $\Delta(b) = \{\delta\}$, where $\delta = |r_2 + r_3 - c_1|$. Thus, S is bland. □

Proposition 4.4 *If* $S = \langle n_1, n_2, n_3 \rangle$ *with* $\mathrm{Betti}(S) = \{b_1, b_2\}$ *and* $b_1 < b_2$, *then* S *is tasty if and only if* $\max \Delta(b_2) > \max \Delta(b_1)$ *and* $b_2 - b_1 \in S$.

Proof For the backward direction, suppose $\max \Delta(b_2) > \max \Delta(b_1)$ and $b_2 - b_1 \in S$. We must have $\max \Delta(S) = \max \Delta(b_2)$ by [7, Theorem 2.5], but since $b_2 - b_1 \in S$, the trade defined at b_1 can be used at b_2 to obtain an element of $\Delta(b_2)$ at most $\max \Delta(b_1)$. As such, $|\Delta(b_2)| \geq 2$, so S is tasty.

Conversely, if $\max \Delta(b_2) \leq \max \Delta(b_1)$, then S is bland since $|\Delta(b_1)| = 1$, and if $b_2 - b_1 \notin S$, then b_1 and b_2 both have singleton delta sets, so again S is bland. This completes the proof. □

Theorem 4.5 *If* $S = \langle n_1, n_2, n_3 \rangle$, *then* $\mathrm{LD}(S)$ *occurs at a Betti element.*

Proof If S has one Betti element, then S is supersymmetric, so apply Theorem 4.2. The claim clearly holds if S is bland, so by Propositions 4.3 and 4.4 it suffices to assume $\mathrm{Betti}(S) = \{b_1, b_2\}$, $b_2 - b_1 \in S$, and $\max \Delta(b_2) > \max \Delta(b_1)$. Write $S = \langle n_1, n_2, n_3 \rangle$,

$$b_1 = c_1 n_1 = c_2 n_2 \quad \text{and} \quad b_2 = c_3 n_3 = r_1 n_1 + r_2 n_2,$$

and let

$$\delta_1 = c_1 - c_2 \quad \text{and} \quad \delta_2 = r_1 + r_2 - c_3.$$

Notice $t_1 = (c_1, -c_2, 0)$ and $t_2 = (r_1, r_2, -c_3)$ form a minimal presentation for S. Assume $c_1 > c_2$, and the two given factorizations of b_2 have extremal lengths in $\mathsf{L}(b_2)$ (or, equivalently, that $|\delta_2|$ is maximal among choices for the trade t_2).

Fix $n \in S$ not uniquely factorable. If $n - b_2 \notin S$, then only the trade t_1 is available, so $\Delta(n) = \{\delta_1\}$ and thus $\mathrm{LD}(n) = \mathrm{LD}(b_1)$. As such, suppose $n - b_2 \in S$. Fix a factorization $z = (z_1, z_2, z_3) \in \mathsf{Z}(n)$ with z_3 maximal, and write $z_3 = qc_3 + r$ with $q, r \in \mathbb{Z}_{\geq 0}$ and $r < c_3$. Since t_2 is the only trade involving n_3, we must have

$$\{y_3 : (y_1, y_2, y_3) \in \mathsf{Z}(n)\} = \{r, r + c_3, \ldots, r + qc_3\}.$$

Moreover, let $\ell = z_1 + z_2 + z_3 - c_3$, and let

$$L = \mathsf{L}(b_2) + \{\ell, \ell + \delta_2, \ldots, \ell + (q-1)\delta_2\}.$$

By repeatedly performing the trade t_2 to z, we see $L \subseteq \mathsf{L}(n)$, and by the maximality of $|\delta_2|$, L is a union of translations of $\mathsf{L}(b_2)$ with matching endpoints. In particular,

$$\mathrm{LD}(L) = \mathrm{LD}(b_2) > \mathrm{LD}(b_1) = \tfrac{1}{\delta_1}.$$

The key observation is now that by the maximality of $|\delta_2|$, every length in $\mathsf{L}(n) \setminus L$ is attained by a factorization obtained from performing the trade t, perhaps multiple times, to some factorization whose length lies in L. As such, letting $m = |\mathsf{L}(n) \setminus L|$, we obtain

$$\mathrm{LD}(n) \geq \frac{|L| + m}{\max L - \min L + m\delta_1} \geq \frac{|L|}{\max L - \min L} = \mathrm{LD}(b_2),$$

as desired. □

Example 4.6 The conclusion of Theorem 4.5 can fail if S has four or more generators. Indeed, the semigroup $S = \langle 20, 28, 42, 73 \rangle$, which appeared as [10, Example 3.3], has Betti element length sets $\mathsf{L}(84) = \{2, 3\}$, $\mathsf{L}(140) = \{4, 5, 7\}$, and $\mathsf{L}(146) = \{2, 4, 5\}$, but a strictly smaller length density results from $\mathsf{L}(202) = \{4, 6, 7, 9\}$.

Maximal Embedding Dimension Numerical Semigroups

We say S is *maximal embedding dimension* (or *MED*) if $e(S) = m(S)$. In this subsection, we write

$$S = \langle m, n_1, \ldots, n_{m-1} \rangle$$

and assume $n_i \equiv i \bmod m$ for each i. It is known in this case that

$$\mathrm{Betti}(S) = \{n_i + n_j : 1 \le i \le j \le m-1\};$$

we refer the reader to [19, Sect. 7.4].

We begin by identifying a class of Betti elements whose delta sets are singletons.

Proposition 4.7 *Suppose $b \in \mathrm{Betti}(S)$ is the smallest Betti element in the equivalence class of $k \in [0, m-1]$ modulo m. If $k = 0$ or k is a unit in $\mathbb{Z}_m$, then $|\mathsf{L}(b)| \le 2$.*

Proof First, suppose $k = 0$, so that b is the smallest Betti element of S divisible by m. We claim any factorization of b involving n_i with $1 \le i \le m-1$ must have length 2 (and thus must be the factorization $n_i + n_{m-i}$). Indeed, by the minimality of b,

$$n_i + n_{m-i} - b = cm$$

for some $c \ge 0$, so any factorization of b of length 3 or more involving n_i would yield a factorization of n_{m-i} of length at least 2, which is impossible since n_{m-i} is a minimal generator of S. This proves $\mathsf{L}(b) = \{2, a\}$, where $b = am$.

In all remaining cases, $b \equiv k \bmod m$ with $1 \le k \le m-1$. Write

$$b = am + n_k = n_i + n_{k-i},$$

where $a \ge 2$ and $1 \le i \le m-2$. By similar reasoning as above, any factorization involving n_i with $i \ne k$ must be the factorization $n_i + n_{k-i}$ by the minimality of b. As such, since k is a unit in $\mathbb{Z}_m$, any factorization involving n_k other than $am + n_k$ must contain $(m+1)n_k$, which is impossible since

$$(m+1)n_k = cn_k + (m+1-c)n_k > n_i + n_{k-i} \ge b$$

for some c. This again yields the desired claim. □

Theorem 4.8 *If $m \ge 3$ is prime, then S is bland.*

Proof Fix $a \ge 1$ so that am is the smallest Betti element that is a multiple of m. By relabeling the generators of S using an appropriately chosen automorphism of $\mathbb{Z}_m$, it suffices to assume $am = n_1 + n_{m-1}$. In what follows, we show

$$\max \Delta(S) = a - 2,$$

which completes the proof since $\Delta(am) = \{a-2\}$ by Proposition 4.7.

First, fix $b' = n_j + n_{m-j} \in \mathrm{Betti}(S)$, and suppose $b' > am$. Let

$$b = n_i + n_{m-i} \in \mathrm{Betti}(S)$$

in such a way that i is maximal such that $i < j$ and $b < b'$ (note $i = 1$ satisfies both constraints, so the set of eligible i is nonempty). By induction on b', we can assume $\max \Delta(b) \le a - 2$. Fix $c, c' \ge 1$ such that

$$b' = cm + n_i + n_{m-i} \qquad \text{and} \qquad n_{i+1} + n_{m-i-1} = c'm + n_i + n_{m-i}.$$

Since n_{i+1} is a minimal generator of S, $n_i + n_1 > n_{i+1}$ and thus $n_i + n_1 \ge n_{i+1} + m$. Similarly, $n_{m-i} + n_{m-1} \ge n_{m-i-1} + m$, and we obtain

$$c'm = n_{i+1} + n_{m-i-1} - (n_i + n_{m-i}) \le n_1 + n_{m-1} - 2m = m(a-2),$$

so by maximality of i, we have $c \le c' \le a - 2$. As such, from

$$\{2\} \cup (c + 2 + \mathsf{L}(b)) \subseteq \mathsf{L}(b'),$$

we conclude $\max \Delta(b') \le a - 2$.

Next, fix $k \in [1, m-1]$. Letting $b = n_1 + n_{k-1} = cm + n_k$, for some c, we see

$$cm = n_1 + n_{k-1} - n_k \le n_1 + n_{m-1} - m = m(a-1).$$

This means $\mathsf{L}(b) \subset [2, c+1] \subset [2, a]$, so $\max \Delta(b) \le a - 2$. More generally, let $b' = n_j + n_{k-j}$ with $1 \le j \le k - j \le k$, and let $b = n_i + n_{j-i}$ with i maximal subject to $i < j$ and $b < b'$. By induction on b', we can assume $\max \Delta(b) \le a - 2$. Fix c, c' so that

$$b' = cm + n_i + n_{k-i} \qquad \text{and} \qquad n_{i+1} + n_{k-i-1} = c'm + n_i + n_{k-i}.$$

We obtain

$$c'm = n_{i+1} + n_{k-i-1} - (n_i + n_{k-i}) \le n_1 + n_{m-1} - 2m = m(a-2),$$

so by the maximality of i, we have $c \le c' \le a - 2$. As before, we have

$$\{2\} \cup (c + 2 + \mathsf{L}(b)) \subseteq \mathsf{L}(b'),$$

and can conclude $\max \Delta(b') \le a - 2$. A similar argument for $b' = n_{k+j} + n_{m-j}$ with $k + 1 \le k + j \le m - j \le m - 1$ completes the proof. □

Proposition 4.9 *There exist both bland and tasty MED numerical semigroups of each composite multiplicity.*

Proof The numerical semigroup $S = \langle m, m+1, \ldots, 2m-1 \rangle$ is bland since it is arithmetical and thus has singleton delta set by [2, Theorem 2.2].

Now, if $m = pq$ with p prime and $q \geq 2$, then we claim $S = \langle m, n_1, \ldots, n_{m-1} \rangle$ with

$$n_i = \begin{cases} m + i & \text{if } p \mid i; \\ 2qm + i & \text{if } p \nmid i \end{cases}$$

is tasty. Indeed, consider a Betti element $b = n_i + n_j$. For convenience in what follows, we write $n_{i+j} = n_{i+j-m}$ if $i + j > m$. If $p \mid i$, then $p \mid j$ if and only if $p \mid (i + j)$, so

$$n_i + n_j = \begin{cases} m + n_{i+j} & \text{if } i + j < m; \\ 2m + n_{i+j-m} & \text{if } i + j > m, \end{cases}$$

and thus $\mathsf{L}(b) \subseteq \{2, 3\}$. Alternatively, if $p \nmid i$ and $p \nmid j$, then $b = n_{i+j} + cm$ with $c \geq 2q$, so the trade $qn_p = (q + 1)m$ can be performed at least once, meaning $\min \Delta(b) = 1$.

The above argument implies each $b \in \text{Betti}(S)$ has either $\Delta(b) = \emptyset$ or $1 \in \Delta(b)$. Lastly, the Betti element $n_1 + n_{m-1} = (2B + 1)m$ has no factorizations of length 3, since any such factorization can use only $m, n_p, \ldots, n_{m-p}$ but $3n_{m-p} < n_1 + n_{m-1}$. Thus, $\max \Delta(S) > 1$ and occurs at a Betti element with non-singleton delta set. □

We close this section by examining multiplicity four MED numerical semigroups, where geometry plays a role in determining whether each semigroup is bland or tasty.

Given $n_1, n_2, n_3 > 4$ with $n_i \equiv i \bmod 4$ for each i, the semigroup $S = \langle 4, n_1, n_2, n_3 \rangle$ is MED if and only if

$$2n_1 > n_2 \qquad n_1 + n_2 > n_3, \qquad n_2 + n_3 > n_1, \qquad \text{and} \qquad 2n_3 > n_2$$

each hold [15]. As such, it is natural to represent each MED numerical semigroup $S = \langle 4, n_1, n_2, n_3 \rangle$ as a point $(n_1, n_2, n_3) \in \mathbb{R}^3$. Examining semigroups with fixed coordinate sum $n_1 + n_2 + n_3$ yields a cross section as depicted on the left in Fig. 2. The two regions labeled "bland" coincide with the semigroups where $\min(n_1, n_2, n_3)$ equals n_1 and n_3, respectively (Proposition 4.10). For the remaining semigroups, if $n_1 + n_3$ is sufficiently larger than n_2, then S is guaranteed tasty by (Theorem 4.12) and lies in the region labeled "tasty" on the left. This phenomenon is also visible in the plot on the right in Fig. 2, which depicts the cross section $n_2 = 18$, placing a large (red) point at (n_1, n_3) if S is bland and a smaller (black) point if S is tasty.

We note that the geometric viewpoint discussed above, first outlined in [15], has proven fruitful in recent years for studying enumerative questions surrounding numerical semigroups; we direct the interested reader to [6, 14].

Proposition 4.10 *If* $m = 4$ *and* $\min(n_1, n_2, n_3) \neq n_2$*, then* S *is bland.*

Proof Throughout this proof, we assume $n_1 = \min(n_1, n_2, n_3)$, as an analogous argument follows upon reversing the roles of n_1 and n_3 throughout.

The Betti elements $n_1 + n_2$ and $n_2 + n_3$ are unique in their respective equivalence classes and have singleton delta set by Proposition 4.7. Consider the Betti elements

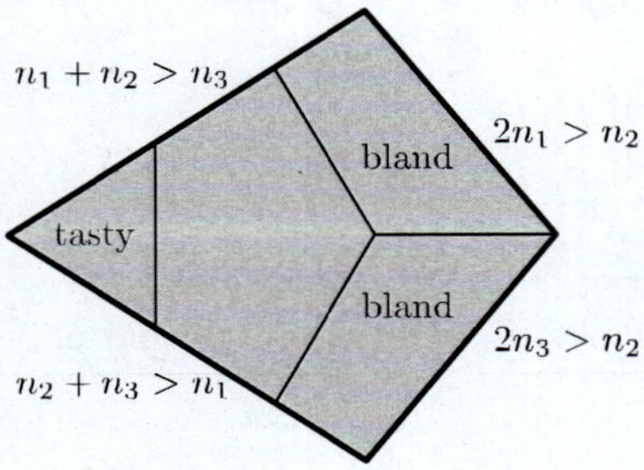

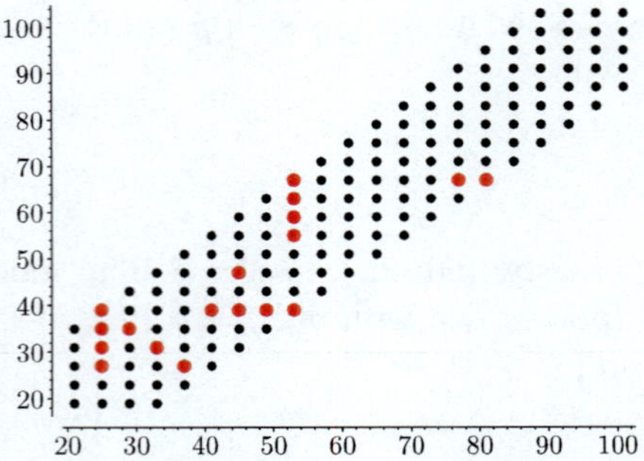

Fig. 2 A diagram (left) of a cross section of MED numerical semigroups $S = \langle 4, n_1, n_2, n_3 \rangle$ with $n_1 + n_2 + n_3$ fixed, and a plot (right) obtained by setting $n_2 = 18$ and placing a large (red) point at (n_1, n_3) if S is bland and a smaller (black) point if S is tasty

$b = 4a$ and $b' = 4(a + c)$ with $a, c \geq 0$. Again by Proposition 4.7, b has singleton delta set, and for b', there are two cases. First, suppose

$$b = 4a = n_1 + n_3 \quad \text{and} \quad b' = 4(a + c) = 4c + n_1 + n_3 = 2n_2.$$

Clearly $\mathsf{L}(b') \subseteq [2, a + c]$, so $\max \Delta(b') \leq \max(c, a - 2)$. We know $\Delta(b) = \{a - 2\}$, and since n_2 is a minimal generator of S,

$$2n_1 \geq n_2 + 4 \quad \text{and} \quad 2n_3 \geq n_2 + 4,$$

so we obtain

$$4c = 2n_2 - (n_1 + n_3) \leq n_1 + n_3 - 8 = 4(a - 2)$$

and thus $c \leq a - 2$. As such, $\max \Delta(b') \leq \max \Delta(b)$. Second, suppose

$$b = 4a = 2n_2 \quad \text{and} \quad b' = 4(a + c) = 4c + 2n_2 = n_1 + n_3.$$

Again, clearly $\mathsf{L}(b') \subseteq [2, a + c]$. Since n_2 and n_3 are minimal generators of S, we have

$$2n_1 \geq n_2 + 4 \quad \text{and} \quad n_1 + n_2 \geq n_3 + 4,$$

from which we obtain $3n_1 \geq n_3 + 8$. From there, $n_1 < n_2$ implies

$$4c = n_1 + n_3 - 2n_2 \leq 4n_1 - 2n_2 - 8 \leq 2n_2 - 8 = 4(a - 2)$$

and we again conclude $c \leq a - 2$ and $\max \Delta(b') \leq \max \Delta(b)$.

This leaves the Betti elements $2n_1$ and $2n_3$. Write

$$2n_1 = n_2 + 4d \quad \text{and} \quad 2n_3 = 2n_1 + 4e$$

for $d, e \geq 1$. Now, since $n_1 < n_2$, we have $4d = 2n_1 - n_2 < n_1 < 4a$, so $d \leq a - 1$. As such, since $\mathsf{L}(2n_1) \subset [2, d+1]$, we have

$$\max \Delta(2n_1) \leq a - 2 = \max \Delta(b).$$

Lastly, from $n_1 + n_2 \geq n_3 + 4$ and $3n_1 \geq n_3 + 8$, we obtain

$$4e = 2n_3 - 2n_1 \leq 2n_2 - 8 \quad \text{and} \quad 4e = 2n_3 - 2n_1 \leq n_1 + n_3 - 8,$$

so $e \leq a - 2$ and thus $\max \Delta(2n_3) \leq a - 2$ as well. □

By examining where in the proof of Proposition 4.10 the hypothesis $n_2 < n_1$ is used, we obtain the following.

Corollary 4.11 *If* $m = 4$, *then* $\max \Delta(S)$ *occurs at* $\max(2n_2, n_1 + n_3)$ *or* $\min(2n_1, 2n_3)$.

Proof By the proof of Proposition 4.10, it suffices to ensure $\max \Delta(S)$ does not occur at

$$b = n_1 + n_2 = 4d + n_3 \quad \text{or} \quad b' = n_2 + n_3 = 4e + n_1.$$

Writing $4a = \min(2n_2, n_1 + n_3)$, summing these equalities above yields

$$4d + 4e = 2n_2 \geq 4a,$$

so $\max \Delta(b) = d - 1 \leq a - 2$ and $\max \Delta(b') = e - 1 \leq a - 2$. □

Theorem 4.12 *Suppose* $m = 4$ *and* $n_2 < n_1, n_3$. *If* $2n_1 + 2n_3 > n_2^2$, *then* S *is tasty. In particular, for fixed* n_2, *there are only finitely many bland numerical semigroups.*

Proof As in the proof of Proposition 4.10, let $2n_2 = 4a$ so that $\mathsf{L}(2n_2) = \{2, a\}$. Let

$$b = n_1 + n_3 = 4c$$

for $c \geq 1$, and write $c = qa + r$ with $0 \leq r < a$. Any factorization of b aside from $n_1 + n_3$ uses only n_2 and 4, the shortest of which is

$$4r + 2qn_2 = 4r + 4aq = 4c = b.$$

This means $\Delta(b) = \{a - 2, 2q + r - 2\}$, so S is tasty if $2q + r \neq a$. This is achieved by

$$4a(2q + r) = 8qa + 4ar > 8qa + 8r = 8c = 2n_1 + 2n_3 > n_2^2 = 4a^2$$

whenever $2n_1 + 2n_3 > n_2^2$.

Supposing $n_1 < n_3$, by Corollary 4.11 it remains to show

$$b' = 2n_1 = 4d + n_2$$

is either non-singleton or does not contain $\max \Delta(S)$. Since $n_1 + n_2 \geq n_3 + 4$, we have

$$8d = 4n_1 - 2n_2 \geq 2n_1 + 2n_3 > n_2^2 > 4n_2 = 8a,$$

meaning some element of $\Delta(b')$ is at most $a - 2$, as desired. An analogous argument when $n_3 < n_1$ completes the proof. □

5 Tasty and Bland Gluings of Numerical Semigroups

Given two numerical semigroups $S_1 = \langle n_1, \ldots, n_r \rangle$ and $S_2 = \langle n_{r+1}, \ldots, n_k \rangle$ and non-atoms $\lambda \in S_1$ and $\mu \in S_2$ such that $\gcd(\lambda, \mu) = 1$, we say the numerical semigroup

$$\mu S_1 + \lambda S_2 = \langle \mu n_1, \ldots, \mu n_r, \lambda n_{r+1}, \ldots, \lambda n_k \rangle$$

is a *gluing* of S_1 and S_2 by λ and μ. It is known that the above generating set for S is minimal, and that

$$\mathrm{Betti}(S) = \mu \mathrm{Betti}(S_1) \cup \lambda \mathrm{Betti}(S_2) \cup \{\lambda\mu\}.$$

For more background on gluings, see [19, Chap. 8].

In this section, we investigate the following question:

If S_1 and S_2 are fixed and μ and λ are allowed to vary, what can be said about how often the gluing $\mu S_1 + \lambda S_2$ is tasty, and how often it is bland?

Our results are extremal in nature. Theorem 5.1 implies there will always be infinitely many tasty gluings $\mu S_1 + \lambda S_2$ for fixed S_1 and S_2, but the same does not hold for bland gluings. In particular, it is possible for two given numerical semigroups S_1 and S_2 to have infinitely many bland gluings (Theorem 5.2), finitely many bland gluings (Theorem 5.3) or no bland gluings (Theorem 5.5).

Theorem 5.1 *Given numerical semigroups S_1 and S_2, there are infinitely many gluings $S = \mu S_1 + \lambda S_2$ that are tasty.*

Proof Let n_k denote the largest atom of S_2, and let $d = \max(\Delta(S_1) \cup \Delta(S_2))$. First, choose any prime $\lambda \in S_1$ such that $|\mathsf{L}_{S_1}(\lambda)| \geq 2$ and $\lambda > n_2$. Next, choose any prime $p > \lambda$ such that

$$p \geq \max \mathsf{L}_{S_1}(\lambda) + d + 1$$

and let $\mu = pn_k$ so that $\min \mathsf{L}_{S_2}(\mu) = p$. Clearly $\gcd(\lambda, \mu) = 1$. Letting $S = \mu S_1 + \lambda S_2$, by [19, Theorem 8.2] we have

$$\mathsf{L}_S(\lambda\mu) = \mathsf{L}_{S_1}(\lambda) \cup \mathsf{L}_{S_2}(\mu)$$

and thus

$$\max \Delta(S) = \max \Delta_S(\lambda\mu) = \min \mathsf{L}_{S_2}(\mu) - \max \mathsf{L}_{S_1}(\lambda) = p - \max \mathsf{L}_{S_1}(\lambda) \geq d + 1,$$

which ensures S is tasty since $|\mathsf{L}_{S_1}(\lambda)| \geq 2$. □

Theorem 5.2 *Fix numerical semigroups S_1 and S_2. If S_1 is bland and*

$$\max \Delta(S_1) \geq \max \Delta(S_2),$$

then there exists infinitely many gluings $\mu S_1 + \lambda S_2$ that are bland.

Proof Let n_k denote the largest atom of S_2. Choose any prime $\lambda \in S_1$ such that

$$\lambda > \max \operatorname{Betti}(S_1), \qquad \lambda > n_2, \qquad \text{and} \qquad \max \mathsf{L}_{S_1}(\lambda) > \max \operatorname{Betti}(S_2),$$

and let $\mu = (\max \mathsf{L}_{S_1}(\lambda))n_k$ so that $\min \mathsf{L}_{S_2}(\mu) = \max \mathsf{L}_{S_1}(\lambda)$. Clearly $\max \mathsf{L}_{S_1}(\lambda) < \lambda$, so $\gcd(\lambda, \mu) = 1$. Letting $S = \mu S_1 + \lambda S_2$, we see $\max \operatorname{Betti}(S) = \lambda\mu$, ensuring

- $\mathsf{L}_S(\mu b) = \mathsf{L}_{S_1}(b)$ for each $b \in \operatorname{Betti}(S_1)$,
- $\mathsf{L}_S(\lambda b) = \mathsf{L}_{S_2}(b)$ for each $b \in \operatorname{Betti}(S_2)$, and
- $\mathsf{L}_S(\lambda\mu) = \mathsf{L}_{S_1}(\lambda) \cup \mathsf{L}_{S_2}(\mu)$ with $\Delta_S(\lambda\mu) \leq \max(\Delta(S_1) \cup \Delta(S_2))$.

We conclude $\max \Delta(S)$ is attained at a Betti element with singleton delta set. □

Theorem 5.3 *Let $S_1 = \langle 2, 3\rangle$ and $S_2 = \langle 6, 9, 20\rangle$. A gluing $S = \mu S_1 + \lambda S_2$ is bland if and only if $\lambda = 4$ and $\mu = 27$.*

Proof Note that $\operatorname{Betti}(S) = \{6\mu, 18\lambda, 60\lambda, \lambda\mu\}$, and that $\Delta_S(6\mu)$, $\Delta_S(18\lambda)$, and $\Delta_S(60\lambda)$ each contain 1.

First, suppose $\mu > 61$. In this case, $\mu\lambda > 60\lambda > 18\lambda$, so $\max \Delta(S) \geq \max \Delta(S_2) = 4$, and $1 \in \Delta_S(\lambda\mu)$ since $1 \in \Delta_{S_2}(\mu)$, so S must be tasty.

Next, suppose $\mu \leq 60$ and $\lambda > 33$. In this case, $1 \in \Delta_{S_1}(\lambda)$ and thus $1 \in \Delta_S(\lambda\mu)$, so S is tasty if and only if $\max \Delta(S) > 1$. We have $\mathsf{L}_{S_2}(\mu) \subseteq [2, 10] \cap \mathbb{Z}$, and since $\lambda > 33$, $\min \mathsf{L}_{S_1}(\lambda) \geq 12$. From this, we conclude $\max \Delta_S(\lambda\mu) \geq 2$ and thus S is tasty.

At this point, only finitely many gluings remain, and an exhaustive computation with the GAP package `numericalsgps` [11] completes the proof. □

Example 5.4 We briefly examine the exceptional case identified in Theorem 5.3. Let $S_1 = \langle 2, 3\rangle$, $S_2 = \langle 6, 9, 20\rangle$, $\lambda = 4$, and $\mu = 27$, and let

$$S = \mu S_1 + \lambda S_2 = \langle 2\mu, 3\mu, 6\lambda, 9\lambda, 20\lambda\rangle.$$

From $\lambda\mu = 6\mu + 6\lambda + 2\mu$, one can readily verify

$$\mathsf{L}_S(60\lambda) = \mathsf{L}_{S_1}(\lambda) \cup (\mathsf{L}_S(6\mu) + 2) = \{3, 7, 8, 9, 10\} \cup \{4, 5, 6, 7\}$$

and

$$\mathsf{L}_S(\lambda\mu) = \mathsf{L}_{S_1}(\lambda) \cup \mathsf{L}_{S_2}(\mu) = \{2, 4\} \cup \{3\},$$

resulting in $\Delta(S) = \{1\}$ and ensuring S is bland.

Theorem 5.5 *Every gluing* $S = \mu S_1 + \lambda S_2$ *of* $S_1 = \langle 2, 3\rangle$ *and* $S_2 = \langle 6, 9, 26\rangle$ *is tasty.*

Proof We begin by noting $\text{Betti}(S_2) = \{18, 78\}$, $\Delta_{S_2}(18) = \{1\}$, and $\Delta_{S_2}(78) = \{1, 6\}$, and that every $n \in S_2$ with $n > 73$ has $|\mathsf{L}_{S_2}(n)| \geq 2$.

First, suppose $\mu > 78$. It is easy to show that 1 lies in $\Delta_S(b)$ for each $b \in \text{Betti}(S)$. Since $1 \in \Delta_{S_2}(\mu)$ implies $1 \in \Delta_S(\mu\lambda)$, and since $\lambda\mu$ is the largest Betti element, 1 lies in the delta set of each $b \in \text{Betti}(S)$. Furthermore, $\max \Delta_S(78\lambda) = 6$, so S is tasty.

Next, suppose $\mu \leq 78$ and $\lambda > 42$. We have $\min \mathsf{L}_{S_1}(\lambda) \geq 15$ and $\mathsf{L}_{S_2}(\mu) \subseteq [2, 13] \cap \mathbb{Z}$, so $\max \Delta_S(\lambda\mu) \geq 2$, and $1 \in \Delta_S(\lambda\mu)$ since $1 \in \Delta_{S_1}(\mu)$. This again implies S is tasty.

Now, all remaining gluings satisfy $\lambda \leq 42$ and $\mu \leq 78$, and an exhaustive computation using [11] verifies each is tasty. □

Our final result of this section concerns gluings $\mu S + \lambda S$ of a numerical semigroup S with itself. In the case $\mathsf{e}(S) = 2$, we identify the asymptotic proportion of gluings $\mu S + \lambda S$ that are tasty (Theorem 5.7). The plots in Fig. 3 depict the choices of λ and μ that are assured tasty or bland by Proposition 5.6.

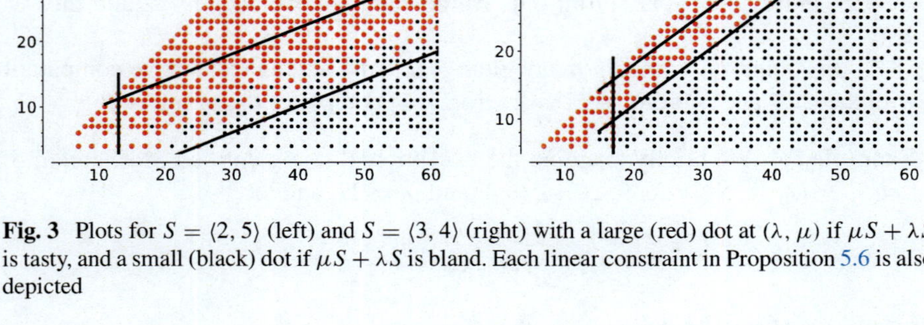

Fig. 3 Plots for $S = \langle 2, 5\rangle$ (left) and $S = \langle 3, 4\rangle$ (right) with a large (red) dot at (λ, μ) if $\mu S + \lambda S$ is tasty, and a small (black) dot if $\mu S + \lambda S$ is bland. Each linear constraint in Proposition 5.6 is also depicted

Proposition 5.6 *Fix* $S = \langle n_1, n_2 \rangle$ *and integers* $\lambda > \mu > 2n_1n_2 - n_1 - n_2$ *with* $\gcd(\lambda, \mu) = 1$, *and denote the gluing* $G = \mu S + \lambda S$.

(a) If $\lambda > \frac{n_2}{n_1}\mu + n_2(n_2 - n_1)$, *then G is tasty.*
(b) If $\lambda < \frac{n_2}{n_1}\mu - n_2(n_2 - n_1)$, *then G is bland.*

Proof The lower bound on λ and μ ensures $|\mathsf{L}_S(\lambda)| \geq 2$ and $|\mathsf{L}_S(\mu)| \geq 2$, and that $\lambda\mu$ is the largest Betti element of G. If $\lambda > \frac{n_2}{n_1}\mu + n_2(n_2 - n_1)$, then

$$\min \mathsf{L}_S(\lambda) \geq \tfrac{1}{n_2}\lambda > \tfrac{1}{n_1}\mu + (n_2 - n_1) \geq \max \mathsf{L}_S(\mu) + \max \Delta(S),$$

and thus $\max \Delta(G) = \max \Delta_G(\lambda\mu) > \max \Delta(S)$. From this, we conclude G is tasty. If, on the other hand, $\lambda < \frac{n_2}{n_1}\mu - n_2(n_2 - n_1)$, then

$$\min \mathsf{L}_S(\lambda) - \max \mathsf{L}_S(\mu) < (\tfrac{1}{n_2}\lambda + (n_2 - n_1)) - (\tfrac{1}{n_1}\mu - (n_2 - n_1)) < n_2 - n_1,$$

and similarly

$$\max \mathsf{L}_S(\lambda) - \min \mathsf{L}_S(\mu) > \tfrac{1}{n_1}\lambda - \tfrac{1}{n_2}\mu - 2(n_2 - n_1) > 0.$$

From these inequalities, we conclude there exists $\delta \in \Delta_G(\lambda\mu)$ with $0 < \delta < n_2 - n_1$, and since $\mathsf{L}_S(\lambda\mu) = \mathsf{L}_{S_1}(\lambda) \cup \mathsf{L}_{S_2}(\mu)$, we must have $\Delta_G(\lambda\mu) = \{\delta, n_2 - n_1\}$. It follows that $\max \Delta(G) = n_2 - n_1$, so $\Delta_G(\lambda n_1 n_2) = \{n_2 - n_1\}$ implies G is bland. □

Theorem 5.7 *Given a numerical semigroup* $S = \langle n_1, n_2 \rangle$, *the proportion of tasty gluings of S with itself is* n_1/n_2. *More precisely,*

$$\lim_{N\to\infty} \frac{\#\,\textit{tasty gluings}\ \mu S + \lambda S\ \textit{with}\ \lambda, \mu < N}{\#\,\textit{gluings}\ \mu S + \lambda S\ \textit{with}\ \lambda, \mu < N} = \frac{n_1}{n_2}.$$

Proof Let L denote the limit on the left-hand side above. Writing

$$A = \{(a, b) \in \mathbb{Z}^2 : \gcd(a, b) = 1\}$$

and $N_0 = 2n_1n_2 - n_1 - n_2$, Proposition 5.6(a) yields a lower bound on L of

$$\begin{aligned}
L &\geq \lim_{N\to\infty} \frac{\#(\lambda, \mu) \in A \cap [N_0, N]^2 \text{ with } \lambda > \mu \text{ and } \lambda > \frac{n_2}{n_1}\mu + n_2(n_2 - n_1)}{\#(\lambda, \mu) \in A \cap [2, N]^2 \text{ with } \lambda > \mu} \\
&= \lim_{N\to\infty} \frac{\#(\lambda, \mu) \in [2, N]^2 \text{ with } \lambda > \mu \text{ and } \lambda > \frac{n_2}{n_1}\mu + n_2(n_2 - n_1)}{\#(\lambda, \mu) \in [2, N]^2 \text{ with } \lambda > \mu} \\
&= \lim_{N\to\infty} \frac{\#(\lambda, \mu) \in [2, N]^2 \text{ with } \frac{n_1}{n_2}\lambda > \mu}{\#(\lambda, \mu) \in [2, N]^2 \text{ with } \lambda > \mu} = \frac{n_1}{n_2},
\end{aligned}$$

where the first equality follows from [16, Chap. IV, Theorem 1]. Proposition 5.6(b) yields an identical upper bound for L, so we conclude $L = n_1/n_2$. □

References

1. A. Assi, M. D'Anna, P. García-Sánchez, *Numerical Semigroups and Applications*, RSME Springer, series 3, Springer, Switzerland, 2020.
2. J. Amos, S. Chapman, N. Hine, J. Paixão, *Sets of lengths do not characterize numerical monoids*, Integers **7** (2007), #A50.
3. T. Barron, C. O'Neill, and R. Pelayo, *On dynamic algorithms for factorization invariants in numerical monoids*, Mathematics of Computation **86** (2017), 2429–2447.
4. T. Barron, C. O'Neill, and R. Pelayo, *On the set of elasticities in numerical monoids*, Semigroup Forum **94** (2017), no. 1, 37–50.
5. C. Bowles, S. Chapman, N. Kaplan, D. Reiser, *On delta sets of numerical monoids*, J. Algebra Appl. **5** (2006) 1–24.
6. W. Bruns, P. García-Sánchez, C. O'Neill, and D. Wilburne, *Wilf's conjecture in fixed multiplicity*, International Journal of Algebra and Computation **30** (2020), no. 4, 861–882.
7. S. Chapman, P. García-Sánchez, D. Llena, A. Malyshev, and D. Steinberg, *On the Delta set and the Betti elements of a BF-monoid*, Arab J. Math 1 (2012), 53–61.
8. S. Chapman, F. Gotti, and R. Pelayo, *On delta sets and their realizable subsets in Krull monoids with cyclic class groups*, Colloq. Math. 137 (2014), no. 1, 137–146.
9. S. Chapman, M. Holden, and T. Moore, *Full elasticity in atomic monoids and integral domains*, Rocky Mountain Journal of Mathematics **36** (2006), no. 5, 1437–1455.
10. S. Chapman, C. O'Neill, and V. Ponomarenko, *On length densities*, preprint. Available at arXiv:2008.06725.
11. M. Delgado, P. García-Sánchez, and J. Morais, *NumericalSgps, A package for numerical semigroups*, Version 1.1.10 (2018), (Refereed GAP package), https://urldefense.com/v3/__https://gap-packages.github.io/numericalsgps/__;!!M2cvx14AM25G2z0FFbZ0Sa4TsktIO1TizEVjdWO_gilyrCCIXrbFEdYPH0xaJpWHQcZ3k0apLtBORzuD9srSUqJ-johHZbvq81t9Ev-9$.
12. J. García-García, M. Moreno-Frías, and A. Vigneron-Tenorio, *Computation of delta sets of numerical monoids*, Monatshefte für Mathematik **178** (2015), no. 3 457–472.
13. P. García-Sánchez, I. Ojeda, and J. Rosales, *Affine semigroups having a unique Betti element*, J. Algebra Appl. **12** (2013), no. 3, 1250177, 11 pp.
14. N. Kaplan and C. O'Neill, *Numerical semigroups, polyhedra, and posets I: the group cone*, to appear, Combinatorial Theory. Available at arXiv:1912.03741.
15. E. Kunz, *Über die Klassifikation numerischer Halbgruppen*, Regensburger Mathematische Schriften **11**, 1987.
16. D. Lehmer, *Asymptotic evaluation of certain totient sums*, American Journal of Mathematics **22** (1900), 293–335.
17. C. O'Neill, *On factorization invariants and Hilbert functions*, Journal of Pure and Applied Algebra **221** (2017), no. 12, 3069–3088.
18. C. O'Neill and R. Pelayo, *Factorization invariants in numerical monoids*, Contemporary Mathematics **685** (2017), 231–249.
19. J. Rosales and P. García-Sánchez, *Numerical Semigroups*, Developments in Mathematics, Vol. 20, Springer-Verlag, New York, 2009.

On a Problem of Cilleruelo and Nathanson, II

Yong-Gao Chen and Jin-Hui Fang

1 Introduction

Let $\mathbb{Z}$ be the set of all integers and let $\mathbb{N}$ be the set of all positive integers. For a set A of integers and an integer n, let

$$d_A(n) = |\{(a, a') : n = a - a', a, a' \in A\}|$$

and

$$s_A(n) = |\{(a, a') : n = a + a', a, a' \in A, a \leq a'\}|.$$

Cilleruelo and Nathanson [3] posed the following problem:

Problem [3, Problem 4.3]. Given two functions $f_1 : \mathbb{N} \to \mathbb{N}$ and $f_2 : \mathbb{Z} \to \mathbb{N}$. Is the condition $\liminf_{u\to\infty} f_1(u) \geq 2$ and $\liminf_{|u|\to\infty} f_2(u) \geq 2$ sufficient to assure that there exists a set A such that $d_A(n) = f_1(n)$ for all $n \in \mathbb{N}$ and $s_A(n) = f_2(n)$ for all $n \in \mathbb{Z}$?

In 2011, the authors [2] answered this problem affirmatively.

Theorem A If two functions $f_1 : \mathbb{N} \to \mathbb{N}$ and $f_2 : \mathbb{Z} \to \mathbb{N}$ satisfy that $\{n \in \mathbb{Z} : f_1(|n|) \geq 2, f_2(n) \geq 2\}$ contains arbitrarily long sequence of consecutive integers,

Y.-G. Chen · J.-H. Fang (✉)
School of Mathematical Sciences and Institute of Mathematics, Nanjing Normal University, Nanjing 210023, P.R., China
e-mail: fangjinhui1114@163.com

Y.-G. Chen
e-mail: ygchen@njnu.edu.cn

M. B. Nathanson (ed.), *Combinatorial and Additive Number Theory V*,
Springer Proceedings in Mathematics & Statistics 395,
https://doi.org/10.1007/978-3-031-10796-2_4

then there exists a set A such that $d_A(n) = f_1(n)$ for all $n \in \mathbb{N}$ and $s_A(n) = f_2(n)$ for all $n \in \mathbb{Z}$.

In this paper, we generalize the above result by considering arithmetic progressions. For a set T of integers and two functions $f_1 : \mathbb{N} \to \mathbb{N}$ and $f_2 : \mathbb{Z} \to \mathbb{N}$, we say that (f_1, f_2) is a *good pairs for* T if $\{n \in \mathbb{Z} : f_1(|n|) \geq 2,\ f_2(n) \geq 2\}$ contains arbitrarily long sequence of consecutive terms in T. A set A of integers is called *difference-sum controllable for* (f_1, f_2, T) if

(C_1) $d_A(n) \leq f_1(n)$ for all $n \in \mathbb{N} \cap T$ and $d_A(n) \leq 1$ for all $n \in \mathbb{N} \setminus T$;
(C_2) $s_A(n) \leq f_2(n)$ for all $n \in T$ and $s_A(n) \leq 1$ for all $n \in \mathbb{Z} \setminus T$.

In this paper, the following result is proved.

Theorem 1 *Let*

$$T = 0 \pmod{m} \cup \pm a_2 \pmod{m} \cup \cdots \cup \pm a_t \pmod{m},$$

where m is a positive integer and $a \pmod{m} = \{a + mk : k \in \mathbb{Z}\}$. If (f_1, f_2) is a good pair for T, then there exists a difference-sum controllable set A for (f_1, f_2, T) such that $d_A(n) = f_1(n)$ for all $n \in \mathbb{N} \cap T$ and $s_A(n) = f_2(n)$ for all $n \in T$.

On the other hand, we pose the following conjecture:

Conjecture Let T be a proper subset of $\mathbb{Z}$ and

$$T = 0 \pmod{m} \cup \pm a_2 \pmod{m} \cup \cdots \cup \pm a_t \pmod{m},$$

where m is a positive integer and $a \pmod{m} = \{a + mk : k \in \mathbb{Z}\}$. Then there exists a good pair (f_1, f_2) for T such that for any set A of integers, $d_A(n) \neq f_1(n)$ or $s_A(n) \neq f_2(n)$ for infinitely many integers n.

Let $\mathbb{Q}$ be the set of all rational numbers and let $\mathbb{Q}^+$ be the set of all positive rational numbers. For any $a < b$, $[a, b] \cap \mathbb{Q}$ is called an interval of $\mathbb{Q}$. For a set $A \subseteq \mathbb{Q}$ and a rational number r, let

$$d_A(r) = |\{(a, a') : r = a - a', a, a' \in A\}|$$

and

$$s_A(r) = |\{(a, a') : r = a + a', a, a' \in A, a \leq a'\}|.$$

The methods in the proof of Theorem 1 (also [2]) can be used to prove the following result.

Theorem 2 *If $f_1 : \mathbb{Q}^+ \to \mathbb{N}$ and $f_2 : \mathbb{Q} \to \mathbb{N}$ are two functions such that $\{\alpha \in \mathbb{Q} : f_1(|\alpha|) \geq 2,\ f_2(\alpha) \geq 2\}$ contains arbitrarily long intervals of $\mathbb{Q}$, then there exists a set $A \subseteq \mathbb{Q}$ such that $d_A(\alpha) = f_1(\alpha)$ for all $\alpha \in \mathbb{Q}^+$ and $s_A(\alpha) = f_2(\alpha)$ for all $\alpha \in \mathbb{Q}$.*

For more related results, one may refer to [1, 4–10] for details.

2 Proof of Theorem 1

Let m, f_1, f_2, and T be as in Theorem 1. Let

$$U = \{n \in \mathbb{Z} : f_1(|n|) \geq 2,\ f_2(n) \geq 2\}.$$

Since (f_1, f_2) is a good pair for T, it follows that U contains arbitrarily long sequence of consecutive terms in T.

We will firstly introduce the following two lemmas.

Lemma 1 *If B is a finite difference-sum controllable set for (f_1, f_2, T) with $B \subseteq T$ and $k \in T$ with $s_B(k) < f_2(k)$, then there exists a finite difference-sum controllable set D for (f_1, f_2, T) with $B \subseteq D \subseteq T$ and $s_D(k) = s_B(k) + 1$.*

Proof Since U contains arbitrarily long sequence of consecutive terms in T and $B \subseteq T$, we can choose an integer d such that

(1) $|d| > 2(\sum_{b \in B} |b| + |k|)$ and $d \equiv 0 \pmod m$;

(2) $\{d + b : b \in B\} \subseteq U \cap T$.

Let $D = B \cup \{-d, d + k\}$. Then $s_D(k) = s_B(k) + 1$. If $s_B(n) \geq 1$ and $n \neq k$, then $n \in (B + B) \setminus \{k\}$ and $s_D(n) = s_B(n)$ by (1). If $s_B(n) = 0$ and $n \neq k$, then by (1) we have $s_D(n) \leq 1$. Since $s_B(n)$ satisfies (C_2), it follows that $s_D(n)$ satisfies (C_2).

Let

$$T_1 = \{|d + b| : b \in B\}, \quad T_2 = \{|d + k - b| : b \in B\} \bigcup \{|2d + k|\}.$$

If $n \in T_1$, then $d_B(n) = 0$, $d_D(n) \leq 2 \leq f_1(n)$ by (1) and (2). If $n \in T_2 \setminus T_1$, then by (1) we have $d_B(n) = 0$ and $d_D(n) = 1$. If n is a positive integer with $n \notin T_1 \cup T_2$, then we have $d_D(n) = d_B(n)$. Since $d_B(n)$ satisfies (C_1), it follows that $d_D(n)$ satisfies (C_1).

Therefore, D is a finite difference-sum controllable set for (f_1, f_2, T) with $B \subseteq D \subseteq T$ and $s_D(k) = s_B(k) + 1$.

This completes the proof of Lemma 1.

Lemma 2 *If B is a finite difference-sum controllable set for (f_1, f_2, T) with $B \subseteq T$ and k is a positive integer of T with $d_B(k) < f_1(k)$, then there exists a finite difference-sum controllable set D for (f_1, f_2, T) with $B \subseteq D \subseteq T$ and $d_D(k) = d_B(k) + 1$.*

Proof Since U contains arbitrarily long sequence of consecutive terms in T and $B \subseteq T$, we can choose an integer d such that

(1) $|d| > 2(\sum_{b \in B} |b| + k)$ and $d \equiv 0 \pmod m$;

(2) $\cup_{b \in B} \{d - b, d + b, d + k - b, d + b + k\} \subseteq U \cap T$.

Let $D = B \cup \{d, d + k\}$ and let

$$S_1 = \{d + b + k : b \in B\} \cup \{d + b : b \in B\}, \quad S_2 = \{2d, 2d + k, 2d + 2k\}.$$

If $n \in S_1$, then $s_B(n) = 0, s_D(n) \le 2 \le f_2(n)$ by (1) and (2). If $n \in S_2$, then by (1) we have $s_B(n) = 0$ and $s_D(n) = 1$. If $n \notin S_1 \cup S_2$, then we have $s_D(n) = s_B(n)$. Since $s_B(n)$ satisfies (C_2), it follows that $s_D(n)$ satisfies (C_2).

Let

$$T_1 = \{|d - b| : b \in B\} \cup \{|d + k - b| : b \in B\}.$$

If $n \in T_1$, then $d_B(n) = 0$, $d_D(n) \le 2 \le f_1(n)$ by (1) and (2). If $n > 0$ and $n \notin T_1$, then we have $d_D(n) = d_B(n)$. Since $d_B(n)$ satisfies (C_1), it follows that $d_D(n)$ satisfies (C_1). It is clear that $d_D(k) = d_B(k) + 1$.

Therefore, D is a finite difference-sum controllable set for (f_1, f_2, T) with $B \subseteq D \subseteq T$ and $d_D(k) = d_B(k) + 1$.

This completes the proof of Lemma 2.

Proof of Theorem 1. Let $t_1 < t_2 < \cdots$ be all positive integers of T. Then $T = \{0, \pm t_1, \pm t_2, \dots\}$.

We will use induction to construct an ascending sequence $A_0 \subseteq A_1 \subseteq \cdots$ of finite difference-sum controllable sets for (f_1, f_2, T) such that for any integer $k \ge 0$, $d_{A_k}(n) = f_1(n)$ for all $n \in T$ with $0 < n \le t_k$ and $s_{A_k}(n) = f_2(n)$ for all $n \in T$ with $|n| \le t_k$.

Let $B_0 = \{0\}$. By using Lemma 1 repeatedly, there exists a finite difference-sum controllable set A_0 for (f_1, f_2, T) with $B_0 \subseteq A_0 \subseteq T$ and $s_{A_0}(0) = f_2(0)$. Suppose that we have constructed $A_0, \dots, A_k$. By using Lemma 1 repeatedly, there exists a finite difference-sum controllable set D_{k+1} for (f_1, f_2, T) such that $A_k \subseteq D_{k+1} \subseteq T$, $s_{D_{k+1}}(t_{k+1}) = f_2(t_{k+1})$ and $s_{D_{k+1}}(-t_{k+1}) = f_2(-t_{k+1})$. By using Lemma 2 repeatedly, there exists a finite difference-sum controllable set A_{k+1} for (f_1, f_2, T) such that $D_{k+1} \subseteq A_{k+1} \subseteq T$ and $d_{A_{k+1}}(t_{k+1}) = f_1(t_{k+1})$.

Let

$$A = \bigcup_{k=0}^{\infty} A_k.$$

Then A is a difference-sum controllable set for (f_1, f_2, T) such that $d_A(n) = f_1(n)$ for all $n \in \mathbb{N} \cap T$ and $s_A(n) = f_2(n)$ for all $n \in T$.

This completes the proof of Theorem 1.

3 Proof of Theorem 2

A proof of Theorem 2 is similar to that of Theorem 1. For convenience of the reader, we give the details here.

Let f_1 and f_2 be as in Theorem 2. Let

$$U = \{\alpha \in \mathbb{Q} : f_1(|\alpha|) \ge 2,\ f_2(\alpha) \ge 2\}.$$

Then U contains arbitrarily long intervals of $\mathbb{Q}$.

A subset A of $\mathbb{Q}$ is called *difference-sum controllable for* $(f_1, f_2, \mathbb{Q})$ if $d_A(\alpha) \le f_1(\alpha)$ for all $\alpha \in \mathbb{Q}^+$ and $s_A(\alpha) \le f_2(\alpha)$ for all $\alpha \in \mathbb{Q}$.

We will firstly introduce the following two lemmas.

Lemma 3 *If B is a finite difference-sum controllable subset for $(f_1, f_2, \mathbb{Q})$ and $\alpha \in \mathbb{Q}$ with $s_B(\alpha) < f_2(\alpha)$, then there exists a finite difference-sum controllable subset D for $(f_1, f_2, \mathbb{Q})$ with $B \subseteq D \subseteq \mathbb{Q}$ and $s_D(\alpha) = s_B(\alpha) + 1$.*

Proof Since U contains arbitrarily long intervals of $\mathbb{Q}$, we can choose a rational number β such that

(1) $|\beta| > 2(\sum_{b\in B} |b| + |\alpha|)$;

(2) $\{\beta + b : b \in B\} \subseteq U$.

Let $D = B \cup \{-\beta, \beta + \alpha\}$. Then $s_D(\alpha) = s_B(\alpha) + 1$. If $s_B(r) \ge 1$ and $r \ne \alpha$, then $r \in (B + B) \setminus \{\alpha\}$ and $s_D(r) = s_B(r)$ by (1). If $s_B(r) = 0$ and $r \ne \alpha$, then by (1) we have $s_D(r) \le 1$. Since $s_B(r) \le f_2(r)$ and $f_2(r) \ge 1$, it follows that $s_D(r) \le f_2(r)$.

Let

$$T_1 = \{|\beta + b| : b \in B\}, \quad T_2 = \{|\beta + \alpha - b| : b \in B\} \bigcup \{|2\beta + \alpha|\}.$$

If $r \in T_1$, then $d_B(r) = 0$, $d_D(r) \le 2 \le f_1(r)$ by (1) and (2). If $r \in T_2 \setminus T_1$, then by (1) we have $d_B(r) = 0$ and $d_D(r) = 1$. If r is a positive rational number with $r \notin T_1 \cup T_2$, then we have $d_D(r) = d_B(r)$. Since $d_B(r) \le f_1(r)$ and $f_1(r) \ge 1$, it follows that $d_D(r) \le f_1(r)$.

Therefore, D is a finite difference-sum controllable set for $(f_1, f_2, \mathbb{Q})$ with $B \subseteq D \subseteq \mathbb{Q}$ and $s_D(\alpha) = s_B(\alpha) + 1$.

This completes the proof of Lemma 3.

Lemma 4 *If B is a finite difference-sum controllable subset for $(f_1, f_2, \mathbb{Q})$ and $\alpha \in \mathbb{Q}^+$ with $d_B(\alpha) < f_1(\alpha)$, then there exists a finite difference-sum controllable subset D for $(f_1, f_2, \mathbb{Q})$ with $B \subseteq D \subseteq \mathbb{Q}$ and $d_D(\alpha) = d_B(\alpha) + 1$.*

Proof Since U contains arbitrarily long intervals of $\mathbb{Q}$, we can choose a rational number β such that

(1) $|\beta| > 2(\sum_{b\in B} |b| + \alpha)$;

(2) $\cup_{b\in B}\{\beta - b, \beta + b, \beta + \alpha - b, \beta + b + \alpha\} \subseteq U$.

Let $D = B \cup \{\beta, \beta + \alpha\}$ and let

$$S_1 = \{\beta + b + \alpha : b \in B\} \cup \{\beta + b : b \in B\}, \quad S_2 = \{2\beta, 2\beta + \alpha. 2\beta + 2\alpha\}.$$

If $r \in S_1$, then $s_B(r) = 0$, $s_D(r) \le 2 \le f_2(r)$ by (1) and (2). If $r \in S_2$, then by (1) we have $s_B(r) = 0$ and $s_D(r) = 1$. If $r \notin S_1 \cup S_2$, then we have $s_D(r) = s_B(r)$. Since $s_B(r) \le f_2(r)$ and $f_2(r) \ge 1$, it follows that $s_D(r) \le f_2(r)$.

Let

$$T_1 = \{|\beta - b| : b \in B\} \cup \{|\beta + \alpha - b| : b \in B\}.$$

If $r \in T_1$, then $d_B(r) = 0$, $d_D(r) \le 2 \le f_1(r)$ by (1) and (2). If $r > 0$ and $r \notin T_1$, then we have $d_D(r) = d_B(r)$. Since $d_B(r) \le f_1(r)$ and $f_1(r) \ge 1$, it follows that $d_D(r) \le f_1(r)$. It is clear that $d_D(\alpha) = d_B(\alpha) + 1$.

Therefore, D is a finite difference-sum controllable set for $(f_1, f_2, \mathbb{Q})$ with $B \subseteq D \subseteq \mathbb{Q}$ and $d_D(\alpha) = d_B(\alpha) + 1$.

This completes the proof of Lemma 4.

Proof of Theorem 2. Since $\mathbb{Q}^+$ is countable, we may write

$$\mathbb{Q}^+ = \{t_1, t_2, t_3, \dots\}.$$

Let $t_0 = 0$ and $t_{-i} = -t_i$ for all $i > 0$.

We will use induction to construct an ascending sequence $A_0 \subseteq A_1 \subseteq \cdots$ of finite difference-sum controllable sets for $(f_1, f_2, \mathbb{Q})$ such that for any integer $k \ge 0$, $d_{A_k}(t_i) = f_1(t_i)$ for all $0 < i \le k$ and $s_{A_k}(t_i) = f_2(t_i)$ for all $|i| \le k$.

Let $B_0 = \{0\}$. By using Lemma 3 repeatedly, there exists a finite difference-sum controllable set A_0 for $(f_1, f_2, \mathbb{Q})$ with $B_0 \subseteq A_0 \subseteq \mathbb{Q}$ and $s_{A_0}(0) = f_2(0)$. Suppose that we have constructed $A_0, \dots, A_k$. By using Lemma 3 repeatedly, there exists a finite difference-sum controllable set D_{k+1} for $(f_1, f_2, \mathbb{Q})$ such that $A_k \subseteq D_{k+1} \subseteq \mathbb{Q}$, $s_{D_{k+1}}(t_{k+1}) = f_2(t_{k+1})$ and $s_{D_{k+1}}(t_{-k-1}) = f_2(t_{-k-1})$. By using Lemma 4 repeatedly, there exists a finite difference-sum controllable set A_{k+1} for $(f_1, f_2, \mathbb{Q})$ such that $D_{k+1} \subseteq A_{k+1} \subseteq \mathbb{Q}$ and $d_{A_{k+1}}(t_{k+1}) = f_1(t_{k+1})$.

Let

$$A = \bigcup_{k=0}^{\infty} A_k.$$

Then A is a difference-sum controllable set for $(f_1, f_2, \mathbb{Q})$ such that $d_A(t_i) = f_1(t_i)$ for all $i > 0$ and $s_A(t_i) = f_2(t_i)$ for all i. Hence, $d_A(\alpha) = f_1(\alpha)$ for all $\alpha \in \mathbb{Q}^+$ and $s_A(\alpha) = f_2(\alpha)$ for all $\alpha \in \mathbb{Q}$.

This completes the proof of Theorem 2.

Acknowledgements This work was supported by the National Natural Science Foundation of China, Grant Nos. 12171243, 12171246 and the Natural Science Foundation of Jiangsu Province, Grant No. BK20211282.

References

1. Chen, Y.G., Ding, Y., On the Erdős-Turán conjecture in the positive rational numbers. Colloq. Math. **152**, 317–323 (2018)
2. Chen, Y.G., Fang, J.H., On a problem of Cilleruelo and Nathanson. Combinatorica **31**, 691–696 (2011)
3. Cilleruelo, J., Nathanson, M.B., Perfect difference sets constructed from Sidon sets. Combinatorica **28**, 401–414 (2008)
4. Cilleruelo, J., Nathanson, M.B., Dense sets of integers with prescribed representation functions. Eur. J. Combin. **34**, 1297–1306 (2013)

5. Lev, V.F., Reconstructing integer sets from their representation functions. Electron. J. Combin. **11**, Research paper 78, 6pp (electronic) (2004)
6. Nathanson, M.B., The inverse problem for representation functions of additive bases. in: Number Theory: New York Seminar 2003, Springer, 253–262 (2004)
7. Nathanson, M.B., Every function is the representation function of an additive basis for the integers. Port. Math. (N.S.) **62**, 55–72 (2005)
8. Pollington, A.D., Vanden, C., The integers as differences of a sequence. Canad. Bull. Math. **24**, 497–499 (1981)
9. Ruzsa, I.Z., An infinite Sidon sequence. J. Number Theory **68**, 63–71 (1998)
10. Tang, M., Unique representation bi-basis for rational numbers field. Period. Math. Hungar. **74**, 250–254 (2017)

Linked Partition Ideals and a Schur-Type Identity of Andrews

Shane Chern

2010MSC. 11P84 · 05A17

1 Introduction

In 1926, Schur [13] proved the following result:

Theorem S. Let $A(n)$ denote the number of partitions of n into parts congruent to ± 1 modulo 6. Let $B(n)$ denote the number of partitions of n into distinct nonmultiples of 3. Let $D(n)$ denote the number of partitions of n of the form $\mu_1 + \mu_2 + \cdots + \mu_s$ where $\mu_i - \mu_{i+1} \geq 3$ with strict inequality if $3 \mid \mu_i$. Then

$$A(n) = B(n) = D(n).$$

This partition theorem has many variants, one of which is due to Andrews [5]:

Theorem A. We consider partitions in which odd parts appear at most once, even parts appear at most twice, and the difference between two parts can never be 1 and can be 2 only if both are odd. Let $E(n)$ denote the weighted count of these partitions with weight $(-1)^{\tau}$ for each partition that has exactly τ parts that appear twice. Then

$$A(n) = B(n) = D(n) = E(n).$$

A combinatorial proof of the fact that $D(n) = E(n)$ was later provided by Yee [14].

S. Chern (✉)
Department of Mathematics and Statistics, Dalhousie University, Halifax, NS B3H 4R2, Canada
e-mail: chenxiaohang92@gmail.com

M. B. Nathanson (ed.), *Combinatorial and Additive Number Theory V*,
Springer Proceedings in Mathematics & Statistics 395,
https://doi.org/10.1007/978-3-031-10796-2_5

In recent years, there are a substantial amount of papers studying generating functions for certain partition sets that can be represented as an *Andrews–Gordon type series* of the form

$$\sum_{n_1,\ldots,n_r\geq 0} \frac{(-1)^{L_1(n_1,\ldots,n_r)} q^{Q(n_1,\ldots,n_r)+L_2(n_1,\ldots,n_r)}}{(q^{A_1};q^{A_1})_{n_1}\cdots(q^{A_r};q^{A_r})_{n_r}}, \tag{1.1}$$

in which L_1 and L_2 are linear forms and Q is a quadratic form in $n_1,\ldots,n_r$, and the q-Pochhammer symbol is defined for $n\in\mathbb{N}\cup\{\infty\}$,

$$(A;q)_n := \prod_{k=0}^{n-1}(1-Aq^k).$$

In particular, in the previous papers of this series [7–10], such representations are associated with the framework of linked partition ideals, with, especially, [7, 9] dealing with identities born out of Schur's Theorem S.

For any partition λ, we denote by $|\lambda|$ the sum of all parts in λ, and by $\sharp(\lambda)$ the number of parts in λ. We also denote by $\tau(\lambda)$ the number of different parts in λ that appear twice.

Let $\mathscr{A}$ denote the set of partitions counted by $E(n)$ for all nonnegative n.

Although it looks like the trivariate generating function for partitions λ in $\mathscr{A}$ that counts both statistics $\sharp(\lambda)$ and $\tau(\lambda)$ does not have a simple representation as an Andrews–Gordon type series, our object is the following non-standard generating function identity.

Theorem 1.1 *We have*

$$\sum_{\lambda\in\mathscr{A}} x^{\sharp(\lambda)} y^{\tau(\lambda)} q^{|\lambda|-\sharp(\lambda)(\sharp(\lambda)-1)} = \frac{(-xq^2;q^2)_\infty}{\prod_{n\geq 0}(1-xq^{2n+1}-x^2yq^{4n+2})}. \tag{1.2}$$

Setting $y=-1$ yields a new proof of Theorem A.

Corollary 1.2 *We have*

$$\sum_{\lambda\in\mathscr{A}}(-1)^{\tau(\lambda)}x^{\sharp(\lambda)}q^{|\lambda|} = \sum_{n_1,n_2\geq 0}\frac{(-1)^{n_2}q^{3\binom{n_1}{2}+18\binom{n_2}{2}+6n_1n_2+n_1+9n_2}x^{n_1+3n_2}}{(q;q)_{n_1}(q^6;q^6)_{n_2}}. \tag{1.3}$$

In particular, for $n\geq 0$,

$$D(n)=E(n).$$

Finally, we recall that the *continuous q-Hermite polynomials* $H_n(x;q)$ are given by

$$H_n(x;q) := e^{in\theta}{}_2\phi_0\left(\begin{matrix}q^{-n},0\\ -\end{matrix};q,q^n e^{-2i\theta}\right) \qquad (\text{with } x=\cos\theta),$$

where the basic hypergeometric series ${}_r\phi_s$ is defined by

$${}_r\phi_s\left(\begin{matrix}a_1,a_2\dots,a_r\\ b_1,b_2,\dots,b_s\end{matrix};q,z\right) := \sum_{n\geq 0}\frac{(a_1;q)_n\cdots(a_r;q)_n}{(q;q)_n(b_1;q)_n\cdots(b_s;q)_n}\left((-1)^n q^{\binom{n}{2}}\right)^{s-r+1} z^n.$$

The continuous q-Hermite polynomials are a family of q-orthogonal polynomials in the basic Askey scheme. See [11, Sect. 3.26] for details. In particular, they satisfy a second-order recurrence for $n\geq 1$,

$$H_{n+1}(x;q) = 2xH_n(x;q) - (1-q^n)H_{n-1}(x;q) \tag{1.4}$$

with $H_0(x;q)=1$ and $H_1(x;q)=2x$.

We show that the generating function for the partition set $\mathscr{A}$ is related to the continuous q-Hermite polynomials.

Corollary 1.3 *We have*

$$\sum_{\lambda\in\mathscr{A}} x^{\sharp(\lambda)}y^{\tau(\lambda)}q^{|\lambda|} = \sum_{M,N\geq 0}\frac{q^{2\binom{M}{2}+4\binom{N}{2}+2MN+M+2N}x^{M+N}t_M(y)}{(q^2;q^2)_M(q^2;q^2)_N}, \tag{1.5}$$

where

$$t_M(y) = (-\mathrm{i})^M y^{M/2} H_M(\tfrac{\mathrm{i}}{2}y^{-1/2};q^2). \tag{1.6}$$

Remark 1.1 By (1.4), we know that as a polynomial in x, $H_n(x;q)$ has degree n. Further, when n is even, the terms in $H_n(x;q)$ with an odd exponent of x vanish; when n is odd, the terms in $H_n(x;q)$ with an even exponent of x vanish. Therefore, $t_M(y)$ is a polynomial in y of degree $\lfloor M/2\rfloor$.

2 Linked Partition Ideals and a Matrix Equation

The general theory of linked partition ideals was proposed by Andrews [1–3] in the 1970s; see [4, Chap. 8] for an introduction. In recent years, a special type of linked partition ideals, called *span one linked partition ideals*, was revisited by Chern and Li [10] and Chern [8] to associate this theory with Andrews–Gordon type series.

Definition 2.1 Assume that we are given

- a finite set $\Pi = \{\pi_1,\pi_2,\dots,\pi_K\}$ of integer partitions with $\pi_1=\emptyset$, the empty partition,

- a *map of linking sets*, $\mathcal{L} : \Pi \to P(\Pi)$, the power set of Π, with especially, $\mathcal{L}(\pi_1) = \mathcal{L}(\emptyset) = \Pi$ and $\pi_1 = \emptyset \in \mathcal{L}(\pi_k)$ for any $1 \leq k \leq K$,
- and a positive integer T, called the *modulus*, which is greater than or equal to the largest part among all partitions in Π.

We say a *span one linked partition ideal* $\mathscr{I} = \mathscr{I}(\langle \Pi, \mathcal{L} \rangle, T)$ is the collection of all partitions of the form

$$\begin{aligned}\lambda &= \phi^0(\lambda_0) \oplus \phi^T(\lambda_1) \oplus \cdots \oplus \phi^{NT}(\lambda_N) \oplus \phi^{(N+1)T}(\pi_1) \oplus \phi^{(N+2)T}(\pi_1) \oplus \cdots \\ &= \phi^0(\lambda_0) \oplus \phi^T(\lambda_1) \oplus \cdots \oplus \phi^{NT}(\lambda_N), \qquad (2.1)\end{aligned}$$

where $\lambda_i \in \mathcal{L}(\lambda_{i-1})$ for each i and λ_N is not the empty partition. We also include in $\mathscr{I}$ the empty partition, which corresponds to $\phi^0(\pi_1) \oplus \phi^T(\pi_1) \oplus \cdots$. Here for any two partitions μ and ν, $\mu \oplus \nu$ gives a partition by collecting all parts in μ and ν, and $\phi^m(\mu)$ gives a partition by adding m to each part of μ.

Lemma 2.1 *$\mathscr{A}$ is the span one linked partition ideal $\mathscr{I}(\langle \Pi, \mathcal{L} \rangle, 2)$, where $\Pi = \{\pi_1 = \emptyset, \pi_2 = (1), \pi_3 = (2), \pi_4 = (2+2)\}$ and*

$$\begin{cases} \mathcal{L}(\pi_1) = \mathcal{L}(\pi_2) = \{\pi_1, \pi_2, \pi_3, \pi_4\}, \\ \mathcal{L}(\pi_3) = \mathcal{L}(\pi_4) = \{\pi_1\}. \end{cases}$$

Proof We decompose each partition in $\mathscr{A}$ into blocks $B_0, B_1, \ldots$ such that all parts between $2i+1$ and $2i+2$ fall into block B_i. By the definition of $\mathscr{A}$, we find that if we apply the operator ϕ^{-2i} to the block B_i, then it is among Π. If $\phi^{-2i}(B_i)$ is π_1 or π_2, then $\phi^{-2(i+1)}(B_{i+1})$ can be any among Π. If $\phi^{-2i}(B_i)$ is π_3 or π_4, then this partition has a part of size $2i+2$ and therefore the next different part is at least $2i+5$ since its difference with $2i+2$ cannot be 1 or 2. Thus, in this case, the block B_{i+1} is empty, that is $\phi^{-2(i+1)}(B_{i+1}) = \pi_1$. Conversely, it is straightforward to verify that all partitions in $\mathscr{I}(\langle \Pi, \mathcal{L} \rangle, 2)$ satisfy the difference conditions defined for $\mathscr{A}$. □

From now on, we always decompose any partition $\lambda \in \mathscr{A} = \mathscr{I}(\langle \Pi, \mathcal{L} \rangle, 2)$ as in (2.1). Further, for each $1 \leq k \leq 4$, we define

$$G_k(x) := \sum_{\substack{\lambda \in \mathscr{A} \\ \lambda_0 = \pi_k}} x^{\sharp(\lambda)} y^{\tau(\lambda)} q^{|\lambda|},$$

the generating function for partitions whose first decomposed block is π_k.

By the definition of span one linked partition ideals, we have

$$G_k(x) = x^{\sharp(\pi_k)} y^{\tau(\pi_k)} q^{|\pi_k|} \sum_{j : \pi_j \in \mathcal{L}(\pi_k)} G_j(xq^2). \qquad (2.2)$$

Therefore,

$$\begin{pmatrix} G_1(x) \\ G_2(x) \\ G_3(x) \\ G_4(x) \end{pmatrix} = \begin{pmatrix} 1 & & & \\ & xq & & \\ & & xq^2 & \\ & & & x^2yq^4 \end{pmatrix} \cdot \begin{pmatrix} 1 & 1 & 1 & 1 \\ 1 & 1 & 1 & 1 \\ 1 & 0 & 0 & 0 \\ 1 & 0 & 0 & 0 \end{pmatrix} \cdot \begin{pmatrix} G_1(xq^2) \\ G_2(xq^2) \\ G_3(xq^2) \\ G_4(xq^2) \end{pmatrix}. \tag{2.3}$$

We then define

$$\begin{pmatrix} F_1(x) \\ F_2(x) \\ F_3(x) \\ F_4(x) \end{pmatrix} = \begin{pmatrix} 1 & 1 & 1 & 1 \\ 1 & 1 & 1 & 1 \\ 1 & 0 & 0 & 0 \\ 1 & 0 & 0 & 0 \end{pmatrix} \cdot \begin{pmatrix} G_1(x) \\ G_2(x) \\ G_3(x) \\ G_4(x) \end{pmatrix}. \tag{2.4}$$

Substituting (2.3) into (2.4) yields the following matrix equation:

$$\begin{pmatrix} F_1(x) \\ F_2(x) \\ F_3(x) \\ F_4(x) \end{pmatrix} = \begin{pmatrix} 1 & 1 & 1 & 1 \\ 1 & 1 & 1 & 1 \\ 1 & 0 & 0 & 0 \\ 1 & 0 & 0 & 0 \end{pmatrix} \cdot \begin{pmatrix} 1 & & & \\ & xq & & \\ & & xq^2 & \\ & & & x^2yq^4 \end{pmatrix} \cdot \begin{pmatrix} F_1(xq^2) \\ F_2(xq^2) \\ F_3(xq^2) \\ F_4(xq^2) \end{pmatrix}. \tag{2.5}$$

3 q-Borel Operators

In this section, following [12, 15], we introduce a family of operators $\mathcal{B}_k$ for integers k, which can be treated as q-analogs of the Borel transformation.

Definition 3.1 Let $\mathbb{K}$ be a field. Let $F(x) = \sum_{n\geq 0} f(n)x^n \in \mathbb{K}(q)[[x]]$. We define the operator $\mathcal{B}_k$ for $k \in \mathbb{Z}$ by

$$\mathcal{B}_k\big(F(x)\big) := \sum_{n\geq 0} f(n)q^{-k\binom{n}{2}}x^n. \tag{3.1}$$

The following property of $\mathcal{B}_k$ will play an important role.

Lemma 3.1 *Let $F(x) \in \mathbb{K}(q)[[x]]$. For any integers k and N, and nonnegative integer M, we have*

$$\mathcal{B}_k\big(x^M F(xq^N)\big) = x^M q^{-k\binom{M}{2}} \mathcal{B}_k\big(F(xq^{N-kM})\big). \tag{3.2}$$

Proof Let us write $F(x) = \sum_{n\geq 0} f(n)x^n$. Then

$$\begin{aligned}\mathcal{B}_k\big(x^M F(xq^N)\big) &= \sum_{n\ge 0} f(n) q^{-k\binom{M+n}{2}+Nn} x^{M+n}\\ &= x^M q^{-k\binom{M}{2}} \sum_{n\ge 0} f(n) q^{-k(Mn+\binom{n}{2})+Nn} x^n\\ &= x^M q^{-k\binom{M}{2}} \sum_{n\ge 0} f(n) q^{-k\binom{n}{2}} (xq^{N-kM})^n\\ &= x^M q^{-k\binom{M}{2}} \mathcal{B}_k\big(F(xq^{N-kM})\big),\end{aligned}$$

which is our desired result. □

4 Non-standard Generating Function

In this section, we solve the matrix equation (2.5) and give a proof of Theorem 1.1. First, we observe from (2.4) that

$$\begin{aligned}\sum_{\lambda\in\mathscr{A}} x^{\sharp(\lambda)} y^{\tau(\lambda)} q^{|\lambda|} &= G_1(x)+G_2(x)+G_3(x)+G_4(x)\\ &= F_1(x). \end{aligned} \tag{4.1}$$

Theorem 4.1 *Let*

$$P(x) := \sum_{\lambda\in\mathscr{A}} x^{\sharp(\lambda)} y^{\tau(\lambda)} q^{|\lambda|}.$$

Then

$$P(x) = (1+xq)P(xq^2) + (xq^2+x^2yq^4)P(xq^4). \tag{4.2}$$

Proof By (2.5), we have

$$F_2(x) = F_1(x) \tag{4.3}$$

and

$$F_3(x) = F_4(x) = F_1(xq^2). \tag{4.4}$$

Also,

$$F_1(x) = F_1(xq^2) + xqF_2(xq^2) + xq^2F_3(xq^2) + x^2yq^4F_4(xq^2).$$

Inserting (4.3) and (4.4) into the above and recalling that $P(x) = F_1(x)$, we arrive at the desired result. □

It looks not easy to solve the q-difference equation in (4.2) directly. Now, we show how to take advantage of the q-Borel operators to prove Theorem 1.1.
Proof of Theorem 1.1 We apply $\mathcal{B}_2$ to both sides of (4.2). Then

$$\begin{aligned}\mathcal{B}_2\big(P(x)\big) = \mathcal{B}_2\big(P(xq^2)\big) + xq\mathcal{B}_2\big(P(x)\big)\\ + xq^2\mathcal{B}_2\big(P(xq^2)\big) + x^2yq^2\mathcal{B}_2\big(P(x)\big).\end{aligned}$$

For convenience, we define

$$Q(x) := \mathcal{B}_2\big(P(x)\big).$$

Then,

$$(1 - xq - x^2yq^2)Q(x) = (1 + xq^2)Q(xq^2).$$

Recalling that $Q(0) = P(0) = 1$, we have

$$Q(x) = \prod_{n\geq 0} \frac{1 + xq^{2n+2}}{1 - xq^{2n+1} - x^2yq^{4n+2}}.$$

Finally, we notice that

$$\begin{aligned}Q(x) &= \mathcal{B}_2\big(P(x)\big)\\ &= \mathcal{B}_2\left(\sum_{\lambda\in\mathscr{A}} x^{\sharp(\lambda)}y^{\tau(\lambda)}q^{|\lambda|}\right)\\ &= \sum_{\lambda\in\mathscr{A}} x^{\sharp(\lambda)}y^{\tau(\lambda)}q^{|\lambda|-2\binom{\sharp(\lambda)}{2}}.\end{aligned}$$

We are therefore led to (1.2). □

5 Theorem A

The object of this section is an alternative proof of Theorem A. Our starting point is (1.3).

Proof of (1.3) Setting $y = -1$ in (1.2) gives

$$\sum_{\lambda\in\mathscr{A}}(-1)^{\tau(\lambda)}x^{\sharp(\lambda)}q^{|\lambda|-\sharp(\lambda)(\sharp(\lambda)-1)} = \frac{(-xq^2;q^2)_\infty}{\prod_{n\geq 0}(1-xq^{2n+1}+x^2q^{4n+2})}$$
$$= (-xq^2;q^2)_\infty\frac{(-xq;q^2)_\infty}{(-x^3q^3;q^6)_\infty}$$
$$= \frac{(-xq;q)_\infty}{(-x^3q^3;q^6)_\infty}.$$

Recall Euler's first and second summations [4, Corollary 2.2, p. 19]:

$$\frac{1}{(t;q)_\infty} = \sum_{m\geq 0}\frac{t^m}{(q;q)_m} \tag{5.1}$$

and

$$(t;q)_\infty = \sum_{m\geq 0}\frac{(-t)^m q^{m(m-1)/2}}{(q;q)_m}. \tag{5.2}$$

We have

$$\sum_{\lambda\in\mathscr{A}}(-1)^{\tau(\lambda)}x^{\sharp(\lambda)}q^{|\lambda|-\sharp(\lambda)(\sharp(\lambda)-1)} = \sum_{n_1\geq 0}\frac{x^{n_1}q^{\binom{n_1}{2}+n_1}}{(q;q)_{n_1}}\sum_{n_2\geq 0}\frac{(-1)^{n_2}x^{3n_2}q^{3n_2}}{(q^6;q^6)_{n_2}}.$$

Therefore,

$$\sum_{\lambda\in\mathscr{A}}(-1)^{\tau(\lambda)}x^{\sharp(\lambda)}q^{|\lambda|} = \sum_{n_1,n_2\geq 0}\frac{(-1)^{n_2}q^{\binom{n_1}{2}+n_1+3n_2}x^{n_1+3n_2}q^{2\binom{n_1+3n_2}{2}}}{(q;q)_{n_1}(q^6;q^6)_{n_2}}$$
$$= \sum_{n_1,n_2\geq 0}\frac{(-1)^{n_2}q^{3\binom{n_1}{2}+18\binom{n_2}{2}+6n_1n_2+n_1+9n_2}x^{n_1+3n_2}}{(q;q)_{n_1}(q^6;q^6)_{n_2}}.$$

This is our desired result. □

Now, we are ready to show that $D(n) = E(n)$.

Let $\mathscr{S}$ denote the set of partitions counted by $D(n)$ in Theorem S for all nonnegative n. Andrews, Bringmann, and Mahlburg [6] proved that the generating function for the partition set $\mathscr{S}$ can be represented as a double series:

$$\sum_{\lambda\in\mathscr{S}}x^{\sharp(\lambda)}q^{|\lambda|} = \sum_{n_1,n_2\geq 0}\frac{(-1)^{n_2}q^{3\binom{n_1}{2}+18\binom{n_2}{2}+6n_1n_2+n_1+9n_2}x^{n_1+2n_2}}{(q;q)_{n_1}(q^6;q^6)_{n_2}}. \tag{5.3}$$

We therefore have

$$\sum_{\lambda\in\mathscr{A}}(-1)^{\tau(\lambda)}q^{|\lambda|}=\sum_{\lambda\in\mathscr{S}}q^{|\lambda|}=\sum_{n_1,n_2\geq 0}\frac{(-1)^{n_2}q^{3\binom{n_1}{2}+18\binom{n_2}{2}+6n_1n_2+n_1+9n_2}}{(q;q)_{n_1}(q^6;q^6)_{n_2}}.$$

This implies that $D(n)=E(n)$.

6 The Continuous q-Hermite Polynomials

Here, we prove the generating function identity in Corollary 1.3. Let

$$S(x)=\sum_{M\geq 0}s_M(y)x^M:=\frac{1}{\prod_{n\geq 0}(1-xq^{2n+1}-x^2yq^{4n+2})}.$$

We may compute that $s_0(y)=1$ and $s_1(y)=q/(1-q^2)$. Also, we have

$$(1-xq-x^2yq^2)S(x)=S(xq^2).$$

Therefore, for $M\geq 1$,

$$(1-q^{2M+2})s_{M+1}(y)=qs_M(y)+yq^2s_{M-1}(y). \tag{6.1}$$

Now, we define, for $M\geq 0$,

$$t_M(y):=(q^2;q^2)_Mq^{-M}s_M(y).$$

Then $t_0(y)=t_1(y)=1$. Also, (6.1) becomes

$$t_{M+1}(y)=t_M(y)+y(1-q^{2M})t_{M-1}(y) \tag{6.2}$$

for $M\geq 1$. To build the connection between $t_M(y)$ and the continuous q-Hermite polynomials, we define, for $M\geq 0$,

$$r_M(y)=(2y)^Mt_M(-\tfrac{1}{4y^2}).$$

Then $r_0(y)=1$ and $r_1(y)=2y$. Further, (6.2) becomes

$$r_{M+1}(y)=2yr_M(y)-(1-q^{2M})r_{M-1}(y) \tag{6.3}$$

for $M \geq 1$. Comparing with (1.4), we have

$$r_M(y) = H_M(y; q^2)$$

for $M \geq 0$. Thus, (1.6) is established.

Finally,

$$\begin{aligned} S(x) &= \sum_{M \geq 0} s_M(y) x^M \\ &= \sum_{M \geq 0} \frac{x^M q^M t_M(y)}{(q^2; q^2)_M}. \end{aligned}$$

Thus, by Euler's second summation (5.2),

$$\begin{aligned} \sum_{\lambda \in \mathscr{A}} x^{\sharp(\lambda)} y^{\tau(\lambda)} q^{|\lambda| - \sharp(\lambda)(\sharp(\lambda)-1)} &= \frac{(-xq^2; q^2)_\infty}{\prod_{n \geq 0} (1 - xq^{2n+1} - x^2 y q^{4n+2})} \\ &= \sum_{M \geq 0} \frac{x^M q^M t_M(y)}{(q^2; q^2)_M} \sum_{N \geq 0} \frac{x^N q^{2\binom{N}{2} + 2N}}{(q^2; q^2)_N}. \end{aligned}$$

We conclude that

$$\sum_{\lambda \in \mathscr{A}} x^{=(\lambda)} y^{\tau(\lambda)} q^{|\lambda|} = \sum_{M, N \geq 0} \frac{t_M(y) q^{2\binom{N}{2} + M + 2N} x^{M+N} q^{2\binom{M+N}{2}}}{(q^2; q^2)_M (q^2; q^2)_N},$$

which yields (1.5).

Acknowledgements The author was supported by a Killam Postdoctoral Fellowship from the Killam Trusts.

References

1. G. E. Andrews, Partition identities, *Advances in Math.* **9** (1972), 10–51.
2. G. E. Andrews, A general theory of identities of the Rogers-Ramanujan type, *Bull. Amer. Math. Soc.* **80** (1974), 1033–1052.
3. G. E. Andrews, Problems and prospects for basic hypergeometric functions, in: *Theory and application of special functions (Proc. Advanced Sem., Math. Res. Center, Univ. Wisconsin, Madison, Wis., 1975)*, 191–224, Math. Res. Center, Univ. Wisconsin, Publ. No. **35**, Academic Press, New York, 1975.
4. G. E. Andrews, *The theory of partitions*, Reprint of the 1976 original. Cambridge Mathematical Library. Cambridge University Press, Cambridge, 1998.

5. G. E. Andrews, Schur's theorem, partitions with odd parts and the Al-Salam–Carlitz polynomials, in: *q-Series from a contemporary perspective (South Hadley, MA, 1998)*, 45–56, Contemp. Math., **254**, Amer. Math. Soc., Providence, RI, 2000.
6. G. E. Andrews, K. Bringmann, and K. Mahlburg, Double series representations for Schur's partition function and related identities, *J. Combin. Theory Ser. A* **132** (2015), 102–119.
7. G. E. Andrews, S. Chern, and Z. Li, Linked partition ideals and the Alladi–Schur theorem, *J. Combin. Theory Ser. A* **189** (2022), Paper No. 105614, 19 pp.
8. S. Chern, Linked partition ideals, directed graphs and q-multi-summations, *Electron. J. Combin.* **27** (2020), no. 3, Paper No. 3.33, 29 pp.
9. S. Chern, Linked partition ideals and Andrews–Gordon type series for Alladi and Gordon's extension of Schur's identity, *Rocky Mountain J. Math.*, accepted.
10. S. Chern and Z. Li, Linked partition ideals and Kanade–Russell conjectures, *Discrete Math.* **343** (2020), no. 7, Paper No. 111876, 24 pp.
11. R. Koekoek and R. F. Swarttouw, *The Askey-scheme of hypergeometric orthogonal polynomials and its q-analogue*, Delft University of Technology, Faculty of Information Technology and Systems, Department of Technical Mathematics and Informatics, Report no. **98-17**, 1998.
12. J.-P. Ramis, About the growth of entire functions solutions of linear algebraic q-difference equations, *Ann. Fac. Sci. Toulouse Math. (6)* **1** (1992), no. 1, 53–94.
13. I. Schur, Zur additiven Zahlentheorie, *S.-B. Preuss. Akad. Wiss. Phys.-Math. Kl.* (1926), 488–495.
14. A. J. Yee, A combinatorial proof of Andrews' partition functions related to Schur's partition theorem, *Proc. Amer. Math. Soc.* 130 (2002), no. 8, 2229–2235.
15. C. Zhang, Une sommation discrète pour des équations aux q-différences linéaires et à coefficients analytiques: théorie générale et exemples (in French), in: *Differential equations and the Stokes phenomenon*, 309–329, World Sci. Publ., River Edge, NJ, 2002.

Semi-magic Matrices for Dihedral Groups

Robert W. Donley

1 Introduction

Just over a century ago, the problem of counting semi-magic squares was initiated by MacMahon [12], who gave formulas for enumerating such squares of size three with fixed line sum. Problems of this type remain an active area of study in combinatorics and, in particular, the subject of permutation polytopes. The theory involved touches on topics including, but not exhaustively, permutation matrices [4], Ehrhart polynomials (see [3] or [14] for general background), Stanley's proof of the Anand-Dumir-Gupta conjecture [1, 13], toric varieties, elliptic curves and zeta functions, and $3j$-symbols in the quantum theory of angular momentum, in the form of Regge symbols and Regge symmetries.

A natural source of semi-magic squares comes from permutation polytopes, which arise from realizations of finite groups as subgroups in a symmetric group S_n, the permutations of which may in turn be realized as permutation matrices. The permutation matrices are the fundamental atoms of the theory. Because these constructions are rooted in group theory, we approach the subject with the tools of representation theory, with an emphasis on intertwining operators, in particular, the homomorphism to permutation matrices. See [10], Theorem 3.2. With this morphism of central interest, we place these sets of semi-magic squares in vector spaces, in fact, semisimple associative algebras, and obtain both old and new results through consideration of the kernel and image.

In our case, we consider semi-magic squares associated with the dihedral groups D_{2n} with $2n$ elements. Combinatorial questions, such as face structure of the permutation polytope, are well understood [2, 6, 15], with several formulas given for the Ehrhart polynomials. Although we obtain the same formulas here, our narrower

R. W. Donley (✉)
Queensborough Community College, Bayside, New York, NY, USA
e-mail: RDonley@qcc.cuny.edu

M. B. Nathanson (ed.), *Combinatorial and Additive Number Theory V*,
Springer Proceedings in Mathematics & Statistics 395,
https://doi.org/10.1007/978-3-031-10796-2_6

concern is counting semi-magic squares obtained from a set of generators. We consider this question using an elementary model ([14], p. 225, Problem 15; p. 561, Problem 53), give a simple form of the generating function, and reconcile some of the several formulas.

On the other hand, because semi-magic matrices are closed under matrix multiplication, the image of the permutation map generalizes the commutative algebra of circulant matrices (for instance, [8, 11]), which has been a subject of traditional, ongoing interest. Again we use methods of representation theory, in particular, characters and projection formulas, to describe this extension both in representation theoretic terms and in terms of the Artin-Wedderburn theorem for semisimple associative algebras. We do not consider properties as a Lie algebra as found in [5], although we give bases with quaternionic properties as a further source of distinguished semi-magic matrices.

We note that, while most results here require only the real numbers, it is helpful to assume all representations are over the complex numbers. The first three sections of [9] cover all required background from representation theory. For a thorough account with consideration of general fields and for Artin-Wedderburn theory of semisimple associative algebras, see [7].

2 Definitions and Notations

Fix $n \geq 3$, and let $G = D_{2n}$ denote the dihedral group with $2n$ elements. We consider four realizations of this group:

- the symmetry group of the regular n-gon,
- a presentation on two generators with three relations,
- a subgroup of the symmetric group on n elements, and
- a subgroup of the group of permutation matrices of size n.

It will be extremely convenient to switch between these realizations, depending on which properties we wish to emphasize. Let e denote the identity in the first three realizations.

In the first realization, we orient the regular n-gon symmetrically about the origin in the xy-plane, labeling the vertices counter-clockwise by 1 through n, with 1 closest to the x-axis in the first quadrant. Vertices lie on the x-axis only when n is odd, in which case the vertex labeled $\frac{n+1}{2}$ lies on the negative x-axis.

Denote by R the counter-clockwise rotation about the origin by $\frac{2\pi}{n}$ and by C the reflection across the x-axis (or complex conjugation). Then D_{2n} contains n rotations of the form

$$R^k \qquad (0 \leq k < n)$$

and n reflections

$$CR^k \qquad (0 \leq k < n).$$

With $|x|$ denoting the order of the element x, the second realization may be given by the relations

$$|R| = n, \qquad |C| = 2, \qquad CRC = R^{-1}. \tag{1}$$

In cycle notation for permutations,

$$R = (12\ldots n), \qquad C = (1n)(2\,n-1)\ldots,$$

and one readily verifies these permutations satisfy the three relations.

Finally, one may assign to each permutation σ its corresponding permutation matrix P_σ: for $1 \le i, j \le n$,

$$if \quad \sigma(i) = j \quad then \quad [P_\sigma]_{ji} = 1, \quad otherwise \quad [P_\sigma]_{ji} = 0.$$

If $\{e_i\}$ denotes the standard basis of $\mathbb{C}^n$, then $P_\sigma(e_i) = e_{\sigma(i)}$ and $P_\tau P_\sigma = P_{\tau\circ\sigma}$, where $\circ$ denotes composition of maps, usually omitted henceforth. The rotations belong to the space of circulant matrices, that is, the corresponding matrices have all entries equal to zero except for one diagonal of ones, continuing through the left-hand side. The diagonal of ones for R^k starts in column one at row $k+1$. On the other hand, with our choice of numbering, the reflections also correspond to (-1)-circulant matrices, now with the diagonal of ones to the left. In particular, the counteridentity matrix P_C has a diagonal of ones along the main diagonal to the left. For example, when $n = 4$,

$$P_R = P_{(1234)} = \begin{vmatrix} 0&0&0&1 \\ 1&0&0&0 \\ 0&1&0&0 \\ 0&0&1&0 \end{vmatrix}, \quad P_C = P_{(14)(23)} = \begin{vmatrix} 0&0&0&1 \\ 0&0&1&0 \\ 0&1&0&0 \\ 1&0&0&0 \end{vmatrix}.$$

Multiplication by P_C on the left inverts columns, while multiplication on the right inverts rows. Thus multiplication by C on either side switches between rotations and reflections or, by P_C, circulant and (-1)-circulant elements of D_{2n}.

3 Structure of D_{2n}

We review the basic structure of D_{2n}, with an emphasis on features needed for character tables. In particular, we consider conjugacy classes and the commutator subgroup. Both items vary based on the parity of n. An efficient calculation of both items follows directly from (1).

Denote the conjugacy class of σ by C_σ, and recall that the commutator subgroup $[G, G]$ of G is generated by the commutators

$$\{xyx^{-1}y^{-1} \mid x, y \in G\}.$$

Example 1 (*n odd*) There are $\frac{n+3}{2}$ conjugacy classes in D_{2n}, given by

- $\{e\}$;
- $\frac{n-1}{2}$ classes of the form $\{R^{\pm k}\}$ for $1 \leq k \leq \frac{n-1}{2}$;
- one single class C_C of size n consisting of all reflections.

The commutator subgroup consists of the rotation subgroup $\langle R \rangle$, so that the abelianization $D_{2n}/[D_{2n}, D_{2n}]$ of D_{2n} has two elements. Thus, the character group consists of two elements

$$D_{2n}^* = \{\chi_{triv}, \ \chi_{det}\},$$

where, for all g in D_{2n},

$$\chi_{triv}(g) = 1, \qquad \chi_{det}(R^k) = 1, \qquad \chi_{det}(CR^k) = -1.$$

With n odd, the sgn character

$$\chi_{sgn}(\sigma) = det(P_\sigma)$$

alternates between χ_{triv} ($\frac{n-1}{2}$ $even$) and χ_{det} ($\frac{n-1}{2}$ odd) based on the number of transpositions in $C = (1n)(2\ n-1)\ldots.$

Example 2 (*n even*) When $n = 2m$, there are $m + 3$ conjugacy classes, given by

- $\{e\}$;
- r^m, the nontrivial central element;
- $m - 1$ classes of the form $\{R^{\pm k}\}$ for $1 \leq k < m$; and
- two reflection classes $C_C = C_{CR^2}$ and C_{CR}, each of size m.

The splitting of the reflections is also seen in permutations; conjugation preserves cycle structure, and reflections have either no fixed points (C) or two fixed points (CR).

The commutator subgroup now consists of the subgroup $\langle R^2 \rangle$, so that the abelianization of D_{2n} has four elements, given by

$$\mathbb{Z}/2 \times \mathbb{Z}/2 \cong \{\bar{e}, \ \bar{R}, \bar{C}, \ \overline{CR}\};$$

the nontrivial elements have order two. Thus, the character group consists of four elements, given by

$$D_{2n}^* = \{\chi_{triv}, \ \chi_{det}, \ \chi_{sgn}, \ \chi_{det} \cdot \chi_{sgn}\}.$$

As before, the values of χ_{sgn} and $\chi_{det} \cdot \chi_{sgn}$ on the reflection classes alternate based on the cycle structure of C.

4 Irreducible Types for ρ

Now we define the permutation representation

$$\rho : D_{2n} \to GL(n, \mathbb{C}),$$

$$\rho(\sigma) = P_\sigma$$

and give its decomposition in terms of irreducible representations of D_{2n}. As a permutation representation, the character of ρ encodes the fixed points for each permutation.

Let $\pi_2 : D_{2n} \to GL(2, \mathbb{C})$ be the two-dimensional representation defined by extending the action of D_{2n} on $\mathbb{R}^2$ to $\mathbb{C}^2$. One verifies from Tables 1 and 2 that π_2 is irreducible.

For an efficient tabulation of all two-dimensional irreducible representations for D_{2n}, the mappings

$$\varphi_j : D_{2n} \to D_{2n},$$

$$\varphi_j(R) = R^j, \qquad \varphi_j(C) = C$$

define homomorphisms for $0 \le j < n$. For $1 \le j < n$, define

$$\pi_2^j = \pi_2 \circ \varphi_j.$$

Table 1 Character table for D_{2n} (n odd)

σ	e	$R^{\pm k}$	C
$\lvert C_\sigma \rvert$	1	2	n
χ_{triv}	1	1	1
χ_{det}	1	1	-1
π_2	2	$2\cos(2k\pi/n)$	0
ρ	n	0	1

Table 2 Character table for D_{2n} ($n = 2m$)

σ	e	R^m	$R^{\pm k}$ (k odd)	$R^{\pm k}$ (k even)	C	CR
$\lvert C_\sigma \rvert$	1	1	2	2	m	m
χ_{triv}	1	1	1	1	1	1
χ_{det}	1	1	1	1	-1	-1
χ_{sgn}	1	$(-1)^m$	-1	1	$(-1)^m$	$(-1)^{m+1}$
$\chi_{det} \cdot \chi_{sgn}$	1	$(-1)^m$	-1	1	$(-1)^{m+1}$	$(-1)^m$
π_2	2	-2	$2\cos(2k\pi/n)$	$2\cos(2k\pi/n)$	0	0
ρ	n	0	0	0	0	2

Then the complete set $\widehat{G}$ of irreducible classes is given by the one-dimensional characters and the two-dimensional representations π_2^j (with $1 \leq j \leq \frac{n-1}{2}$ for n odd, and with $1 \leq j < m-1$ for $n = 2m$).

In Table 1, the character table for n odd, we append each character for π_2^j, which equals that of π_2, but with the replacement $2\cos(\frac{2jk\pi}{n})$. Of course, since the characters agree, π_2^j and π_2^{-j} are equivalent as representations.

To see the row orthogonality relations, one uses basic trigonometric identities with the following lemma and the corresponding identity for sine.

Lemma 1 *Suppose $n \geq 2$ and j is not a multiple of n. Then*

$$\sum_{k=0}^{n-1} \cos(2jk\pi/n) = 0.$$

Proof Suppose $\omega \neq 1$ is an n-th root of unity. Then

$$1 + \omega + \omega^2 + \cdots + \omega^{n-1} = 0.$$

Taking the real part of each term with $\omega = e^{2j\pi i/n}$, the lemma follows. □

Now orthogonality relations for characters immediately give the following.

Proposition 1 *Suppose n is odd. As representations of D_{2n},*

$$\rho \cong \chi_{triv} \oplus \bigoplus_{j=1}^{\frac{n-1}{2}} \pi_2^j.$$

Table 2 shows the character table when $n = 2m$. Here the character for π_2^j requires the change to cosine as before and additionally the -2 becomes $(-1)^j 2$. One also sees immediately that

$$\pi_2^m \cong \chi_{sgn} \oplus \chi_{det} \cdot \chi_{sgn}.$$

Proposition 2 *Suppose $n = 2m$. As representations of D_{2n},*

$$\rho \cong \chi_{triv} \oplus \chi' \oplus \bigoplus_{j=1}^{m-1} \pi_2^j,$$

where

$$\chi' = \chi_{sgn}\ (m\ odd) \quad or \quad \chi' = \chi_{det} \cdot \chi_{sgn}\ (m\ even).$$

5 The Intertwining Operator Φ_ρ

We recall the definition of the group algebra for a finite group G.

Definition 1 Let G be a finite group, and suppose $g_1, \ldots, g_n$ are the elements of G. Define $\mathbb{C}[G]$ to be the vector space of linear combinations over the formal basis $\{e_{g_1}, \ldots, e_{g_n}\}$. We define multiplication by extending the multiplication of basis elements on indices:

$$e_g \cdot e_{g'} = e_{g \cdot g'}.$$

Note that $\dim_{\mathbb{C}} \mathbb{C}[G] = |G|$.

If (π, V_π) is a representation of G, then one extends π to a homomorphism of associative algebras by

$$\Phi_\pi : \mathbb{C}[G] \to Hom_{\mathbb{C}}(V_\pi, V_\pi),$$

$$\Phi_\pi\left(\sum x_i e_{g_i}\right) = \sum x_i \pi(g_i).$$

Both the kernel and image of Φ_π are subrepresentations of $G \times G$ in the domain and range, respectively, and one has the isomorphism of algebras

$$Im(\Phi_\pi) \cong \mathbb{C}[G]/Ker\ \Phi_\pi.$$

Here the corresponding actions of $G \times G$ are given by extending

$$(L \otimes R)(g_1, g_2)e_x = e_{g_1 x g_2^{-1}},$$

$$(Hom(\pi, \pi)(g_1, g_2))M = \pi(g_1)M\pi(g_2^{-1}).$$

Additionally, we have the analogue of the Peter-Weyl theorem for finite groups:

Proposition 3 *As representations of $G \times G$,*

$$\mathbb{C}[G] \cong \bigoplus_{\pi'} Hom_{\mathbb{C}}(V_{\pi'}, V_{\pi'}),$$

where the sum ranges over the set $\widehat{G}$ of irreducible classes of representations for G. Each summand on the right-hand side is irreducible for $G \times G$.

Each irreducible subrepresentation for $G \times G$ corresponds to a simple two-sided ideal of the algebra structure, and the one-dimensional ideals in $\mathbb{C}[G]$ are directly obtained as follows.

Proposition 4 *Suppose $\chi : G \to \mathbb{C}^*$ is a (one-dimensional) character of G. Then*

$$|G|e_\chi = \sum_{g \in G} \chi(g^{-1})e_g$$

spans the subspace of $\mathbb{C}[G]$ corresponding to $Hom_{\mathbb{C}}(\mathbb{C}_\chi, \mathbb{C}_\chi)$.

Next Schur's lemma now implies that there exists a subset X_π of the set $\widehat{G}$ of irreducible classes such that

$$Ker(\Phi_\pi) = \bigoplus_{\pi' \in X_\pi} Hom_{\mathbb{C}}(V_{\pi'}, V_{\pi'}),$$

$$Im(\Phi_\pi) = \bigoplus_{\pi' \notin X_\pi} Hom_{\mathbb{C}}(V_{\pi'}, V_{\pi'}).$$

From Propositions 1 and 2, we immediately obtain for the permutation representation Φ_ρ for D_{2n}:

Proposition 5 *(a) If n is odd, then $X_\rho = \{\chi_{det}\}$. That is,*

$$Ker(\Phi_\rho) = \mathbb{C}e_{\chi_{det}}.$$

(b) If $n = 2m$, then $X_\rho = \{\chi_{det}, \chi''\}$, where

$$\chi'' = \chi_{sgn}\ (m\ even) \quad or \quad \chi'' = \chi_{det} \cdot \chi_{sgn}\ (m\ odd).$$

That is,

$$Ker(\Phi_\rho) = \mathbb{C}e_{\chi_{det}} \oplus \mathbb{C}e_{\chi''}.$$

6 Semi-magic Matrices

Closely related to permutation representations are semi-magic matrices. We recall basic properties about these matrices, with attention to properties from the previous section.

Definition 2 An element M of $M(n, \mathbb{C})$ is called a **semi-magic matrix** if the sums along each row and column are equal. This common sum r_M is called the **line sum** of M. Denote the vector space of semi-magic matrices of size n by $MM(n)$.

Example 3 Every permutation matrix is a semi-magic matrix with line sum 1, and in fact the set of all permutation matrices form a spanning set of $MM(n)$. The line sum of a linear combination of permutation matrices is evident:

$$M = \sum x_\sigma P_\sigma \quad \mapsto \quad r_M = \sum x_\sigma .$$

Example 4 Consider the element J in $MM(n)$ with all entries equal to one. Then one sees immediately that the line sum of J is n, and, for all M in $MM(n)$ with line sum r,

$$JM = MJ = rJ,$$

so that J is in the center of $MM(n)$. In particular, $J^2 = nJ$.

In general,

Proposition 6 *If M_1 and M_2 are in $MM(n)$ with line sums r_1 and r_2 then so is $M_1 M_2$ with line sum $r_1 r_2$.*

Proof Let $e = (1, \ldots, 1)$ be in $\mathbb{C}^n$. That M has the magic property with line sum r is equivalent to the eigenvector condition

$$Me = M^T e = re.$$

Then

$$M_1 M_2 e = r_2 M_1 e = r_1 r_2 e$$

and

$$(M_1 M_2)^T e = M_2^T M_1^T e = r_1 M_2^T e = r_2 r_1 e.$$

□

Thus $MM(n)$ is an associative algebra over $\mathbb{C}$, and the line sum functional is a linear character with respect to the algebra structure. Furthermore, the corresponding map

$$\Phi : \mathbb{C}[S_n] \to MM(n)$$

is a surjective intertwining operator between representations of $S_n \times S_n$ as before, and it is not difficult to see that the decomposition into irreducible subrepresentations is given by

$$MM(n) \cong \mathbb{C}J \oplus MM_0(n),$$

where

$$MM_0(n) = \{M \in MM(n) \mid JM = 0\}$$

is the simple two-sided ideal in $MM(n)$ where all elements have line sum equal to zero. Thus, for the center of $MM(n)$,

$$Z(MM(n)) = Span_{\mathbb{C}}(I_n, J),$$

where I_n is the identity matrix of size n.

More useful for what follows, the weaker statement that products of linear combinations are closed follows from the closure property of group multiplication:

$$\left(\sum x_\sigma P_\sigma\right)\left(\sum y_\tau P_\tau\right) = \sum x_\sigma y_\tau P_{\sigma\tau}.$$

With this note, the next two definitions merely recast the permutation mapping from the previous section.

Definition 3 Suppose G is a subgroup of S_n. Denote by $MM(G) \subseteq MM(n)$ the associative algebra over $\mathbb{C}$ generated by the elements of G as permutation matrices. We call $MM(G)$ the **permutation algebra** associated to G. This algebra depends on the embedding of G in S_n.

Definition 4 The intertwining operator

$$\Phi_\rho : \mathbb{C}[G] \to MM(G) \subset MM(n),$$

$$\sum x_\sigma e_\sigma \quad \mapsto \quad \sum x_\sigma P_\sigma$$

is a surjection between both $G \times G$ representations and associative algebras over $\mathbb{C}$. With respect to either structure,

$$MM(G) \; \cong \mathbb{C}[G]/Ker(\Phi_\rho).$$

Additionally, Φ_ρ carries coefficient sums to line sums.

7 Semi-magic Squares for $MM(D_{2n})$

We now consider the problem of enumerating certain types of semi-magic squares.

Definition 5 A semi-magic matrix M is called a **semi-magic square** if it has entries in the non-negative integers. Alternatively, M may be written as

$$M = \sum n_\sigma P_\sigma,$$

where each n_σ is a non-negative integer. In this case, the line sum of M is $\sum n_\sigma$.

Definition 6 Let $H_G(r)$ denote the number of semi-magic squares in the permutation algebra $MM(G)$ with line sum equal to r.

Traditionally $H_n(r)$ is used to denote the number of all semi-magic squares of size n with line sum equal to r. Problems of this type go back to MacMahon for the case of size three. We base our results for $H_{D_{2n}}(r)$ on the model found in [14], p. 225, Problem 15 for the case of $n = 3$.

Theorem 1 *Suppose n is odd and $H_{D_{2n}}(r)$ counts the number of semi-magic squares in $MM(D_{2n})$ with line sum r. Then*

$$H_{D_{2n}}(r) = \binom{r+2n-1}{2n-1} - \binom{r+n-1}{2n-1},$$

which has generating function

$$F_n(x) = \sum_{r\geq 0} H_{D_{2n}}(r)x^r = \frac{1-x^n}{(1-x)^{2n}}.$$

Proof Since the permutation matrices form a spanning set for $MM(D_{2n})$, we may identify each linear combination in $MM(D_{2n})$ with a $2n$-tuple of complex numbers, ordered so that the rotation elements precede the reflection elements. That is,

$$\sum x_i P_{\sigma_i} \mapsto (x_1, \ldots, x_{2n}).$$

By Proposition 5, the non-uniqueness of this representation is captured by the single dependence relation from χ_{det}:

$$\sum_{\sigma} \chi_{det}(\sigma)P_\sigma = 0 \qquad or \qquad \sum_k P_{R^k} = \sum_k P_{CR^k} \ \ (= J).$$

In terms of $2n$-tuples, this relation presents as

$$(x_1+1, \ \ldots, \ x_n+1, \ x_{n+1}, \ \ldots, \ x_{2n}) = (x_1, \ \ldots, \ x_n, \ x_{n+1}+1, \ \ldots, \ x_{2n}+1).$$

Thus each semi-magic square with line sum r is uniquely represented by a $2n$-tuple of non-negative integers such that

- the entries sum to r and
- at least one value in the last n entries is equal to zero.

The first term of $H_{D_{2n}}(r)$ counts the number of way to place r balls in $2n$ boxes (weak compositions of r into $2n$ parts); to guarantee a vanishing entry, we discard $2n$-tuples in which the last n entries are non-zero, or of the form

$$(x_1, \ \ldots, \ x_n, \ 1+x_{n+1}, \ \ldots, \ 1+x_{2n}),$$

where the x_i are non-negative integers and

$$\sum_{i=1}^{2n} x_i = r - n.$$

Finally, with $s \geq 1$, the generating function follows from the binomial series

$$\frac{1}{(1-x)^s} = \sum_{r \geq 0} \binom{r+s-1}{s-1} x^r.$$

□

Theorem 2 *Suppose $n = 2m$ and $H_{D_{2n}}(r)$ counts the number of semi-magic squares in $MM(D_{2n})$ with line sum r. Then*

$$H_{D_{2n}}(r) = \binom{r+4m-1}{2n-1} - 2\binom{r+3m-1}{2n-1} + \binom{r+2m-1}{2n-1},$$

which has generating function

$$F_n(x) = \sum_{r \geq 0} H_{D_{2n}}(r) x^r = \frac{(1-x^m)^2}{(1-x)^{2n}}.$$

Proof The approach is essentially the same as the previous theorem, except now Proposition 5 yields two relations

$$\sum \chi_{det}(\sigma) P_\sigma = 0 \quad \text{and} \quad \sum \chi''(\sigma) P_\sigma = 0,$$

which, noting the sign for r^m, reduce to the equalities

$$\sum P_{R^{2k}} = \sum P_{CR^{2k+1}} \;\; (= J_1), \qquad \sum P_{R^{2k+1}} = \sum P_{CR^{2k}} \;\; (= J_2).$$

Here J_1 is a matrix of alternating ones and zeros, starting with a one in the upper left corner; a similar pattern holds for J_2 but begins with a zero in this corner.

If we order the group elements as above with four sections, we are now counting as before but with at least one zero entry in both the second and fourth sections. If $r = r_1 + r_2$, then the distribution of r_1 to the first n entries and r_2 to the last n entries is independent, so that

$$H_{D_{2n}}(r) = \sum_{r_1=0}^{r} H_{D_n}(r_1) H_{D_n}(r - r_1).$$

Noting that this sum is a discrete convolution, we obtain the generating function by squaring that of $H_{D_n}(r)$, from which we obtain the counting formula. □

The equalities with J_1 and J_2 are directly seen using circulant matrices for rotations and reflections, recalling that multiplication by P_C on the left inverts columns. Also note that

$$J = J_1 + J_2, \quad J_1^2 = mJ_1, \quad J_2^2 = mJ_1, \quad J_1 J_2 = J_2 J_1 = mJ_2.$$

Finally, the following corollary gives the analogue of the Anand-Dumir-Gupta conjecture for D_{2n} after Stanley.

Corollary 1 *Suppose $G = D_{2n} \subset S_n$. The following properties of $H_G(r)$ hold:*

- *for n odd (resp., even), $H_G(r)$ is a polynomial of degree $2n-2$ (resp. $2n-3$);*
- *$H_G(-r) = (-1)^{n-1} H_G(r-n)$;*
- *$H_G(-1) = H_G(-2) = \cdots = H_G(-n+1) = 0$; and*
- *the coefficients of the numerator $h^*(x)$ in the reduced generating function are positive, symmetric, and unimodal.*

Proof The counting formula shows that $H_G(r)$ is a polynomial in r, with degree given by reducing $F_n(x)$ and applying the binomial series. Once reduced, the numerator is a polynomial in x with coefficients as noted. The remaining properties follow by expanding the binomial coefficients above into factorials. □

8 Alternative Counting Formulas

Alternative formulas for $H_{D_{2n}}(r)$ arise by either reducing the generating function or changing counting method. We consider when n is odd; for the even case, a similar formula follows with alteration for the discrete convolution. See [2] for the h^*-vector and generating function in the context of permutation polytopes.

Corollary 2 *Suppose n is odd, and $G = D_{2n}$. Then*

$$H_G(r) = \sum_{i=0}^{n-1} \binom{r + 2n - 2 - i}{2n - 2}.$$

Proof Using the proof of Theorem 1, we give an alternative counting scheme for the unique representations with sum r as follows. First, let X_0 denote all $2n$-tuples with a zero in the $2n$-th entry, and let X_i denote all $2n$-tuples with the last i entries non-zero and a zero in the $2n-i$-th entry. That is, all X_i have elements of the form

$$(0, 0, \ldots, 0, 1, \ldots, 1) + (x_1, x_2, \ldots, x_{2n-i-1}, 0, x_{2n-i+1}, \ldots),$$

where the first vector ends with i ones and all $x_k \geq 0$. Then, for $0 \leq i \leq n-1$, the number of elements in X_i is the i-th term of the sum. □

Effectively, we have reduced the generating function and applied the binomial series. When n is odd, the numerator of the reduced generating function is

$$h^*(x) = 1 + x + x^2 + \cdots + x^{n-1};$$

when n is even, this numerator is

$$h^*(x) = (1 + x + x^2 + \cdots + x^{m-1})^2 = 1 + 2x + 3x^2 + \cdots + 3x^{n-4} + 2x^{n-3} + x^{n-2}.$$

Another formula follows from the Principle of Inclusion-Exclusion ([3], p. 39); here we sum directly over the choice of i positions for zeros in the last n entries. See Theorem 1.2 of [6] for a similar formula.

Corollary 3 *Suppose n is odd, and $G = D_{2n}$. Then*

$$H_G(r) = \sum_{i=1}^{n} (-1)^{i+1} \binom{n}{i} \binom{r + 2n - 1 - i}{2n - 1 - i}.$$

To reconcile this formula with the generating function, if we start the sum at $i = 0$, then the binomial series gives

$$\sum_{r \geq 0} H_G(r) x^r = \sum_{i=0}^{n} (-1)^{i+1} \binom{n}{i} \frac{(1-x)^i}{(1-x)^{2n}} = \frac{-x^n}{(1-x)^{2n}}.$$

Now we subtract the term at $i = 0$ to get the result.

9 Orthogonal Idempotents for $MM(D_{2n})$

Let $N = \langle R \rangle$ be the normal subgroup of rotations in D_{2n}. The algebra $MM(D_{2n})$ extends the commutative algebra $Circ(n) = MM(N)$ of circulant matrices of size n. As a commutative algebra generated by a permutation matrix, this algebra, of dimension n, may be simultaneously diagonalized. In fact, with respect to the action of $N \times N$, we have

$$\mathbb{C}[N] \cong Im(\Phi_{\rho|_N}) \cong \bigoplus_{\chi \in N^*} Hom_{\mathbb{C}}(\mathbb{C}_\chi, \mathbb{C}_\chi),$$

where N^* is the set of n characters $(\chi, \mathbb{C}_\chi)$ on N. As a decomposition into simple two-sided ideals of $\mathbb{C}[N]$ and $Circ(n)$, the elements

$$e_\chi = \frac{1}{n} \sum \chi(R^{-k}) e_{R^k} \quad \textit{and} \quad U_\chi = \Phi_\rho(e_\chi) = \frac{1}{n} \sum \chi(R^{-k}) P_{R^k},$$

respectively, span the isotypic component $Hom(\chi, \chi)$ for $N \times N$. In particular, if $\omega = e^{2\pi i/n}$, $0 \leq j \leq n-1$, and $\chi_j(R) = \omega^j$, then

$$U_{\chi_j} = \frac{1}{n}\sum_{k=0}^{n-1} \omega^{-jk}\, P_{R^k} = \frac{1}{n}\begin{vmatrix} 1 & \omega^{j} & \omega^{2j} & \omega^{3j} & \dots \\ \omega^{-j} & 1 & \omega^{j} & \omega^{2j} & \dots \\ \omega^{-2j} & \omega^{-j} & 1 & \omega^{j} & \dots \\ \dots & \dots & \dots & \dots & \dots \end{vmatrix}$$

is evidently circulant and semi-magic with line sum equal to one if χ is trivial, zero otherwise.

In addition to giving diagonalizing bases for these algebras, we obtain a complete set of orthogonal idempotents corresponding to each simple two-sided ideal, that is, for instance, on $Circ(n)$,

$$U_\chi^2 = U_\chi, \qquad U_\chi U_{\chi'} = 0 \ (\chi \neq \chi'), \quad \sum U_\chi = I.$$

Then, as a decomposition of simple two-sided ideals,

$$Circ(n) = \sum_\chi Circ(n)U_\chi = \sum_\chi \mathbb{C}U_\chi.$$

As seen in Propositions 1 and 2, a similar decomposition for $MM(D_{2n})$ consists of simple two-sided ideals with diagonal blocks of size 1 and 2. A full description of this decomposition is given by the corresponding complete set of orthogonal idempotents.

Theorem 3 *Suppose n is odd, and define $c_j = \cos(2j\pi/n)$. The complete set of $\frac{n+1}{2}$ orthogonal idempotents for $MM(D_{2n})$ is given by*

$$U_{triv} = \frac{1}{n}\, J,$$

and, for each $1 \le j \le \frac{n-1}{2}$,

$$U_{\pi_2^j} = \frac{2}{n}\sum_{k=0}^{n-1} \cos(2kj\pi/n)\, P_{R^k} = \frac{2}{n}\begin{vmatrix} 1 & c_j & c_{2j} & c_{3j} & \dots \\ c_j & 1 & c_j & c_{2j} & \dots \\ c_{2j} & c_j & 1 & c_j & \dots \\ \dots & \dots & \dots & \dots & \dots \end{vmatrix}.$$

Proof Propositions 1 and 2 identify the isotypic components for D_{2n} in ρ. To obtain each U, we note the general projection formula onto the isotypic component for π:

$$P_\pi(v) = \frac{d_\pi}{|G|}\sum_g \chi_\pi(g^{-1})\rho(g)v,$$

where d_π is the dimension of the irreducible representation (π, V_π). Since the blocks in the group algebra may be identified uniquely by the left or right action alone, we use the left action applied to the identity element to obtain

$$U_\pi = P_\pi(I) = \frac{d_\pi}{2n} \sum_\sigma \chi_\pi(\sigma^{-1}) P_\sigma.$$

The idempotent, orthogonality, and completeness properties for U are inherited from the P_π. □

Theorem 4 *Suppose $n = 2m$, and preserve the notation of the previous theorem. The complete set of $m+1$ orthogonal idempotents for $MM(D_{2n})$ is given by*

$$U_{triv} = \frac{1}{n} J,$$

$$U_{\chi'} = \frac{1}{n} \left| \begin{array}{ccccc} 1 & -1 & 1 & -1 & \dots \\ -1 & 1 & -1 & 1 & \dots \\ 1 & -1 & 1 & -1 & \dots \\ \dots & \dots & \dots & \dots & \dots \end{array} \right|$$

and, as defined in Theorem 3, each $U_{\pi_2^j}$ for $1 \le j \le m-1$.

Proof The proof is unchanged from the previous theorem. For the second idempotent, it is helpful to note

$$P_C J_2 = P_C \sum P_{R^{2k+1}} = \sum P_{CR^{2k+1}} = \sum P_{R^{2k}} = J_1,$$

$$P_C J_1 = P_C \sum P_{R^{2k}} = \sum P_{CR^{2k}} = \sum P_{R^{2k+1}} = J_2.$$

□

Several items are worth noting in both theorems:

- each U is symmetric and circulant,
- U_{triv} has line sum equal to 1,
- U has line sum equal to 0 otherwise,
- although true from general theory, that these U form a complete set of orthogonal idempotents is directly checked,
- the set of all U gives a basis for the center of $MM(D_{2n})$, and
- the two-dimensional idempotents are twice the real parts of the idempotents U_{χ_j} in $Circ(n)$.

Corollary 4 *As a sum of simple two-sided ideals,*

$$MM(D_{2n}) = \mathbb{C}J \oplus \bigoplus_{j=1}^{\frac{n-1}{2}} MM(D_{2n})U_{\pi_2^j} \quad (n \text{ odd}), \quad and$$

$$MM(D_{2n}) = \mathbb{C}J \oplus \mathbb{C}U_{\chi'} \oplus \bigoplus_{j=1}^{m-1} MM(D_{2n})U_{\pi_2^j} \quad (n = 2m).$$

When n is odd (resp., $n = 2m$), the dimension of $MM(D_{2n})$ is $2n - 1$ (resp., $2n - 2$).

Now Propositions 1 and 2 are given explicitly by the following.

Corollary 5 *Fix $n \geq 3$, define c_j as before, and let $s_j = \sin(2j\pi/n)$. The orthogonal decomposition of $\mathbb{C}^n$ into D_{2n}-invariant subspaces under ρ is given by*

$$\mathbb{C}^n = \mathbb{C}(1, 1, \ldots, 1) \oplus \bigoplus_{j=1}^{\frac{n-1}{2}} V_{\pi_2^j} \quad (n \text{ odd}), \quad and$$

$$\mathbb{C}^n = \mathbb{C}(1, 1, \ldots, 1) \oplus \mathbb{C}(1, -1, 1, \ldots, -1) \oplus \bigoplus_{j=1}^{m-1} V_{\pi_2^j} \quad (n = 2m).$$

Here $V_{\pi_2^j} = U_{\pi_2^j}\mathbb{C}^n$ with an orthonormal basis given by

$$u_j = \sqrt{\frac{2}{n}}\,(1, c_j, c_{2j}, \ldots, c_{(n-1)j}), \qquad v_j = \sqrt{\frac{2}{n}}\,(0, s_j, s_{2j}, \ldots, s_{(n-1)j}).$$

Proof If

$$w_j = \frac{1}{\sqrt{n}}\,(1, \omega^j, \omega^{2j}, \ldots, \omega^{(n-1)j}),$$

a unit eigenvector for R with eigenvalue ω^{-j}, then the vectors

$$u_j = \frac{w_j + w_{-j}}{\sqrt{2}}, \qquad v_j = \frac{w_j - w_{-j}}{\sqrt{2}i}$$

span a subspace invariant and irreducible under D_{2n}. That is, the span is invariant under all $\rho(R^k)$, and

$$\rho(RC)u_j = u_j, \quad \rho(RC)v_j = -v_j.$$

Since $\rho(R)$ is an orthogonal matrix, these eigenvectors are orthogonal with respect to the usual Hermitian inner product on $\mathbb{C}^n$, and orthonormality in the corollary follows immediately.

Alternatively, we may simply apply $U_{\pi_2^j}$ to the standard basis vectors e_1 and e_2 and use Gram-Schmidt orthogonalization, or we may unravel the multiplication of basis vectors in the next section. □

10 Quaternionic Bases

Finally, we use the results of the previous section to give bases for each component of type π_2^j in Corollary 4. In addition to obtaining further examples of special semi-magic matrices, we exhibit the quaternionic structure for such components.

Now we consider the imaginary part of U_{χ_j}. With $s_j = \sin(2j\pi/n)$, we define

$$U'_{\pi_2^j} = \frac{2}{n}\begin{vmatrix} 0 & s_j & s_{2j} & s_{3j} & \cdots \\ -s_j & 0 & s_j & s_{2j} & \cdots \\ -s_{2j} & -s_j & 0 & s_j & \cdots \\ \cdots & \cdots & \cdots & \cdots & \cdots \end{vmatrix}.$$

From the idempotent properties of U_{χ_j} and $U_{\pi_2^j}$, one has

$$U_{\pi_2^j}U'_{\pi_2^j} = U'_{\pi_2^j}U_{\pi_2^j} = U'_{\pi_2^j}, \qquad (U'_{\pi_2^j})^2 = -U_{\pi_2^j}.$$

Furthermore, we have

$$P_C U_{\pi_2^j} P_C = U_{\pi_2^j}, \qquad P_C U'_{\pi_2^j} P_C = -U'_{\pi_2^j}.$$

We immediately obtain

Theorem 5 *The component $MM(D_{2n})_{\pi_2^j}$ associated to π_2^j in $MM(D_{2n})$ obtains a quaternionic structure as follows: define a basis*

$$q_1 = U_{\pi_2^j}, \quad q_2 = iP_C U_{\pi_2^j}, \quad q_3 = U'_{\pi_2^j}, \quad q_4 = iP_C U'_{\pi_2^j}.$$

Then, for all $1 \le t \le 4$, $q_1 q_t = q_t q_1 = q_t$, and

$$q_2^2 = q_3^2 = q_4^2 = -q_1, \quad q_2 q_3 q_4 = -q_1.$$

Example 5 When $n = 3$, there is only one two-dimensional type, with corresponding basis in $MM(D_6)$, the space of semi-magic squares of size 3:

$$q_1 = \frac{1}{3}\begin{vmatrix} 2 & -1 & -1 \\ -1 & 2 & -1 \\ -1 & -1 & 2 \end{vmatrix}, \quad q_2 = \frac{i}{3}\begin{vmatrix} -1 & -1 & 2 \\ -1 & 2 & -1 \\ 2 & -1 & -1 \end{vmatrix},$$

$$q_3 = \frac{\sqrt{3}}{3}\begin{vmatrix} 0 & 1 & -1 \\ -1 & 0 & 1 \\ 1 & -1 & 0 \end{vmatrix}, \quad q_4 = \frac{i\sqrt{3}}{3}\begin{vmatrix} 1 & -1 & 0 \\ -1 & 0 & 1 \\ 0 & 1 & -1 \end{vmatrix}.$$

We obtain a basis for $MM(D_6)$ by including J.

The case $n = 4$ also has a single two-dimensional type. We have

$$q_1 = \frac{1}{2}\begin{vmatrix} 1 & 0 & -1 & 0 \\ 0 & 1 & 0 & -1 \\ -1 & 0 & 1 & 0 \\ 0 & -1 & 0 & 1 \end{vmatrix}, \quad q_2 = \frac{i}{2}\begin{vmatrix} 0 & -1 & 0 & 1 \\ -1 & 0 & 1 & 0 \\ 0 & 1 & 0 & -1 \\ 1 & 0 & -1 & 0 \end{vmatrix},$$

$$q_3 = \frac{1}{2}\begin{vmatrix} 0 & 1 & 0 & -1 \\ -1 & 0 & 1 & 0 \\ 0 & -1 & 0 & 1 \\ 1 & 0 & -1 & 0 \end{vmatrix}, \quad q_4 = \frac{i}{2}\begin{vmatrix} 1 & 0 & -1 & 0 \\ 0 & -1 & 0 & 1 \\ -1 & 0 & 1 & 0 \\ 0 & 1 & 0 & -1 \end{vmatrix}$$

Again we obtain a basis for $MM(D_8)$ by including J_1 and J_2.

References

1. Anand, H., Dumir, V. C., and H. Gupta: A combinatorial distribution problem. Duke Math. J. **33**, 757–769 (1966)
2. Baumeister, B, Haase, C., Nill, B., and A. Paffenholz: Polytopes associated to dihedral groups. Ars Math. Contemp. **7**(1), 30–38 (2014)
3. Beck, M. and R. Sanyal: Combinatorial Reciprocity Theorems. An invitation to enumerative geometric combinatorics. Graduate Studies in Mathematics **195**. American Mathematical Society, Providence, RI. (2018)
4. Birkhoff, G.: Tres observaciones sobre el álgebra lineal. Univ. Nac. Tucumán Rev. Ser. A **5**, 147–151 (1946)
5. Boukas, A., Feinsilver, P., and Fellouris, A.: Structure and decompositions of the linear span of generalized stochastic matrices. Commun. Stoch. Anal. **9**(2), 239–250 (2015)
6. Burggraf, K., De Loera, J., and M. Omar: On volumes of permutation polytopes. Discrete geometry and optimization, 55–77, Fields Inst. Commun. **69**, Springer, New York (2013)
7. Curtis, C. W. and I. Reiner: Representation theory of finite groups and and associative algebras. Pure and Applied Mathemtaics, Vol. XI, Interscience Publishers, New York-London (1962)
8. Davis, P. J.: Circulant matrices. AMS Chelsea Publishing, Providence, R. I. (1994)
9. Fulton, W. and J. Harris: Representation Theory. A first course. Graduate Texts in Mathematics **129**. Reading in Mathematics. Springer-Verlag, New York (1991)
10. Guralnick, R. M. and D. Perkinson: Permutation polytopes and indecomposable elements in permutation groups. J. Comb. Theory A **113**(7), 1243–1256 (2006)
11. Kra, I. and S. Simanca: On circulant matrices. Notices Amer. Math. Soc. **59**(3), 368–377 (2012)

12. MacMahon, P. A.: Combinatory Analysis, vols. 1 and 2. Cambridge University Press, (1915, 1916); reprinted by Chelsea, New York (1960) and Dover, New York (2004)
13. Stanley, R. P.: Combinatorics and commutative algebra. Second edition. Progress in Mathematics **41**. Birkhäuser, Boston, MA (1996)
14. Stanley, R. P.: Enumerative combinatorics, Volume 1. Second edition. Cambridge Studies in Advanced Mathematics **49**, Cambridge University Press, Cambridge (2012)
15. Steinkamp, H.: Convex polytopes and permutation matrices. Master's thesis, Reed College, Portland, Oregon (1999)

Is the Syracuse Falling Time Bounded by 12?

Shalom Eliahou, Jean Fromentin, and Rénald Simonetto

1 Introduction

We denote by $\mathbb{N}$ the set of positive integers. Let $T : \mathbb{N} \to \mathbb{N}$ be the notorious $3x + 1$ function, defined by $T(n) = n/2$ if n is even, $T(n) = (3n + 1)/2$ if n is odd. For $k \geq 0$, denote by $T^{(k)}$ the kth iterate of T. The *orbit* of n under T is the sequence

$$\mathcal{O}_T(n) = (n, T(n), T^{(2)}(n), \ldots).$$

The famous Collatz conjecture states that for all $n \geq 1$, there exists $r \geq 1$ such that $T^{(r)}(n) = 1$. The least such r is denoted $\sigma_\infty(n)$ and called the *total stopping time* of n. An equivalent version of the Collatz conjecture states that for all $n \geq 2$, there exists $s \geq 1$ such that $T^{(s)}(n) < n$. The least such s is denoted by $\sigma(n)$ and called the *stopping time* of n. For instance, we have

$$\sigma(n) = \begin{cases} 1 & \text{if } n \text{ is even,} \\ 2 & \text{if } n \equiv 1 \bmod 4, \end{cases} \tag{1}$$

S. Eliahou (✉) · J. Fromentin · R. Simonetto
UR 2597 - LMPA - Laboratoire de Mathématiques Pures et Appliquées Joseph Liouville, Univ. Littoral Côte d'Opale, F-62100 Calais, France
e-mail: eliahou@univ-littoral.fr

J. Fromentin
e-mail: fromentin@univ-littoral.fr

R. Simonetto
e-mail: renalds@microsoft.com

CNRS, FR2037, Calais, France

R. Simonetto
Microsoft France, 37 Quai du Président Roosevelt, 92130 Issy-les-Moulineaux, France

M. B. Nathanson (ed.), *Combinatorial and Additive Number Theory V*,
Springer Proceedings in Mathematics & Statistics 395,
https://doi.org/10.1007/978-3-031-10796-2_7

as is well known and easy to check. A *stopping time record* is an integer $n \geq 2$ such that $\sigma(m) < \sigma(n)$ for all $2 \leq m \leq n-1$.

For the original slower version $C\colon \mathbb{N} \to \mathbb{N}$, where $C(n) = n/2$ or $3n+1$ according as n is even or odd, the analog of the stopping time is called the *glide* in [10]. The list of all currently known *glide records*, complete up to at least 2^{60}, is maintained in [11]. It is quite likely that glide records and stopping time records coincide; we have verified it by computer up to 2^{32}.

It is well known that $\sigma(n)$ is unbounded as n grows. For instance, since

$$T^{(\ell)}(2^\ell - 1) = 3^\ell - 1, \tag{2}$$

as follows from the formula $T(2^a 3^b - 1) = 2^{a-1}3^{b+1} - 1$ for $a \geq 1$, we have $\sigma(2^\ell - 1) \geq \ell$ for all $\ell \geq 2$.

In this paper, we propose an accelerated version of the function T. The idea, somewhat as in [13], is to apply an iterate of T to n depending on the number of digits of n in base 2. Accordingly, we introduce the following function.

Definition 1 The *jump function* $\mathrm{jp}\colon \mathbb{N} \to \mathbb{N}$ is defined for $n \in \mathbb{N}$ by

$$\mathrm{jp}(n) = T^{(\ell)}(n),$$

where $\ell = \lfloor \log_2(n) + 1 \rfloor$ is the number of digits of n in base 2.

Example 1 We have $\mathrm{jp}(1) = T^{(1)}(1) = 2$, and $\mathrm{jp}(2) = T^{(2)}(2) = 2$ since 2 is of length $\ell = 2$ in base 2. For $n = 27$, written 11011 in base 2, hence of length $\ell = 5$, we have $\mathrm{jp}(27) = T^{(5)}(27) = 71$. In turn, 71 is of length $\ell = 7$ in base 2 since $2^6 \leq 71 < 2^7$, whence $\mathrm{jp}(71) = T^{(7)}(71) = 137$. The orbit of 27 under jumps is displayed below in (4).

Example 2 A single jump at $n = 2^\ell - 1$ with $\ell \geq 1$ yields

$$\mathrm{jp}(2^\ell - 1) = 3^\ell - 1. \tag{3}$$

This follows from the equalities $\ell = \lfloor \log_2(2^\ell - 1) + 1 \rfloor$ and (2).

Example 3 We have $\mathrm{jp}(2n) = \mathrm{jp}(n)$ for all $n \geq 1$. Indeed, $2n$ is of length one more than n in base 2.

In analogy with the stopping time relative to T, we now introduce the falling time relative to jumps. As $\mathrm{jp}(1) = \mathrm{jp}(2) = 2$, we only consider $n \geq 3$.

Definition 2 Let $n \geq 3$. The *falling time* of n, denoted $\mathrm{ft}(n)$, is the least $k \geq 1$ such that $\mathrm{jp}^{(k)}(n) < n$, or ∞ if there is no such k.

Note that, for a presumed cyclic orbit under T with minimum $m \geq 3$, we would have $\mathrm{jp}(m) = \infty$.

There is no tight comparison between stopping time and falling time. It may happen that $\sigma(a) < \sigma(b)$ whereas $\text{ft}(a) > \text{ft}(b)$. For instance, for $a = 41$ and $b = 43$, we have

$$\sigma(41) = 2 < \sigma(43) = 5,$$
$$\text{ft}(41) = 8 > \text{ft}(43) = 2.$$

It may also happen that $\text{ft}(n) > \sigma(n)$, as shown by the case $n = 41$.

Of course, the Collatz conjecture is equivalent to $\text{ft}(n) < \infty$ for all $n \geq 3$. In Sect. 2, we provide computational evidence leading us to a stronger conjecture, namely that $\text{ft}(n)$ is in fact bounded for all $n \geq 3$. Specifically, all integers n we have tested so far satisfy $\text{ft}(n) \leq 16$. See Conjecture 1. In Sect. 3, in analogy with the falling time, we introduce the *Syracuse falling time* $\text{sft}(n)$, and corresponding conjectures, by only considering the odd terms in the orbits $\mathcal{O}_T(n)$. In Sect. 4, we report surprising computational results on $\text{ft}(2^\ell - 1)$ and $\text{sft}(2^\ell - 1)$ for $\ell \leq 500\,000$, and we formulate corresponding conjectures. In the last Sect. 5, inspired by the case $n = 2^\ell - 1$, we formulate still stronger conjectures on $\text{ft}(n)$ and $\text{sft}(n)$ for very large integers n. We conclude the paper with some supporting heuristics.

For a wealth of information, developments and commented references related to the $3x + 1$ problem, see the webpage and book of J. C. Lagarias [8, 9]. To date, the Collatz conjecture has been verified by computer up to 2^{68} by D. Barina [1]. Using this bound, it follows from [4] that any non-trivial cycle of T must have length at least 114 208 327 604.

2 Falling Time Records

In this section, we only consider those positive integers n satisfying $\sigma(n) \geq 3$, i.e. such that $n \equiv 3 \bmod 4$ by (1). Let us denote by $4\mathbb{N} + 3$ the set of those integers. Here is our first computational evidence that the falling time remains small.

Proposition 1 *We have* $\text{ft}(n) \leq 14$ *for all* $n \in [1, 2^{44} - 1]$ *such that* $n \equiv 3 \bmod 4$.

Proof In a few days of computing time with CALCULCO [2].

As shown in Table 1, the smallest $n \in 4\mathbb{N} + 3$ such that $\text{ft}(n) \geq 14$, namely $n = 12\,235\,060\,455$, actually satisfies $\text{ft}(n) = 14$ and $n > 2^{33}$.

Definition 3 A *falling time record* is an integer $n \in 4\mathbb{N} + 3$ such that $\text{ft}(m) < \text{ft}(n)$ for all $m \in 4\mathbb{N} + 3$ with $m < n$.

The list of falling time records up to 2^{35} is given in Table 1. It was built while establishing Proposition 1. For instance, we have $\text{ft}(3) = 2$, $\text{ft}(7) = 3$ and $\text{ft}(n) \leq 3$ for all $3 \leq n < 27$ such that $n \equiv 3 \bmod 4$. The value $\text{ft}(27) = 8$ follows from the

Table 1 Falling time records up to 2^{35}

$n \equiv 3 \bmod 4$	$\lfloor \log_2(n) + 1 \rfloor$	$\mathrm{ft}(n)$
3	2	2
7	3	3
27	5	8
60 975	16	9
1 394 431	21	10
6 649 279	23	11
63 728 127	26	13
12 235 060 455	34	14

Table 2 Some new falling times

n	111	103	71	55	217 740 015
$\mathrm{ft}(n)$	4	5	6	7	12

fact that 8 jumps are needed from 27 to fall below it, as shown by the orbit of 27 under jumps:

$$\mathcal{O}_{\mathrm{jp}}(27) = (27, 71, 137, 395, 566, 3\,644, 650, 53, 8, 2, 2, \dots). \tag{4}$$

Interestingly, five of the falling time records in Table 1 are also glide records, namely, 3, 7, 27, 63 728 127 and 12 235 060 455, as seen by consulting [11].

Table 1 shows that the number 12 and a few smaller ones fail to occur as falling time records. One may then wonder about the smallest $n \in 4\mathbb{N} + 3$ reaching $\mathrm{ft}(n) = 12$.

The answer is to be found in Table 2. Let us define a *new falling time* as an integer $n \in 4\mathbb{N} + 3$ such that $\mathrm{ft}(n)$ is distinct from $\mathrm{ft}(m)$ for all smaller $m \in 4\mathbb{N} + 3$. Of course, every falling time record is a new falling time. The list of new falling times we know so far, which are not already falling time records, is given in Table 2.

Integers Satisfying $\mathrm{ft}(n) > 14$

For $a, b \in \mathbb{Z}$, we denote by $[a, b] = \{n \in \mathbb{Z} \mid a \leq n \leq b\}$ the integer interval they span. Recalling Example 3, implying that $\mathrm{ft}(2m) = \mathrm{ft}(m)$ for all $m \geq 3$, a single integer n satisfying $\mathrm{ft}(n) > 14$ yields *infinitely many* integers N satisfying $\mathrm{ft}(N) > 14$, namely $N = 2^r n$ for all $r \geq 1$. However, the latter numbers have stopping time equal to 1, and hence are not particularly interesting.

Only those integers n satisfying $\mathrm{ft}(n) > 14$ and having a reasonably large stopping time are really interesting, for their apparent rarity and their relevance to the Collatz

Table 3 24-persistent integers satisfying $\text{ft}(n) = 15$

1 513 398 373 944 347,	1 702 573 170 687 391,	2 017 864 498 592 463,
2 553 859 756 031 087,	3 405 146 341 374 783,	3 830 789 634 046 631,
5107719512062175,	5 746 184 451 069 947,	6 464 457 507 453 691,
7 272 514 695 885 403,	22 370 169 558 105 279	

conjecture. Hence, we shall restrict our search to 24-*persistent* integers, i.e. to those n satisfying

$$\sigma(n) \geq 24.$$

The property for n of being 24-persistent only depends on its class mod 2^{24}. See [14] for more details on the description of the condition $\sigma(n) \geq k$ by classes mod 2^k. See also [5]. For $k = 24$, the number of 24-persistent classes mod 2^{24} is exactly 286 581.

As shown below, the occurrence of $\text{ft}(n) > 14$ among 24-persistent numbers seems to be very rare. Here is our first computational result.

Proposition 2 *The smallest* 24-*persistent integer* n *such that* $\text{ft}(n) > 14$ *is*

$$n_0 = 1\,008\,932\,249\,296\,231.$$

It satisfies $\text{ft}(n_0) = 15$, $\sigma(n_0) = 886$ *and* $\lfloor \log_2(n_0) + 1 \rfloor = 50$.

Proof In a few days of computing time with CALCULCO [2].

The neighborhood of g_{30} It turns out that n_0 is a glide record, and as such is listed under the name

$$n_0 = g_{30}$$

in Table 5 below. We have verified by computer that g_{30} *is the smallest* 24-*persistent integer* n *satisfying* $\text{ft}(n) > 14$. However, because of the restriction $\sigma(n) \geq 24$, we do not know whether g_{30} is an actual falling time record.

In a small neighborhood of g_{30} in the Collatz tree, we found 11 more 24-persistent integers n satisfying $\text{ft}(n) > 14$. By *small neighborhood* of g_{30}, we mean here integers m such that

$$T^{(i)}(m) = T^{(j)}(g_{30})$$

for some $i \in [0, 40]$, $j \in [0, 30]$. It turns out that these 11 integers all satisfy $\text{ft}(n) = 15$, as g_{30} itself. They are displayed in Table 3.

The Neighborhood of g_{32} There is another glide record in Table 5 with falling time 15, namely

$$g_{32} = 180\,352\,746\,940\,718\,527.$$

Table 4 24-persistent integers satisfying $\mathrm{ft}(n) = 16$

49 312 2600 554 790 303,	739 683 900 832 185 455,
986 245 201 109 580 607,	1 479 367 801 664 370 911

Table 5 Top ten known glide records

n		$\lfloor \log_2(n) + 1 \rfloor$	glide of n	$\sigma(n)$	$\mathrm{ft}(n)$
g_{34}	2 602 714 556 700 227 743	62	1 639	1005	13
g_{33}	1 236 472 189 813 512 351	61	1 614	990	14
g_{32}	180 352 746 940 718 527	58	1 575	966	**15**
g_{31}	118 303 688 851 791 519	57	1 471	902	12
g_{30}	1 008 932 249 296 231	50	1 445	886	**15**
g_{29}	739 448 869 367 967	50	1 187	728	12
g_{28}	70 665 924 117 439	47	1 177	722	13
g_{27}	31 835 572 457 967	45	1 161	712	13
g_{26}	13 179 928 405 231	44	1 122	688	14
g_{25}	2 081 751 768 559	41	988	606	12

In this case, looking at a somewhat larger neighborhood of g_{32} in the Collatz tree, namely at all m such that

$$T^{(i)}(m) = T^{(j)}(g_{32})$$

for some $i \in [0, 50]$, $j \in [0, 30]$, we found four 24-persistent integers n reaching $\mathrm{ft}(n) = 16$. These four integers are displayed in Table 4.

However, these four integers n have a small stopping time. They all satisfy $\sigma(n) \in [35, 48]$, as compared to $\sigma(g_{32}) = 966$. Hence again, they are not particularly interesting.

This leads us to the following conjecture.

Conjecture 1 There exists $B \geq 16$ such that $\mathrm{ft}(n) \leq B$ for all $n \geq 3$.

An even bolder conjecture, based on the data we currently have, would be to take $B = 16$ above. Anyway, with whatever value of B, Conjecture 1 constitutes a strong form of the Collatz conjecture.

Glide Records

Eric Roosendaal maintains the list of all currently known glide records [11], complete up to at least 2^{60}. At the time of writing, there are 34 of them, denoted $g_1, \ldots, g_{34}$ below. As noted in [11], only the first 32 ones have been independently checked. The ten biggest are displayed in descending order in Table 5. It turns out that

$$\mathrm{ft}(g_1), \ldots, \mathrm{ft}(g_{34}) \leq 15.$$

Moreover, among them, the highest value $\mathrm{ft}(n) = 15$ is only reached by g_{30} and g_{32}. Table 4 displays four 24-persistent integers n satisfying $\mathrm{ft}(n) = 16$ in the neighborhood of g_{32}. We do not know whether $\mathrm{ft}(n) \geq 17$ is at all reachable.

Falling Time Distribution

In three distinct graphics, we display the distribution of the values taken by the falling time function in large integer intervals. These graphics show that the proportion of the case $\mathrm{ft}(n) \geq 3$ in the integer intervals $[2^\ell, 2^{\ell+1} - 1]$ tends to 0 as ℓ grows.

- Figure 1 displays the proportion of the occurrence of $\mathrm{ft}(n) = 1$, $\mathrm{ft}(n) = 2$ and $\mathrm{ft}(n) \geq 3$, respectively, among all odd integers in the integer intervals $[2^\ell, 2^{\ell+1} - 1]$ for $2 \leq \ell \leq 40$.
- Figure 2 does the same but separates the cases $n \equiv 1 \bmod 4$ and $n \equiv 3 \bmod 4$. The purpose is to show that the former case, with stopping time 2, behaves like the more interesting latter case, and so may be safely ignored.
- Finally, Fig. 3 is restricted to 24-persistent integers in the integer intervals $[2^\ell, 2^{\ell+1} - 1]$ for $24 \leq \ell \leq 50$.

A Variant of Jumps

Let $h \in \mathbb{N}$. For all $n \in \mathbb{N}$, we define

$$\mathrm{jp}_h(n) = T^{(h\ell)}(n)$$

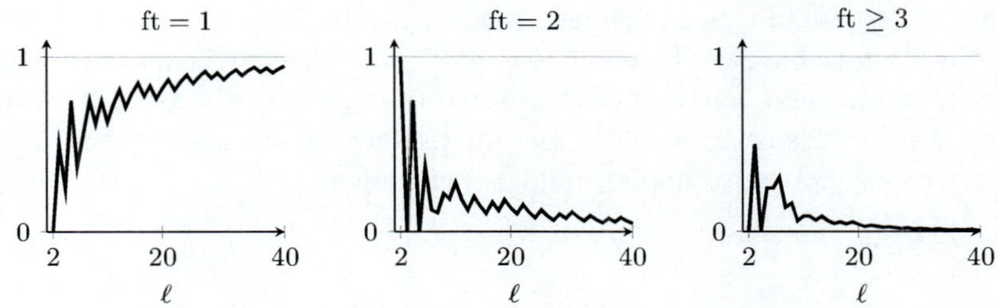

Fig. 1 Proportion of odd integers in $[2^\ell, 2^{\ell+1} - 1]$ with falling time equal to 1, 2 and greater than 2, respectively. The integer ℓ goes from 2 to 40

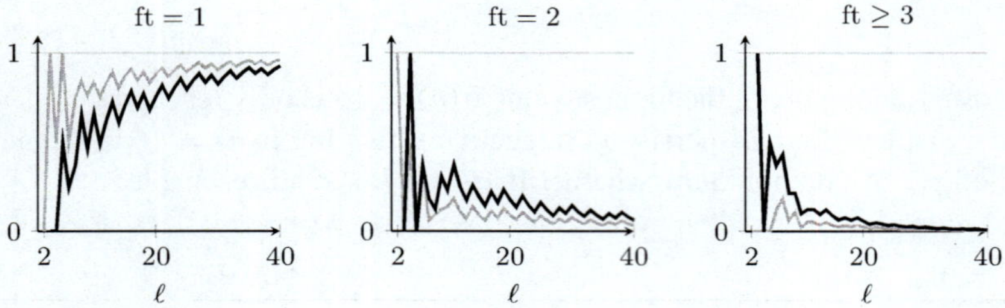

Fig. 2 Same plot as for Fig. 1 except that integers are separated with respect to their class 1 or 3 modulo 4. Gray curves are for integers congruent to 1 modulo 4 while black ones are for those congruent to 3 modulo 4

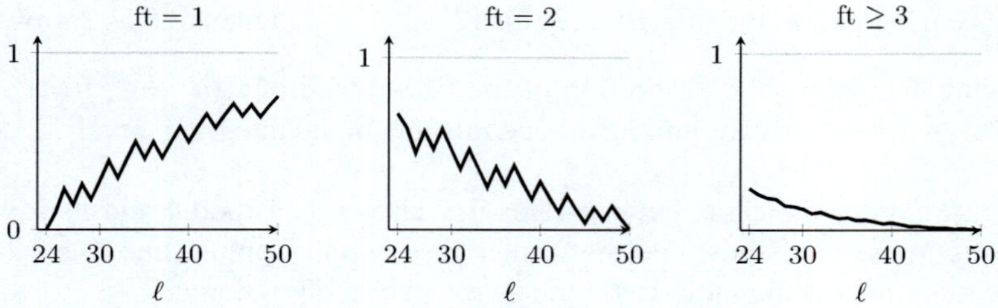

Fig. 3 Proportion of 24-persistent integers in $[2^\ell, 2^{\ell+1} - 1]$ with falling time equal to 1, 2 and greater than 2, respectively. The integer ℓ goes from 24 to 50

where, as before, ℓ is the number of digits of n in base 2. This is not the same, of course, as the h-iterate $\mathrm{jp}^{(h)}(n)$. Note also that for $h = 1$, we recover jumps, i.e. $\mathrm{jp}_1(n) = \mathrm{jp}(n)$. For $n \geq 3$, the *h-falling time* $\mathrm{ft}_h(n)$ is defined correspondingly, as the smallest $k \geq 1$, if any, such that $\mathrm{ft}_h^{(k)}(n) < n$.

It turns out that for $h = 18$, and for the glide records $g_1, \ldots, g_{34}$, we have

$$\mathrm{ft}_{18}(g_i) = 1$$

for all $1 \leq i \leq 34$. In view of that fact, is it true that $\mathrm{ft}_{18}(n) = 1$ for all $n \geq 3$? We do not know. But we have verified it up to $n \leq 2^{30}$, and it cannot be outright dismissed, given the conjectural behavior of $\mathrm{ft}(n)$ for very large n as discussed in Sect. 5. Of course, a positive answer would imply the Collatz conjecture. On the other hand, uncovering any counterexample would be quite a feat.

3 The Syracuse Version

Let $\mathbb{O} = 2\mathbb{N} + 1$ denote the set of odd positive integers. Another well-studied version of the $3x + 1$ function is $\mathrm{syr}\colon\ \mathbb{O} \to \mathbb{O}$, defined on any $x \in \mathbb{O}$ by

$$\mathrm{syr}(x) = (3x + 1)/2^{\nu},$$

where $\nu \geq 1$ is the largest integer such that 2^{ν} divides $3x + 1$. That is, $\mathrm{syr}(x)$ is the largest odd factor of $3x + 1$. This specific version is called the *Syracuse function* in [13]. It has been amply investigated in the past, though under different notation or names. For instance in [3], where lower bounds on the length of presumed nontrivial cycles of $\mathrm{syr}(x)$ are given in terms of the convergents p_n/q_n to $\log_2(3)$; or in [6, 7, 12], where statistical properties of $\mathrm{syr}(x)$ and related maps are studied using the Structure theorem of Sinai, which we briefly recall below.

In analogy with the functions $\mathrm{jp}(n)$ and $\mathrm{ft}(n)$ related to the $3x + 1$ function $T(x)$, we now introduce the corresponding functions $\mathrm{sjp}(n)$ and $\mathrm{sft}(n)$ related to the Syracuse version $\mathrm{syr}(x)$.

Definition 4 We define the *Syracuse jump function* $\mathrm{sjp}\colon \mathbb{O} \to \mathbb{O}$ by

$$\mathrm{sjp}(n) = \mathrm{syr}^{(\ell)}(n), \text{ where } \ell = \lfloor \log_2(n) + 1 \rfloor.$$

Example 4 We have $\mathrm{sjp}(1) = 1$, $\mathrm{sjp}(3) = 1$ and $\mathrm{sjp}(27) = \mathrm{syr}^{(5)}(27) = 107$.

Here is the corresponding Syracuse falling time.

Definition 5 Let $n \in \mathbb{O}\setminus\{1\}$. The *Syracuse falling time* of n, denoted $\mathrm{sft}(n)$, is the least $k \geq 1$ such that $\mathrm{sjp}^{(k)}(n) < n$, or ∞ if there is no such k.

Example 5 We have $\mathrm{sft}(27) = 6$, as witnessed by the orbit of 27 under Syracuse jumps, namely

$$\mathcal{O}_{\mathrm{sjp}}(27) = (27, 107, 233, 377, 911, 53, 1, 1, \ldots).$$

As one may expect, the inequality $\mathrm{sft}(n) \leq \mathrm{ft}(n)$ holds very often, but not always. For instance, for $n = 199$, we have $\mathrm{ft}(199) = 1$ but $\mathrm{sft}(199) = 5$. The former equality follows from the orbit

$$\mathcal{O}_T(199) = (199, 299, 449, 674, 337, 506, 253, 380, 190, \ldots)$$

and the value $\lfloor \log_2(199) + 1 \rfloor = 8$, yielding $\mathrm{jp}(199) = 190$, while the latter one follows from the orbit

$$\mathcal{O}_{\mathrm{syr}}(199) = (199, 323, 395, 479, 577, 1, \ldots).$$

Definition 6 A *Syracuse falling time record* is an integer $n \in 4\mathbb{N} + 3$ such that $n \geq 7$ and $\mathrm{sft}(m) < \mathrm{sft}(n)$ for all $m \in 4\mathbb{N} + 3$ with $m < n$.

The complete list of Syracuse falling time records up to 2^{35} is displayed in Table 6. Compared with Table 1, it turns out that all current Syracuse falling time records are also falling time records. The converse does not hold, as shown by the falling time records 60 975 and 1 394 431 in Table 1.

Table 6 Syracuse falling time records up to 2^{35}

$n \equiv 3 \bmod 4$	$\lfloor \log_2(n) + 1 \rfloor$	$\mathrm{sft}(n)$
7	3	2
27	5	6
6 649 279	23	7
63 728 127	26	9

Current Maximum

The Collatz conjecture is equivalent to the statement $\mathrm{sft}(n) < \infty$ for all $n \in \mathbb{O} \setminus \{1\}$. Again, it is likely that a stronger form holds, namely, that $\mathrm{sft}(n)$ is *bounded* on $\mathbb{O} \setminus \{1\}$. Besides the computational evidence above and below, some heuristics point to that possibility in Sect. 5. Similarly to Proposition 1, here is a computational result in that direction.

Proposition 3 *We have* $\mathrm{sft}(n) \leq 9$ *for all* $n \in [3, 2^{35} - 1]$ *such that* $n \equiv 3 \bmod 4$.

Proof By computer with CALCULCO [2].

As yet another hint pointing to the same direction, it turns out that

$$\mathrm{sft}(g_1), \ldots, \mathrm{sft}(g_{34}) \leq 10 \tag{5}$$

for the 34 currently known glide records. For definiteness, Table 7 displays the Syracuse falling times of the top ten glide records as listed in Table 5.

Among the g_i, and as in Sect. 2 for the falling time, only g_{30} and g_{32} reach the current maximum of the Syracuse falling time, namely $\mathrm{sft}(n) = 10$. Interestingly, the biggest currently known glide record, namely $n = g_{34}$, only satisfies $\mathrm{sft}(n) = 8$. With Proposition 3 and (5) in the background, here is our formal conjecture.

Conjecture 2 There exists $C \geq 10$ such that $\mathrm{sft}(n) \leq C$ for all $n \equiv 3 \bmod 4$.

Again, the truth of this conjecture would yield a strong positive solution of the Collatz conjecture. At the time of writing, no single positive integer $n \equiv 3 \bmod 4$ is known to satisfy $\mathrm{sft}(n) \geq 11$. Thus, a still bolder conjecture would be to take $C = 10$ in Conjecture 2, or $C = 12$ to be on a safer side. Whence the title of this paper.

Table 7 Syracuse falling times of top ten glide records

n	g_{25}	g_{26}	g_{27}	g_{28}	g_{29}	g_{30}	g_{31}	g_{32}	g_{33}	g_{34}
$\mathrm{sft}(n)$	9	8	8	8	8	**10**	8	**10**	9	8

A Variant of Syracuse Jumps

As in Sect. 2 for jumps, we propose here an accelerated variant of Syracuse jumps. Let $h \in \mathbb{N}$. For all $n \in \mathbb{O}$, we define

$$\mathrm{sjp}_h(n) = \mathrm{syr}^{(h\ell)}(n),$$

where ℓ is the number of digits of n in base 2. Of course, $\mathrm{sjp}_1(n) = \mathrm{sjp}(n)$. For $n \geq 3$, the *Syracuse h-falling time* $\mathrm{sft}_h(n)$ is defined correspondingly, as the smallest $k \geq 1$, if any, such that $\mathrm{sft}_h^{(k)}(n) < n$. It turns out that for $h = 12$, and for the glide records $g_1, \ldots, g_{34}$, we have

$$\mathrm{sft}_{12}(g_i) = 1$$

for all $1 \leq i \leq 34$. Again, we may ask whether $\mathrm{sft}_{12}(n) = 1$ holds for all odd $n \geq 3$. A positive answer would imply the Collatz conjecture. We have verified it up to $n \leq 2^{30}$, and our semi-random search did not yield any counterexample.

Anyway, the occurrence of $\mathrm{sft}_{12}(n) = 1$ as n grows to infinity is probably overwhelming; and, just possibly, tools such as Sinai's structure theorem and its applications [6, 7, 12] might help prove that this is indeed the case. But we shall not pursue here this line of investigation.

For convenience, let us briefly recall the statement of that theorem. Given $x \in 6\mathbb{N} + \{1, 5\}$, let $x_i = \mathrm{syr}^{(i)}(x)$ for all $i \geq 0$, and let $k_i \geq 1$ be such that $x_i = (3x_{i-1} + 1)/2^{k_i}$ for all $i \geq 1$. Moreover, for $m \geq 1$, set

$$\gamma_m(x) = (k_1, \ldots, k_m).$$

Sinai's structure theorem states that *given any* $(k_1, \ldots, k_m) \in \mathbb{N}^m$, *the set of all* $x \in 6\mathbb{N} + \{1, 5\}$ *such that* $\gamma_m(x) = (k_1, \ldots, k_m)$ *consists of a unique and full congruence class mod* $6 \cdot 2^{k_1+\cdots+k_m}$ *in* $\mathbb{N}$.

4 The Case $2^\ell - 1$

In sharp contrast with the stopping time of $2^\ell - 1$, for which $\sigma(2^\ell - 1) \geq \ell$ for all $\ell \geq 2$, the falling time and the Syracuse falling time of $2^\ell - 1$ seem to remain very small as ℓ grows. Here is some strong computational evidence.

Proposition 4 *Besides* $\mathrm{ft}(2^5 - 1) = \mathrm{ft}(2^6 - 1) = 8$, *we have* $\mathrm{ft}(2^\ell - 1) \leq 5$ *for all* $2 \leq \ell \leq 500\,000$ *with* $\ell \notin \{5, 6\}$.

Proof In a few days of computing time with CALCULCO [2].

Moreover, the value $\mathrm{ft}(2^\ell - 1) = 5$ seems to occur finitely many times only, the last one being presumably at $\ell = 132$. In turn, the value $\mathrm{ft}(2^\ell - 1) = 4$ seems to occur infinitely often. Whence the following conjecture, verified by computer up to $\ell = 500\,000$.

Conjecture 3 We have $\mathrm{ft}(2^\ell - 1) \leq 4$ for all $\ell \geq 133$.

Here are the analogous statement and conjecture for the Syracuse falling time.

Proposition 5 *Besides* $\mathrm{sft}(2^5 - 1) = \mathrm{sft}(2^6 - 1) = 5$, *and* $\mathrm{sft}(2^{24} - 1) = 4$, *we have*

$$\mathrm{sft}(2^\ell - 1) \in \{2, 3\} \textit{ for all } \ell \in [2, 4\,624] \setminus \{5, 6, 24\},$$
$$\mathrm{sft}(2^\ell - 1) = 2 \textit{ for all } \ell \in [4\,625, 500\,000].$$

Proof In a few days of computing time with CALCULCO [2].

This leads us to the following conjecture, true up to $\ell \leq 500\,000$.

Conjecture 4 We have $\mathrm{sft}(2^\ell - 1) = 2$ for all $\ell \geq 4\,625$.

5 For Very Large *n*

As hinted by the computational evidence and conjectures of Sect. 4 on the case $n = 2^\ell - 1$, by intensive semi-random search, and by the heuristics below, it appears to be increasingly difficult for integers n to satisfy $\mathrm{ft}(n) \geq 5$ or $\mathrm{sft}(n) \geq 3$ as they grow very large. Here then are still bolder conjectures.

Conjecture 5 We have $\mathrm{ft}(n) \leq 4$ for all $n \geq 2^{500}$.

This threshold of 2^{500} is inspired by Conjecture 3, of course with a margin for safety. It cannot be significantly lowered, since $\mathrm{ft}(2^{132} - 1) = 5$ as noted before Proposition 4. Moreover, intensive random search revealed one integer $n \in [2^{70}, 2^{71} - 1]$ satisfying $\mathrm{ft}(n) = 5$, namely

$$n = 1\,884\,032\,044\,420\,885\,877\,201\,579\,449\,071\,924\,925\,072\,300\,117\,065\,411.$$

However, this integer is congruent to 3 mod 16 and hence has stopping time equal to 4 only.

Here is the analogous conjecture for the Syracuse falling time. Its threshold of 2^{5000} is similarly inspired by Proposition 5 and Conjecture 4.

Conjecture 6 We have $\mathrm{sft}(n) \leq 2$ for all odd $n \geq 2^{5000}$.

Heuristics

Besides the computational evidence leading to Conjectures 1, 2, 3, 4, 5 and 6, a heuristic argument would run as follows. It is well known that the Collatz conjecture is equivalent to the statement that, starting with any integer $n \geq 1$, the probability for $T^{(k)}(n)$ to be even or odd tends to $1/2$ as k grows to infinity. Thus, even if n written in base 2 is a highly structured binary string, as e.g. for $n = 2^\ell - 1$, one may expect that for $\ell =$ the length of that string, then $T^{(\ell)}(n)$ in base 2 will already look more random. That is, a single jump or Syracuse jump at $n \geq 3$ should already introduce a good dose of randomness, all the more so as n grows very large. And therefore, a bounded number of jumps or Syracuse jumps at n might well suffice to fall below n.

A Challenge

We hope that the experts in highly efficient computation of the $3x + 1$ function will tackle the challenge of probing these conjectures to much higher levels than the ones reported here. For instance, as both a challenge and a request to the reader, and in view of Conjecture 5, if you do find any $n \geq 2^{500}$ satisfying $\mathrm{ft}(n) \geq 5$, please e-mail it to the authors. Your solution will be duly recorded on a dedicated webpage.

Acknowledgements The computations performed for this paper were carried out on the CALCULCO high performance computing platform provided by SCoSI/ULCO (Service COmmun du Système d'Information de l'Université du Littoral Côte d'Opale).

References

1. D. Barina, *Convergence verification of the Collatz problem*, The Journal of Supercomputing 77 (2021), 2681–2688.
2. CALCULCO, a computing platform at Université du Littoral Côte d'Opale.
3. R. E. Crandall, *On the " $3x + 1$" Problem*, Math. Comp. 32 (1978), 1281–1292.
4. S. Eliahou, *The 3x+1 problem: new lower bounds on nontrivial cycle lengths*, Discrete Math. 11 (1993), 45–56.
5. C. J. Everett, *Iteration of the Number-Theoretic Function* $f(2n) = n$, $f(2n + 1) = 3n + 2$, Adv. Math. 25 (1977), 42–45.
6. A. V. Kontorovich and S. J. Miller, *Benford's law, values of L-functions, and the* 3x + 1 *problem*, Acta Arith. 120 (2005), 269–297.
7. A. V. Kontorovich and Ya. G. Sinai, *Structure Theorem for* (d, g, h)*-maps*, Bull. Braz. Math. Soc. (N.S.) 33 (2002), 213–224.
8. J. C. Lagarias, $3x + 1$ problem and related problems, https://dept.math.lsa.umich.edu/~lagarias//3x+1.html
9. The Ultimate Challenge: The $3x + 1$ Problem. J. C. Lagarias, Editor. Amer. Math. Soc., Providence, RI, 2010.
10. E. Roosendaal, www.ericr.nl/wondrous
11. E. Roosendaal, www.ericr.nl/wondrous/glidrecs.html

12. Ya. G. Sinai, *Statistical* (3x+1)-*Problem*, Comm. Pure Appl. Math. 56 (2003), 1016–1028.
13. T. Tao, *Almost all orbits of the Collatz map attain almost bounded values* (2019) arXiv:1909.03562
14. R. Terras, *A stopping time problem on the positive integers*, Acta Arith. 30 (1976), 241–252.

Genera of Numerical Semigroups and Polynomial Identities for Degrees of Syzygies

Leonid G. Fel

2010 Mathematics Subject Classification: Primary—20M14, Secondary—11P81.

1 Introduction

Two sets of polynomial and quasi-polynomial identities for degrees of syzygies in numerical semigroups $S_m = \langle d_1, \ldots, d_m \rangle$ were derived recently [7] when studying the rational representation (Rep) of the Hilbert series of S_m and the quasi-polynomial Rep of the restricted partition function. A part of polynomial identities of degrees $1 \leq n \leq m-2$ were coincided with Herzog-Kühl's equations [13] for the Betti numbers of graded Cohen-Macaulay modules of codimension $m-1$, but a new additional polynomial identity of degree $n = m-1$ turned out to be an important tool in a study of symmetric (not complete intersection) semigroups with small embedding dimension (*edim*) $m = 4, 5, 6$ [6, 8, 9].

A further application [10] of polynomial identities for higher degrees of syzygies, $n \geq m$, involves power sums $G_{n-m} = \sum_{s \in \Delta_m} s^{n-m}$, which are called *genera* of numerical semigroups, where $\Delta_m = \mathbb{Z}_> \setminus S_m$ and $G_0 = \#\Delta_m$ denote a set of semigroup gaps and its cardinality (*genus*), respectively. A set Δ_m is uniquely defined by semigroup generators d_j. Albeit there are $\mu - 1$ explicitly known gaps $1 \leq s \leq \mu - 1$, where $\mu = \min\{d_1, \ldots, d_m\}$ denotes a semigroup multiplicity, a most of gaps $s > \mu$ (including the largest gap F_m which called the Frobenius number) cannot be determined explicitly.

L. G. Fel (✉)
Department of Civil Engineering, Technion—Israel Institute of Technology, Haifa 32000, Israel
e-mail: lfel@technion.ac.il

M. B. Nathanson (ed.), *Combinatorial and Additive Number Theory V*,
Springer Proceedings in Mathematics & Statistics 395,
https://doi.org/10.1007/978-3-031-10796-2_8

By a fundamental theorem of symmetric polynomials [14], there exists a finite number g_m of algebraically independent genera G_r. On the other hand, an involvement of G_r into polynomial identities for syzygies degrees may decrease this number. In the present paper, we study this question for arbitrary semigroup S_m and find how does the number g_m depend on special characteristics of a semigroup (e.g., *edim*) and its properties (non-symmetric, symmetric, complete intersection). For this purpose, we derive polynomial identities of higher degrees $n \geq m$ and find algebraic equalities related a finite number $g_m + 1$ of genera.

The paper is organized in six sections. In Sect. 2, we obtain polynomial identities (9) of higher degrees $n \geq m$ following an approach, suggested in [13]. The rest of this section is completely technical: we determine necessary expressions for all entries that appeared in formula (9). In Sect. 3, we derive linear equations (19) for alternating power sums $\mathbb{C}_k(S_m)$ and put forward a conjecture on the linear Rep of coefficients K_p in (22) by genera G_r of a semigroup and special polynomials T_r defined in (25). In Sect. 4, we prove the existence of a polynomial equation $\mathcal{R}_G(G_0, \ldots, G_{g_m}) = 0$ for arbitrary non-symmetric semigroups, where $g_m = B_m - m + 1$, and $B_m = \sum_{k=1}^{m-1} \beta_k$ and β_k denote the total and partial Betti numbers of S_m, and find such equation for S_3. In Sect. 5, we discuss supplementary relations for G_k in symmetric semigroups and complete intersection (CI), and give formulas for g_m in both cases, e.g., in the latter case, it looks much simple, $g_m = m - 2$. In Sect. 6, we give concluding remarks about coefficients K_p.

2 Polynomial Identities for Numerical Semigroups

Recall the basic facts on numerical semigroups and polynomial identities following [7]. Let a numerical semigroup S_m be minimally generated by a set of natural numbers $\{d_1, \ldots, d_m\}$, where $\mu \geq m$ and $\gcd(d_1, \ldots, d_m) = 1$. Its generating function $H(S_m; z)$,

$$H(S_m; z) = \sum_{s \in S_m} z^s, \qquad z < 1, \qquad 0 \in S_m, \tag{1}$$

is referred to as *the Hilbert series* of S_m and has a rational Rep,

$$H(S_m; z) = \frac{Q(S_m; z)}{\prod_{i=1}^{m}(1 - z^{d_i})}, \qquad C_{k,j} \in \mathbb{Z}_>, \quad 1 \leq k \leq m-1, \quad 1 \leq j \leq \beta_k, \tag{2}$$

$$Q(S_m; z) = 1 - \sum_{j=1}^{\beta_1} z^{C_{1,j}} + \sum_{j=1}^{\beta_2} z^{C_{2,j}} - \cdots \pm \sum_{j=1}^{\beta_{m-1}} z^{C_{m-1,j}}, \qquad \sum_{k=0}^{m-1} (-1)^k \beta_k = 0, \quad \beta_0 = 1, \tag{3}$$

where $C_{k,j}$ and β_k stand for degrees of syzygies and partial Betti's numbers, respectively. The largest degree Q_m of the $(m-1)$-th syzygy is related to the Frobenius number F_m of S_m,

$$Q_m = F_m + \sigma_1, \quad Q_m = \max \mathsf{PF}(S_m), \quad \mathsf{PF}(S_m) = \{C_{m-1,1}, \ldots, C_{m-1,\beta_{m-1}}\}, \quad \sigma_1 = \sum_{j=1}^{m} d_j, \tag{4}$$

where $\mathsf{PF}(S_m)$ is called a set of pseudo-Frobenius numbers. Denote by $\mathbb{C}_k(S_m)$ the alternating power sum of syzygies degrees,

$$\mathbb{C}_k(S_m) = \sum_{j=1}^{\beta_1} C_{1,j}^k - \sum_{j=1}^{\beta_2} C_{2,j}^k + \ldots - (-1)^{m-1} \sum_{j=1}^{\beta_{m-1}} C_{m-1,j}^k, \tag{5}$$

and write the polynomial identities (Theorem 1 in [7]) for a semigroup S_m,

$$\mathbb{C}_0(S_m) = 1, \qquad \mathbb{C}_r(S_m) = 0, \quad 1 \le r \le m-2, \qquad \mathbb{C}_{m-1}(S_m) = (-1)^m (m-1)! \pi_m, \tag{6}$$

where $\pi_m = \prod_{i=1}^m d_i$.

Polynomial Identities of Arbitrary Degree

Start with relation for the Hilbert series $H(S_m; z)$ and a generating function $\Phi(S_m; z)$ for the semigroup gaps $s \in \Delta_m$,

$$\Phi(S_m; z) + H(S_m; z) = \frac{1}{1-z}, \qquad \Phi(S_m; z) = \sum_{s \in \Delta_m} z^s, \tag{7}$$

and present the numerator $Q(S_m; z)$ in (2) as follows,

$$Q(S_m; z) = (1-z)^{m-1} \Pi(S_m; z), \qquad \Pi(S_m; z) = \Psi(S_m; z)[1 - (1-z)\Phi(S_m; z)], \tag{8}$$

where $\Psi(S_m; z)$ is a product of cyclotomic polynomials $\Psi_j(S_m; z)$,

$$\Psi(S_m; z) = \prod_{j=1}^{m} \Psi_j(z), \qquad \Psi_j(z) = \sum_{k=0}^{d_j - 1} z^k, \quad \Psi_j(1) = d_j, \quad \Psi(S_m; 1) = \pi_m.$$

Differentiating r times the first equality in (8), we obtain an infinite set of algebraic equations related syzygies degrees $C_{k,j}$ of a semigroup S_m with its generators d_j and gaps $s \in \Delta_m$,

$$Q_z^{(r)}(z) = \sum_{k=0}^{r} (-1)^k \frac{(m-1)!}{(m-k-1)!} \binom{r}{k} (1-z)^{m-k-1} \Pi_z^{(r-k)}(z), \quad r \geq 1, \tag{9}$$

where

$$Q_z^{(r)}(z) = \frac{d^r Q(S_m; z)}{dz^r}, \qquad \Pi_z^{(r)}(z) = \frac{d^r \Pi(S_m; z)}{dz^r}.$$

Calculate separately derivatives $Q_z^{(r)}(z)$ and $\Pi_z^{(r-k)}(z)$. According to expression (3), we obtain,

$$Q_z^{(r)}(z) = -\sum_{j=1}^{\beta_1} (C_{1,j})_r z^{C_{1,j}-r} + \sum_{j=1}^{\beta_2} (C_{2,j})_r z^{C_{2,j}-r} - \ldots + (-1)^{m-1} \sum_{j=1}^{\beta_{m-1}} (C_{m-1,j})_r z^{C_{m-1,j}-r}, \tag{10}$$

where

$$(C_{i,j})_r = C_{i,j}(C_{i,j}-1) \times \ldots \times (C_{i,j}-r+1), \quad \text{if } r \leq C_{i,j} \quad \text{and} \quad (C_{i,j})_r = 0, \quad \text{if } r > C_{i,j},$$

and $(x)_r = x(x-1) \times \ldots \times (x-r+1)$ denotes the falling factorial.

Making use of alternating sums $\mathbb{C}_k(S_m)$ in (5), present the polynomial expansion (10) as follows,

$$Q_z^{(r)}(1) = -\sum_{k=0}^{r} \mathfrak{S}_k^r \mathbb{C}_k(S_m), \qquad \mathfrak{S}_k^n = (-1)^{n-k} \left[{n \atop k} \right], \qquad \mathfrak{S}_n^n = 1, \tag{11}$$

where $\mathfrak{S}_k^n$ denote Stirling's numbers of the 1st kind and symbols $\left[{n \atop k} \right]$ satisfy the recurrence relation,

$$\left[{n+1 \atop k} \right] = n \left[{n \atop k} \right] + \left[{n \atop k-1} \right], \quad 1 \leq k \leq n, \qquad \left[{n \atop n} \right] = 1. \tag{12}$$

In Appendix 6 we present the first expressions for $\left[{n \atop n-k} \right]$ up to $k = 9$.

A straightforward calculation of the derivative $\Pi_z^{(r)}(z)$ gives

$$\Pi_z^{(r)}(z) = \Psi_z^{(r)}(z) + \sum_{k=1}^{r} k \binom{r}{k} \Psi_z^{(r-k)}(z) \Phi_z^{(k-1)}(z) - (1-z) \sum_{k=0}^{r} \binom{r}{k} \Psi_z^{(r-k)}(z) \Phi_z^{(k)}(z), \tag{13}$$

where

$$\Psi_z^{(r)}(z) = \sum_{k_1,\ldots,k_m \geq 0}^{r=k_1+\ldots+k_m} \frac{r}{k_1! \cdots k_m!} \prod_{j=1}^{m} \Psi_{j,z}^{(k_j)}(z), \quad \Psi_{j,z}^{(k)}(z) = \sum_{l \geq k}^{d_j-1} (l)_k z^{l-k}, \quad \Phi_z^{(k)}(z) = \sum_{\substack{s \in \Delta_m \\ s \geq k}} (s)_k z^{s-k},$$

and

$$\Psi_z^{(r)}(z) = \frac{d^r \Psi\,(S_m;z)}{dz^r}, \qquad \Psi_{j,z}^{(r)}(z) = \frac{d^r \Psi_j\,(S_m;z)}{dz^r}, \qquad \Phi_z^{(r)}(z) = \frac{d^r \Phi\,(S_m;z)}{dz^r}. \tag{14}$$

Thus, we obtain an expression for the derivative $\Pi_z^{(r)}(z)$ at $z=1$,

$$\Pi_{z=1}^{(r)} = \Psi_{z=1}^{(r)} + \sum_{k=1}^{r} k\binom{r}{k}\Psi_{z=1}^{(r-k)}\Phi_{z=1}^{(k-1)} = \Psi_{z=1}^{(r)} + r\sum_{k=0}^{r-1}\binom{r-1}{k}\Psi_{z=1}^{(r-k-1)}\Phi_{z=1}^{(k)}, \tag{15}$$

$$\Pi_{z=1}^{(r)} = \Pi_z^{(r)}(z=1),\quad \Psi_{z=1}^{(r)} = \Psi_z^{(r)}(z=1),\quad \Psi_{j,z=1}^{(k)} = \Psi_{j,z}^{(k)}(z=1),\quad \Phi_{z=1}^{(k)} = \Phi_z^{(k)}(z=1).$$

Derivatives $\Phi_{z=1}^{(r)}$, $\Psi_{z=1}^{(r)}$ *and* $\Pi_{z=1}^{(r)}$

Derivatives $\Phi_{z=1}^{(r)}$ may be calculated separately in accordance with (14),

$$\Phi_{z=1}^{(r)} = \sum_{k=0}^{r} \mathfrak{S}_k^r G_k, \qquad G_k = \sum_{s\in\Delta_m} s^k, \qquad \text{e.g.} \tag{16}$$

$$\Phi_{z=1}^{(0)} = G_0,\qquad \Phi_{z=1}^{(1)} = G_1,\qquad \Phi_{z=1}^{(2)} = G_2 - G_1,\qquad \Phi_{z=1}^{(3)} = G_3 - 3G_2 + 2G_1,$$

$$\Phi_{z=1}^{(4)} = G_4 - 6G_3 + 11G_2 - 6G_1,\qquad \Phi_{z=1}^{(5)} = G_5 - 10G_4 + 35G_3 - 50G_2 + 24G_1.$$

The sums G_k are known as *genera* of numerical semigroup S_m and G_0 denotes a semigroup *genus*.

General formulas for derivatives $\Psi_{z=1}^{(r)}$ and $\Pi_{z=1}^{(r)}$ are given in (13) and (15) and cannot be simplified essentially for arbitrary r. Here, we present the expressions for $\Psi_{z=1}^{(r)}$ and $\Pi_{z=1}^{(r)}$ for small $r \le 4$. All necessary calculations are given in Appendix 6.

$$\frac{\Psi_{z=1}^{(0)}}{\pi_m}=1,\quad \frac{\Psi_{z=1}^{(1)}}{\pi_m} = \frac{\sigma_1 - m}{2},\quad \frac{\Psi_{z=1}^{(2)}}{\pi_m} = \left(\frac{\sigma_1 - m}{2}\right)^2 + \frac{\sigma_2 - 6\sigma_1 + 5m}{12}, \tag{17}$$

$$\frac{\Psi_{z=1}^{(3)}}{\pi_m}=\left(\frac{\sigma_1 - m}{2} - 1\right)\left[\left(\frac{\sigma_1 - m}{2}\right)^2 + \frac{\sigma_2 - 4\sigma_1 + 3m}{4}\right],$$

$$\frac{\Psi_{z=1}^{(4)}}{\pi_m}=\frac{1}{3}\left(\frac{\sigma_2 - 6\sigma_1 + 5m}{4}\right)^2 + \left(\frac{\sigma_1 - m}{2}\right)^2 \frac{\sigma_2 - 6\sigma_1 + 5m}{2} + \left(\frac{\sigma_1 - m}{2}\right)^4 -$$
$$\frac{\sigma_1 - m}{2}(\sigma_2 - 4\sigma_1 + 3m) - \frac{\sigma_4 - 110\sigma_2 + 360\sigma_1 - 251m}{120},$$

where $\sigma_k = \sum_{j=1}^m d_j^k$ are power sums of generators d_j. Substituting (16, 17) into formulas (15), we arrive at expressions for $\Pi_{z=1}^{(r)}$,

$$
\begin{aligned}
&\frac{\Pi_{z=1}^{(0)}}{\pi_m}=1, \quad \frac{\Pi_{z=1}^{(1)}}{\pi_m}=\frac{\sigma_1-m}{2}+G_0, \qquad (18)\\
&\frac{\Pi_{z=1}^{(2)}}{\pi_m}=\left(\frac{\sigma_1-m}{2}\right)^2+\frac{\sigma_2-6\sigma_1+5m}{12}+(\sigma_1-m)G_0+2G_1,\\
&\frac{\Pi_{z=1}^{(3)}}{\pi_m}=\left(\frac{\sigma_1-m}{2}-1\right)\left[\left(\frac{\sigma_1-m}{2}\right)^2+\frac{\sigma_2-4\sigma_1+3m}{4}\right]+\\
&\quad\left[3\left(\frac{\sigma_1-m}{2}\right)^2+\frac{\sigma_2-6\sigma_1+5m}{4}\right]G_0+3(\sigma_1-m)G_1+3(G_2-G_1),\\
&\frac{\Pi_{z=1}^{(4)}}{\pi_m}=\frac{1}{3}\left(\frac{\sigma_2-6\sigma_1+5m}{4}\right)^2+\left(\frac{\sigma_1-m}{2}\right)^2\frac{\sigma_2-6\sigma_1+5m}{2}+\left(\frac{\sigma_1-m}{2}\right)^4-\\
&\quad\frac{\sigma_1-m}{2}(\sigma_2-4\sigma_1+3m)-\frac{\sigma_4-110\sigma_2+360\sigma_1-251m}{120}+\\
&\quad\left(\frac{\sigma_1-m}{2}-1\right)\left[(\sigma_1-m)^2+\sigma_2-4\sigma_1+3m\right]G_0+\\
&\quad\left[3(\sigma_1-m)^2+\sigma_2-6\sigma_1+5m\right]G_1+6(\sigma_1-m)(G_2-G_1)+4(G_3-3G_2+2G_1).
\end{aligned}
$$

3 Linear Equations for Alternating Power Sums $\mathbb{C}_k(S_m)$

In the right-hand side of expression (9) for $Q_z^{(r)}(1)$, there survives a solely one term, namely, when $k = m - 1$. Combining the resulting expression in (9) with (11), we obtain,

$$
(-1)^m(m-1)!\binom{r}{m-1}\Pi_{z=1}^{(r-m+1)} = \sum_{k=m-1}^{r} \mathfrak{S}_k^r \mathbb{C}_k(S_m), \qquad r \geq m-1.
$$

Represent the last equation in a more convenient way by shifting the variable r, i.e., $r = m + p$,

$$
\sum_{k=m-1}^{m+p} \mathfrak{S}_k^{m+p}\mathbb{C}_k(S_m) = (-1)^m \frac{(m+p)!}{(1+p)!}\,\Pi_{z=1}^{(p+1)}, \qquad p \geq -1. \qquad (19)
$$

Thus, we arrive at the matrix equation with $p + 2$ variables $\mathbb{C}_k(S_m)$, where $k = m - 1, \ldots, m + p$,

$$
\begin{pmatrix}
\mathfrak{S}_{m-1}^{m-1} & 0 & 0 & \ldots & 0\\
\mathfrak{S}_{m-1}^{m} & \mathfrak{S}_{m}^{m} & 0 & \ldots & 0\\
\mathfrak{S}_{m-1}^{m+1} & \mathfrak{S}_{m}^{m+1} & \mathfrak{S}_{m+1}^{m+1} & \ldots & 0\\
\ldots & \ldots & \ldots & \ldots & 0\\
\mathfrak{S}_{m-1}^{m+p} & \mathfrak{S}_{m}^{m+p} & \mathfrak{S}_{m+1}^{m+p} & \ldots & \mathfrak{S}_{m+p}^{m+p}
\end{pmatrix}
\begin{pmatrix}
\mathbb{C}_{m-1}(S_m)\\
\mathbb{C}_{m}(S_m)\\
\mathbb{C}_{m+1}(S_m)\\
\ldots\\
\mathbb{C}_{m+p}(S_m)
\end{pmatrix}
= (-1)^m
\begin{pmatrix}
(m-1)!\,\Pi_{z=1}^{(0)}\\
m!\,\Pi_{z=1}^{(1)}\\
\frac{(m+1)!}{2!}\,\Pi_{z=1}^{(2)}\\
\ldots\\
\frac{(m+p)!}{(p+1)!}\,\Pi_{z=1}^{(p+1)}
\end{pmatrix},
\qquad (20)
$$

where, according to definition of the Stirling numbers (12), we have in a diagonal $\mathfrak{S}_r^r = 1, r \geq 0$.

The general solution of matrix equation (20) may be written as follows,

$$\mathbb{C}_{m+p}(S_m) = (-1)^m \frac{(m+p)!}{(p+1)!} \Pi_{z=1}^{(p+1)} - \sum_{j=1}^{p+1} (-1)^j \begin{bmatrix} m+p \\ m+p-j \end{bmatrix} \mathbb{C}_{m+p-j}(S_m), \quad \text{e.g.,} \tag{21}$$

$$\mathbb{C}_{m-1}(S_m) = (-1)^m (m-1)!\, \Pi_{z=1}^{(0)},$$

$$\mathbb{C}_m(S_m) = (-1)^m m!\, \Pi_{z=1}^{(1)} + \begin{bmatrix} m \\ m-1 \end{bmatrix} \mathbb{C}_{m-1}(S_m),$$

$$\mathbb{C}_{m+1}(S_m) = (-1)^m \frac{(m+1)!}{2} \Pi_{z=1}^{(2)} + \begin{bmatrix} m+1 \\ m \end{bmatrix} \mathbb{C}_m(S_m) - \begin{bmatrix} m+1 \\ m-1 \end{bmatrix} \mathbb{C}_{m-1}(S_m).$$

Calculating $\mathbb{C}_{m+p}(S_m)$ in (21) by consecutive substitution of $\mathbb{C}_{m+q-1}(S_m)$ into $\mathbb{C}_{m+q}(S_m)$, where $q = 0, \ldots, p$, we arrive at the final expression,

$$\mathbb{C}_n(S_m) = \frac{(-1)^m n!}{(n-m)!} \pi_m K_{n-m}, \qquad \mathbb{C}_{m-1}(S_m) = (-1)^m (m-1)!\, \pi_m, \tag{22}$$

where a coefficient K_p is a linear combination of genera $G_0, \ldots, G_p$. We present here expressions[1] for K_p when $p \leq 6$,

$$K_0 = G_0 + \delta_1, \qquad \delta_p = \frac{\sigma_p - 1}{2^p}, \tag{23}$$

$$K_1 = G_1 + \frac{\sigma_1}{2} G_0 + \frac{3\delta_1^2 + \delta_2}{6},$$

$$K_2 = G_2 + \sigma_1 G_1 + \frac{3\sigma_1^2 + \sigma_2}{12} G_0 + \frac{\delta_1\left(\delta_1^2 + \delta_2\right)}{3},$$

$$K_3 = G_3 + \frac{3}{2}\sigma_1 G_2 + \frac{3\sigma_1^2 + \sigma_2}{4} G_1 + \frac{\sigma_1(\sigma_1^2 + \sigma_2)}{8} G_0 + \frac{15\delta_1^4 + 30\delta_1^2\delta_2 + 5\delta_2^2 - 2\delta_4}{60}.$$

$$K_4 = G_4 + 2\sigma_1 G_3 + \frac{3\sigma_1^2 + \sigma_2}{2} G_2 + \frac{\sigma_1(\sigma_1^2 + \sigma_2)}{2} G_1 +$$
$$\frac{15\sigma_1^4 + 30\sigma_1^2\sigma_2 + 5\sigma_2^2 - 2\sigma_4}{240} G_0 + \delta_1 \frac{3\delta_1^4 + 10\delta_1^2\delta_2 + 5\delta_2^2 - 2\delta_4}{15},$$

[1] Formulas for K_0, K_1, K_2, K_3 were calculated by consecutive substitution of expressions (22) and (18) into (21). The other three formulas for K_4, K_5, K_6 were found in two steps: (1) by analytic derivations (with help of Mathematica software) of expressions for $\Psi_{z=1}^{(r)}$ and $\Pi_{z=1}^{(r)}$, $r = 5, 6, 7$, which are extremely lengthy to be disposed in the paper, (2) by consecutive substitution of the found expressions for $\Pi_{z=1}^{(r)}$ and $\mathbb{C}_{m+r-1}(S_m)$ into (21).

$$
\begin{aligned}
K_5 = {} & G_5 + \frac{5}{2}\sigma_1 G_4 + \frac{5}{6}\left(3\sigma_1^2 + \sigma_2\right) G_3 + \frac{5}{4}\sigma_1(\sigma_1^2 + \sigma_2)G_2 + \\
& \frac{15\sigma_1^4 + 30\sigma_1^2\sigma_2 + 5\sigma_2^2 - 2\sigma_4}{48} G_1 + \sigma_1 \frac{3\sigma_1^4 + 10\sigma_1^2\sigma_2 + 5\sigma_2^2 - 2\sigma_4}{96} G_0 + \\
& \frac{63\delta_1^6 + 315\delta_1^4\delta_2 + 315\delta_1^2\delta_2^2 - 126\delta_1^2\delta_4 + 35\delta_2\delta_4 - 42\delta_2^3 + 16\delta_6}{378} \\
K_6 = {} & G_6 + 3\sigma_1 G_5 + \frac{5}{4}\left(3\sigma_1^2 + \sigma_2\right) G_4 + \frac{5}{2}\sigma_1(\sigma_1^2 + \sigma_2)G_3 + \\
& \frac{15\sigma_1^4 + 30\sigma_1^2\sigma_2 + 5\sigma_2^2 - 2\sigma_4}{16} G_2 + \sigma_1 \frac{3\sigma_1^4 + 10\sigma_1^2\sigma_2 + 5\sigma_2^2 - 2\sigma_4}{16} G_1 + \\
& \frac{63\sigma_1^6 + 315\sigma_1^4\sigma_2 + 315\sigma_1^2\sigma_2^2 - 126\sigma_1^2\sigma_4 + 35\sigma_2\sigma_4 - 42\sigma_2^3 + 16\sigma_6}{4032} G_0 + \\
& \delta_1 \frac{9\delta_1^6 + 63\delta_1^4\delta_2 + 105\delta_1^2\delta_2^2 - 42\delta_1^2\delta_4 + 35\delta_2^3 - 42\delta_2\delta_4 + 16\delta_6}{63} .
\end{aligned}
$$

A straightforward calculation of higher K_p (even with help of Mathematica software) encounters with enormous technical difficulties. On the other hand, a careful observation of formulas (23) allows to put forward a conjecture about a general formula for K_p for arbitrary p, which is related to a special kind of symmetric polynomials $P_n = P_n(x_1, \ldots, x_m)$ of degree n in m variables, discussed in [11],

$$
P_n = \sum_{j=1}^{m} x_j^n - \sum_{j>r=1}^{m} (x_j + x_r)^n + \sum_{j>r>i=1}^{m} (x_j + x_r + x_i)^n - \ldots - (-1)^m \left(\sum_{j=1}^{m} x_j\right)^n . \tag{24}
$$

In what follows, we make use of a remarkable property of polynomials P_n: its factorization reads [11],

$$
P_n(x_1, \ldots, x_m) = \frac{(-1)^{m+1} n!}{(n-m)!} \chi_m T_{n-m}(x_1, \ldots, x_m), \qquad \chi_m = \prod_{j=1}^{m} x_j, \quad T_0 = 1, \tag{25}
$$

where $T_r(x_1, \ldots, x_m)$ is a symmetric polynomial of degree r in m variables. According to [11], this polynomial satisfies inequality,

$$
T_r(x_1, \ldots, x_m) > 0, \qquad x_1, \ldots, x_m > 0. \tag{26}
$$

Denote $X_k = \sum_{j=1}^{m} x_j^k$ and, by a fundamental theorem of symmetric polynomials [14], use such power sums as a basis for algebraic Rep of polynomials T_r. In other words, instead of $T_r(x_1, \ldots, x_m)$, we make use in (25) of polynomials $T_r(X) = T_r(X_1, \ldots, X_r)$, which were derived in [11] and presented in Appendix 6. To pose a conjecture, define two polynomials $T_r(\sigma) = T_r(\sigma_1, \ldots, \sigma_r)$ and $T_r(\delta) = T_r(\delta_1, \ldots, \delta_r)$ by replacing $X_k \to \sigma_k$ and $X_k \to \delta_k$ in $T_r(X_1, \ldots, X_r)$, where σ_k and δ_k are defined in (17) and (23), respectively.

Conjecture 1 Let a semigroup $S_m = \langle d_1, \ldots, d_m \rangle$ be given and G_r denote its genera according to (16). Then the alternating power sums $\mathbb{C}_k(S_m)$ in (5) are given by (22) with K_p as follows

$$K_p = \sum_{r=0}^{p} \binom{p}{r} T_{p-r}(\sigma) G_r + \frac{2^{p+1}}{p+1} T_{p+1}(\delta), \qquad (27)$$

If (27) holds for any p, then, combining it with (26) and keeping in mind $\sigma_i, \delta_i > 0$, we get $K_p > 0$.

4 Algebraic Equations for K_r in Numerical Semigroups

Consider a numerical semigroup S_m and write equalities (6) and (22) for alternating sums $\mathbb{C}_k(S_m)$, defined in (5), as a system of non-homogeneous polynomial equations for positive integer variables $C_{k,j}$. For convenience, rename $C_{k,j}$ by one-index variable z_i in such a way, that i runs through the two-index (k, j) table. enumerating elements of the first and following rows successively,

$$z_1 = C_{1,1}, \quad z_2 = C_{1,2}, \quad \ldots, \quad z_{\beta_1} = C_{1,\beta_1}, \quad z_{\beta_1+1} = C_{2,1}, \quad z_{\beta_1+2} = C_{2,2}, \quad \ldots, \quad z_{\zeta_m} = C_{m-1,\beta_{m-1}},$$

where a total number $\zeta_m = \#\{z_i\}$ of independent variables z_i is dependent on inner properties of S_m, e.g., non-symmetric, symmetric (not CI), symmetric CI and others more sophisticated (Weierstrass', Arf's, hyperelliptic *etc.*). We consider here the three basic kinds of semigroups mentioned above.

A study of $n + 1$ *non-homogeneous* polynomial equations $f_j(z_1, \ldots, z_n) = 0$ in n variables z_i goes back to classical works of Bézout, Sylvester, Caley and Macaulay [15], who has mostly considered an equivalent problem with $n + 1$ *homogeneous* polynomial equations in $n + 1$ variables. The use of a multivariate resultant $\mathcal{R}es\,\{f_0, f_1, \ldots, f_n\}$, which is an irreducible polynomial over a ring $A[f_0, f_1, \ldots, f_n]$, generated by coefficients of f_j, and vanishes whenever all polynomials f_j have a common root, is a standard computational tool in the elimination theory. An interest in finding explicit formulas for resultants, extending Macaulay's formulas as a quotient of two determinants, has been renewed in the 1990th (see [2] and references therein).

Bearing in mind a special form of Eqs. (6) and (22), we consider here the most general properties of these equations: the existence of an algebraic relation among K_p, which entered in (22), and its rescaled version, given below. Rewrite polynomial equations (6) and (22) in new notations,

$$\Gamma_k(z_1, \ldots, z_{\zeta_m}, L_k) = 0, \quad k \geq 1, \qquad \zeta_m = B_m, \tag{28}$$
$$L_k = \begin{cases} 0, & if\ 1 \leq k \leq m-2, \\ (-1)^m (m-1)! \pi_m, & if\ k = m-1, \\ (-1)^m \frac{k!}{(k-m)!} \pi_m K_{k-m}, & if\ k \geq m, \end{cases}$$

where $\Gamma_k(z_1, \ldots, z_{\zeta_m}, L_k)$ is a homogeneous polynomial of degree k with respect to all variables z_i and linear in L_k, and $B_m = \sum_{k=1}^{m-1} \beta_k$ denotes the total Betti number of non-symmetric semigroups.

Making use of a homogeneity of the polynomial $\Gamma_k(\xi_1, \ldots, \xi_{\zeta_m}, \ell_k)$, rescale the variables and the whole Eq. (28) as follows,

$$\Gamma_k(\xi_1, \ldots, \xi_{\zeta_m}, \ell_k) = 0, \quad k \geq 1, \qquad \xi_j = \frac{z_j}{\upsilon_m}, \quad \upsilon_m = \pi_m^{1/(m-1)}, \tag{29}$$

$$\ell_k = \begin{cases} 0, & if\ 1 \leq k \leq m-2, \\ (-1)^m (m-1)! & if\ k = m-1, \\ (-1)^m \frac{k!}{(k-m)!} \varkappa_{k-m}, & if\ k \geq m, \qquad \varkappa_p = K_p \upsilon_m^{-(p+1)} \end{cases} \tag{30}$$

where according to (5), the polynomial $\Gamma_k(\xi_1, \ldots, \xi_{\zeta_m}, \ell_k)$ for the arbitrary non-symmetric semigroup S_m reads,

$$\Gamma_k(\xi_1, \ldots, \xi_{\zeta_m}, \ell_k) = \sum_{j=1}^{\beta_1} \xi_j^k - \sum_{j=\beta_1+1}^{\beta_1+\beta_2} \xi_j^k + \ldots + (-1)^m \sum_{j=\zeta_m-\beta_{m-1}}^{\zeta_m} \xi_j^k - \ell_k. \tag{31}$$

Theorem 1 *Let S_m be a non-symmetric semigroup with the Hilbert series given in (1). Then there exists an algebraic equation in $g_m + 1$ variables, $\varkappa_0, \varkappa_1, \ldots, \varkappa_{g_m}$,*

$$\mathcal{R}_K(\varkappa_0, \varkappa_1, \ldots, \varkappa_{g_m-1}, \varkappa_{g_m}) = 0, \qquad g_m = B_m - m + 1, \tag{32}$$

and the polynomial $\mathcal{R}_K$ is irreducible over a ring $A[\varkappa_0, \varkappa_1, \ldots, \varkappa_{g_m-1}, \varkappa_{g_m}]$.

Proof Choose[2] the first $B_m + 1$ polynomial equations (29) in B_m variables $\xi_1, \ldots, \xi_{B_m}$ and build a new system of B_m equations in $B_m - 1$ variables $\xi_2, \ldots, \xi_{B_m}$ by eliminating ξ_1 in resultants $\mathcal{R}es_1\{\Gamma_1, \Gamma_j\}$,

$$\Gamma_j^1(\xi_2, \ldots, \xi_{B_m}, \ell_1, \ell_j) = 0, \qquad j = 2, \ldots, B_m + 1, \tag{33}$$
$$\Gamma_j^1(\xi_2, \ldots, \xi_{B_m}, \ell_1, \ell_j) = \mathcal{R}es_1\{\Gamma_1(\xi_1, \ldots, \xi_{B_m}, \ell_1), \Gamma_j(\xi_1, \ldots, \xi_{B_m}, \ell_j)\}.$$

The polynomial Γ_j^1 in (33) is irreducible [2] over a ring $A[\xi_2, \ldots, \xi_{B_m}, \ell_1, \ell_j]$ (see a detailed proof of a resultant irreducibility for two polynomials in [12]).

[2] Note that inequality $B_m > m - 1$ holds, since, according to [18], we have $\beta_1 \geq m - 1$ while the other β_k are positive.

At the second step, choose the first B_m polynomial equations (33) in $B_m - 1$ variables $\xi_2, \ldots, \xi_{B_m}$ and build $B_m - 1$ equations in $B_m - 2$ variables $\xi_3, \ldots, \xi_{B_m}$ by eliminating ξ_2 in resultants $\mathcal{R}es_2\left\{\Gamma_j^1, \Gamma_2^1\right\}$,

$$\Gamma_j^{1,2}(\xi_3, \ldots, \xi_{B_m}, \ell_1, \ell_2, \ell_j) = 0, \qquad j = 3, \ldots, B_m + 1,$$
$$\Gamma_j^{1,2}(\xi_3, \ldots, \xi_{B_m}, \ell_1, \ell_2, \ell_j) = \mathcal{R}es_2\left\{\Gamma_2^1(\xi_2, \ldots, \xi_{B_m}, \ell_1, \ell_2), \Gamma_j^1(\xi_2, \ldots, \xi_{B_m}, \ell_1, \ell_j)\right\}.$$

The polynomial $\Gamma_j^{1,2}$ is irreducible [2] over a ring $A[\xi_3, \ldots, \xi_{B_m}, \ell_1, \ell_2, \ell_j]$ by reasons mentioned above.

Continuing to eliminate the variables ξ_k successively and constructing the families of resultants,

$$\Gamma_j^{1,\ldots,k}(\xi_{k+1}, \ldots, \xi_{B_m}, \ell_1, \ldots, \ell_k, \ell_j) = 0, \qquad j = k + 1, \ldots, B_m + 1,$$
$$\Gamma_j^{1,\ldots,k}(\xi_{k+1}, \ldots, \xi_{B_m}, \ell_1, \ldots, \ell_k, \ell_j) = \mathcal{R}es_k\left\{\Gamma_k^{1,\ldots,k-1}, \Gamma_j^{1,\ldots,k-1}\right\},$$

we arrive at the B_mth step at one resultant equation

$$\mathcal{R}es_{Bm}\left\{\Gamma_{Bm}^{1,\ldots,Bm-1}\left(\xi_{Bm}, \ell_1, \ldots, \ell_{Bm-1}, \ell_{Bm}\right), \Gamma_{Bm+1}^{1,\ldots,Bm-1}\left(\xi_{Bm}, \ell_1, \ldots, \ell_{Bm-1}, \ell_{Bm+1}\right)\right\} = 0. \quad (34)$$

The polynomial $\mathcal{R}es_{B_m}$ in the left hand side of (34) is irreducible [2] over a ring $A[\ell_1, \ldots, \ell_{B_m-1}, \ell_{B_m}, \ell_{B_m+1}]$ as well as all resultants of two polynomials at previous steps.

Equation (34) is free of any variable ξ_i and involves only $B_m + 1$ coefficients ℓ_n, $1 \leq n \leq B_m + 1$. Keeping in mind the two first relations in (30), namely, $\ell_n = 0$ if $1 \leq n < m - 1$, and an independence of ℓ_{m-1} on $\varkappa_j$, we conclude that Eq. (34) is algebraic in $B_m - m + 2$ variables $\ell_m, \ell_{m+1}, \ldots, \ell_{B_m}, \ell_{B_m+1}$. However, by the third relation in (30), such equation can be represented in $\varkappa_0, \varkappa_1, \ldots, \varkappa_{B_m-m}, \varkappa_{B_m-m+1}$ as given in (32). □

Corollary 1 *Let S_m be a non-symmetric semigroup with the Hilbert series given in (1). Then there exists g_m algebraically independent genera. The set of such genera reads,*

$$\{G_0, G_1, \ldots, G_{g_m-1}\}.$$

Proof Combining Theorem 1 and formulas (23) as well as (27), by assumption that Conjecture 1 is true, we arrive at algebraic equation

$$\mathcal{R}_G(G_0, G_1, \ldots, G_{g_m-1}, G_{g_m}) = 0,$$

with an irreducible polynomial $\mathcal{R}_G$ over a ring $A[G_0, G_1, \ldots, G_{g_m-1}, G_{g_m}]$. Resolving the last equation with respect to G_{g_m} and keeping in mind an irreduciblity of $\mathcal{R}_G$, we arrive at the algebraic function $G_{g_m} = \mathcal{F}(G_0, \ldots, G_{g_m-1})$, where the set

$\{G_0, G_1, \ldots, G_{g_m-1}\}$ comprises genera for any numerical semigroup S_m, which are algebraically independent. □

Theorem 1 may be extended on algebraic equations included $\varkappa_n$, $n > g_m$, with a similar proof.

Theorem 2 *There exists an algebraic equation in $g_m + 1$ variables $\varkappa_0, \varkappa_1, \ldots, \varkappa_{g_m-1}$ and $\varkappa_n$,*

$$\mathcal{R}_K(\varkappa_0, \varkappa_1, \ldots, \varkappa_{g_m-1}, \varkappa_n) = 0, \qquad n > g_m, \tag{35}$$

and the polynomial $\mathcal{R}_K$ is irreducible over a ring $A[\varkappa_0, \varkappa_1, \ldots, \varkappa_{g_m-1}, \varkappa_n]$.

Numerical Semigroups S_3

In this section, we consider the most simple case of non-symmetric numerical semigroups S_3 generated by three integers. The numerator in the rational Rep (3) of its Hilbert series $H(S_3; z)$ reads,

$$Q(S_3; z) = 1 - \left(z^{x_1} + z^{x_2} + z^{x_3}\right) + z^{y_1} + z^{y_2}, \qquad \beta_1 = 3, \quad \beta_2 = 2, \quad g_3 = 3.$$

Six polynomial equations (6) and (22) for five symmetric polynomials, $X_k = \sum_{j=1}^{3} x_j^k$, $k = 1, 2, 3$, and $Y_r = \sum_{j=1}^{2} y_j^r$, $r = 1, 2$, are given below,

$$\begin{array}{lll} Y_1 = X_1, & Y_2 = X_2 + 2\pi_3, & Y_3 = X_3 + 6\pi_3 K_0, \\ Y_4 = X_4 + 24\pi_3 K_1, & Y_5 = X_5 + 60\pi_3 K_2, & Y_6 = X_6 + 120\pi_3 K_3, \end{array} \tag{36}$$

where K_i are given in (23). Bearing in mind the Newton identities [14] related symmetric polynomials,

$$\begin{aligned} &Y_3 = \frac{1}{2}\left(3Y_2 - Y_1^2\right) Y_1, \quad Y_4 = \frac{1}{2}\left(Y_2^2 + 2Y_1^2 Y_2 - Y_1^4\right), \\ &Y_5 = \frac{1}{4}\left(5Y_2^2 - Y_1^4\right) Y_1, \quad Y_6 = \frac{1}{4}\left(Y_2^2 + 6Y_1^2 Y_2 - 3Y_1^4\right) Y_2, \\ &X_4 = \frac{1}{6}\left(X_1^4 - 6X_1^2 X_2 + 8X_1 X_3 + 3X_2^2\right), \quad X_5 = \frac{1}{6}\left(X_1^5 - 5X_1^3 X_2 + 5X_1^2 X_3 + 5X_2 X_3\right), \\ &X_6 = \frac{1}{12}\left(X_1^6 - 3X_1^4 X_2 - 9X_1^2 X_2^2 + 3X_2^3 + 4X_1^3 X_3 + 12X_1 X_2 X_3 + 4X_3^2\right), \end{aligned} \tag{37}$$

we present six equations (36) as follows

$$
\begin{aligned}
\frac{1}{2}Y_1\left(3Y_2-Y_1^2\right)&=X_3+6\pi_3K_0, \qquad Y_1=X_1, \qquad Y_2=X_2+2\pi_3, \qquad (38)\\
3\left(Y_2^2+2Y_1^2Y_2-Y_1^4\right)&=X_1^4-6X_1^2X_2+8X_1X_3+3X_2^2+144\pi_3K_1,\\
\frac{3}{2}Y_1\left(5Y_2^2-Y_1^4\right)&=X_1^5-5X_1^3X_2+5X_1^2X_3+5X_2X_3+360\pi_3K_2,\\
3Y_2\left(Y_2^2+6Y_1^2Y_2-3Y_1^4\right)&=X_1^6-3X_1^4X_2-9X_1^2X_2^2+3X_2^3+4X_1^3X_3+12X_1X_2X_3+\\
&\quad 4X_3^2+1440\pi_3K_3.
\end{aligned}
$$

Substituting three first relations of (38) into the three last and simplifying the final expressions, we obtain

$$
\begin{aligned}
&Y_1^2-4K_0Y_1+12K_1+\pi_3=Y_2, \qquad Y_1^3-2K_0Y_1^2+4\pi_3K_0+24K_2=Y_2(Y_1+2K_0),\\
&Y_1^4-2\pi_3Y_1^2+8\pi_3K_0Y_1+8\pi_3K_0^2-\frac{4}{3}\pi_3^2+80K_3=Y_2^2-2Y_2(\pi_3-4K_0Y_1). \qquad (39)
\end{aligned}
$$

Combining separately the 1st relation in (39) with the second and third relations, we get, respectively,

$$
\begin{aligned}
\left(3K_1-2K_0^2+\frac{\pi_3}{4}\right)Y_1&=6K_2-6K_0K_1+\frac{\pi_3}{2}K_0, \qquad (40)\\
\left(3K_1-2K_0^2+\frac{\pi_3}{4}\right)Y_1^2&=10K_3-18K_1^2+\pi_3K_0^2-\frac{\pi_3^2}{24},
\end{aligned}
$$

and further, due to (36, 37, 39),

$$
\begin{aligned}
X_2&=4\left(\frac{3K_2-6K_0K_1+2K_0^3}{3K_1-2K_0^2+\frac{\pi_3}{4}}\right)^2-4\left(K_0^2-3K_1\right)-\pi_3, \qquad X_1=Y_1,\\
X_3&=Y_1^3-6K_0Y_1^2+3\left(6K_1+\frac{\pi_3}{2}\right)Y_1-6K_0\pi_3, \qquad Y_2=X_2+2\pi_3.
\end{aligned}
$$

Rescaling $K_r=\varkappa_r\sqrt{\pi_3^{r+1}}$, write a necessary condition to have non-trivial solutions for Eq. (40),

$$
3\varkappa_1\neq 2\varkappa_0^2-\frac{1}{4},
$$

otherwise, there exist three equalities

$$
\varkappa_1=\frac{2}{3}\varkappa_0^2-\frac{1}{12}, \qquad \varkappa_2=\varkappa_0\left(\varkappa_1-\frac{1}{12}\right), \qquad \varkappa_3=\frac{9}{5}\varkappa_1^2-\frac{1}{10}\varkappa_0^2+\frac{1}{240}, \qquad (41)
$$

which define a special class of semigroups S_3. In Sect. 5, we show that the last three formulas are related to symmetric 3-generated semigroups.

Combining two equalities (40), we obtain, in accordance with Theorem 1, equation (32) in rescaled variables $\varkappa_0, \varkappa_1, \varkappa_2, \varkappa_3$,

$$\left(10\varkappa_3 - 18\varkappa_1^2 + \varkappa_0^2 - \frac{1}{24}\right)\left(3\varkappa_1 - 2\varkappa_0^2 + \frac{1}{4}\right) = \left(6\varkappa_2 - 6\varkappa_0\varkappa_1 + \frac{\varkappa_0}{2}\right)^2, \quad (42)$$

that manifests three algebraically independent genera G_0, G_1, G_2.

Extension on Higher $\varkappa_n$ in S_3

To derive the Eq. (35) for $\varkappa_4$ let us consider the power sums X_7, Y_7,

$$X_7 = \frac{X_1^7 - 21X_1^3X_2^2 + 7X_1^4X_3 + 21X_2^2X_3 + 28X_1X_3^2}{36}, \qquad Y_7 = \frac{(Y_1^6 - 7Y_1^4Y_2 + 7Y_1^2Y_2^2 + 7Y_2^3)Y_1}{8},$$

and substitute them into equality $Y_7 = X_7 + 840\pi_3 K_4$. Making use of the last equality and five first relations in (38) and performing necessary calculations, we arrive at three equations for Y_1 and Y_2.

$$Y_1^2 - 4K_0Y_1 + 12K_1 + \pi_3 = Y_2, \qquad Y_1^3 - 2K_0Y_1^2 + 4\pi_3K_0 + 24K_2 = Y_2(Y_1 + 2K_0),$$
$$8K_0^2\pi_3Y_1 + Y_1(Y_1^2 - Y_2)(Y_2 - \pi_3) + K_0(Y_1^4 - 4\pi_3^2 + 4\pi_3Y_2 - 4Y_1^2Y_2 - Y_2^2) = -60K_4, \quad (43)$$

which is similar to (39) by exception of the last one. Equations (43) can be resolved in Y_1 as follows,

$$\left(3K_1 - 2K_0^2 + \frac{\pi_3}{4}\right)Y_1 = 6K_2 - 6K_0K_1 + \frac{\pi_3}{2}K_0,$$
$$\left(Y_1^2 - 2K_0Y_1 + 12K_1\right)\left(3K_1 - 2K_0^2 + \frac{\pi_3}{4}\right)Y_1 = 15K_4 - K_0\left(\frac{\pi_3}{2} - 6K_1\right)^2, \quad (44)$$

Combining two equations in (44) and rescaling the coefficients $K_r = \varkappa_r\sqrt{\pi_3^{r+1}}$, we obtain, in accordance with Theorem 1, Eq. (35) in $\varkappa_0, \varkappa_1, \varkappa_2, \varkappa_4$ variables,

$$\left(15\varkappa_4 - \varkappa_0\left(6\varkappa_1 - \frac{1}{2}\right)^2\right)\left(3\varkappa_1 - 2\varkappa_0^2 + \frac{1}{4}\right)^2 = \left(6\varkappa_2 - 6\varkappa_1\varkappa_0 + \frac{\varkappa_0}{2}\right)\mathcal{K}_4(\varkappa_0, \varkappa_1, \varkappa_2), \quad (45)$$

where $\mathcal{K}_4(\varkappa_0, \varkappa_1, \varkappa_2)$ is a positive definite function

$$\mathcal{K}_4 = 2(1+12\varkappa_1)\varkappa_0^4 + 24\varkappa_2\varkappa_0^3 - 18\varkappa_1(1+4\varkappa_1)\varkappa_0^2 + 3\varkappa_2(1-36\varkappa_1)\varkappa_0$$
$$+ 36\varkappa_2^2 + \frac{3\varkappa_1}{4}(1+12\varkappa_1)^2$$

in the positive octant $\varkappa_0, \varkappa_1, \varkappa_2 > 0$. In order to prove that, we suppose, by way of contradiction, that $\mathcal{K}_4(\varkappa_0, \varkappa_1, \varkappa_2) = 0$. The last equation may be resolved as quadratic in $\varkappa_2$,

$$\varkappa_2^{\pm} = \pm\left(1 - 8\varkappa_0^2 + 12\varkappa_1\right)\sqrt{\varkappa_0^2 - 12\varkappa_1} + \varkappa_0\left(36\varkappa_1 - 1 - 8\varkappa_0^2\right), \quad \varkappa_2^- \le \varkappa_2^+. \tag{46}$$

Consider the largest real root $\varkappa_2^+ \ge \varkappa_2^-$ and require $12\varkappa_1 \le \varkappa_0^2$. Combining (46) with the last inequality, we arrive for $\varkappa_0, \varkappa_1 > 0$ at the upper bound,

$$\varkappa_2^+ \le \left(1 - 7\varkappa_0^2\right)\sqrt{\varkappa_0^2 - 12\varkappa_1} - \varkappa_0\left(1 + 5\varkappa_0^2\right) \le \varkappa_0\left(1 - 7\varkappa_0^2 - 1 - 5\varkappa_0^2\right) \le -12\varkappa_0^3.$$

Thus, the both roots $\varkappa_2^{\pm}$ are never positive. In other words, in the positive octant $\varkappa_0, \varkappa_1, \varkappa_2 > 0$ the function $\mathcal{K}_4(\varkappa_0, \varkappa_1, \varkappa_2)$ is never vanished. Since $\mathcal{K}_4(\varkappa_0, 0, 0) = 2\varkappa_0^4$, we conclude that the function $\mathcal{K}_4(\varkappa_0, \varkappa_1, \varkappa_2)$ is always positive.

Coming back to relation (44) and Eqs. (45), we conclude that there exists one special case, when all (44, 45) are satisfied identically,

$$3\varkappa_1 - 2\varkappa_0^2 + \frac{1}{4} = 0, \quad 6\varkappa_2 - 6\varkappa_1\varkappa_0 + \frac{\varkappa_0}{2} = 0, \quad 15\varkappa_4 - \varkappa_0\left(6\varkappa_1 - \frac{1}{2}\right)^2 = 0, \tag{47}$$

This case is related to symmetric 3-generated semigroups and equalities (47) are coincided with the three corresponding formulas for K_1, K_2, K_4 in (67).

5 Supplementary Relations for K_r and G_r in Symmetric Semigroups

If a numerical semigroup S_m is symmetric, then degrees $C_{k,j}$ of syzygies and Betti's numbers β_k in the rational Rep (2) of the Hilbert series are related as follows,

$$\beta_k = \beta_{m-k-1}, \quad \beta_{m-1} = 1, \qquad C_{k,j} + C_{m-k-1,j} = Q_m, \tag{48}$$

while the number of gaps and non-gaps of S_m are equal to G_0. Therefore, according to (4, 23), we have

$$F_m = 2G_0 - 1, \quad 2K_0 = Q_m = F_m + \sigma_1, \qquad \sum_{j=1}^{G_0} s_j^n + \sum_{j=1}^{G_0}(F_m - s_j)^n = \sum_{j=0}^{F_m} j^n. \tag{49}$$

The last identity in (49) may be represented as follows

$$2G_{2r} + \sum_{q=1}^{2r}(-1)^q\binom{2r}{q}F_m^q G_{2r-q} = \frac{F_m^{2r+1}}{2r+1} + \frac{F_m^{2r}}{2} + \sum_{q=1}^{2r-1}\binom{2r}{q-1}\mathcal{B}_{2r-q+1}\frac{F_m^q}{q}, \tag{50}$$

where $\mathcal{B}_k$ denotes the Bernoulli number. Equality (50) reduces the number of independent genera G_k twice, making G_{2r} dependent on genera with odd indices, G_{2j-1}, $j = 1, \ldots, r$,

$$\frac{G_{2r}}{F_m} = \frac{1}{2}\sum_{p=0}^{2r-2}\left[\binom{2r}{p}\frac{\mathcal{B}_{2r-p}}{p+1} + (-1)^p\binom{2r}{p+1}G_{2r-p-1}\right]F_m^p - \frac{2r-1}{2r+1}\frac{F_m^{2r}}{4}, \quad (51)$$

e.g.,

$$\frac{G_2}{F_m} = G_1 - \frac{F_m^2-1}{12}, \qquad \frac{F_m^2-1}{12} = \frac{G_0(G_0-1)}{3}, \quad (52)$$
$$\frac{G_4}{F_m} = 2G_3 - F_m^2 G_1 + \frac{F_m^2-1}{12}\frac{6F_m^2+1}{5},$$
$$\frac{G_6}{F_m} = 3G_5 - 5F_m^2 G_3 + 3F_m^4 G_1 - \frac{F_m^2-1}{12}\frac{51F_m^4+9F_m^2+2}{14},$$
$$\frac{G_8}{F_m} = 4G_7 - 14F_m^2 G_5 + 28F_m^4 G_3 - 17F_m^6 G_1 + \frac{F_m^2-1}{12}\frac{310F_m^6+55F_m^4+13F_m^2+3}{15}.$$

To find a number g_m of algebraically independent genera of symmetric semigroups, we apply Theorem 1 with a new number ζ_m of independent variables, which differs from (31). For this purpose, represent formula (3) for the numerator $Q(S_m; t)$ when $m = 0, 1 (\mathrm{mod}\ 2)$ separately and account for independent variables $x_j, y_j, \ldots, z_j, Q_m$ in every of two cases,

- $m = 2q,\ q \geq 2, \qquad \zeta_{2q} = \sum_{j=0}^{q-1}\beta_j$

$$1-t^{Q_m} - \sum_{j=1}^{\beta_1}\left(t^{x_j} - t^{Q_m-x_j}\right) + \sum_{j=1}^{\beta_2}\left(t^{y_j} - t^{Q_m-y_j}\right) - \ldots - (-1)^q\sum_{j=1}^{\beta_{q-1}}\left(t^{z_j} - t^{Q_m-z_j}\right), \quad (53)$$

- $m = 2q+1\ \ q \geq 2, \qquad \zeta_{2q+1} = \sum_{j=0}^{q}\beta_j \qquad (-1)^q\beta_q = \sum_{j=0}^{q-1}(-1)^{j+1}\beta_j$

$$1+t^{Q_m} - \sum_{j=1}^{\beta_1}\left(t^{x_j} + t^{Q_m-x_j}\right) + \sum_{j=1}^{\beta_2}\left(t^{y_j} + t^{Q_m-y_j}\right) - \ldots + (-1)^q\sum_{j=1}^{\beta_q}\left(t^{z_j} + t^{Q_m-z_j}\right). \quad (54)$$

Note, that $\zeta_{2q+1} = 0\ (\mathrm{mod}\ 2)$ since ζ_{2q+1} may be presented for $m = 1, 3 (\mathrm{mod}\ 4)$ as follows

$$\zeta_{4q+1} = 2\sum_{j=1}^{q}\beta_{2j-1}, \qquad \zeta_{4q+3} = 2\left(\sum_{j=1}^{q}\beta_{2j} + 1\right). \quad (55)$$

Theorem 3 *Let S_m be a symmetric (not CI) semigroup, then there are g_m independent genera*

$$g_{2q} = \zeta_{2q} - q, \quad G_0, G_1, G_3, \ldots G_{2g_{2q}-3}, \tag{56}$$

$$g_{2q+1} = \zeta_{2q+1} - q, \quad G_0, G_1, G_3, \ldots G_{2g_{2q+1}-3}. \tag{57}$$

Proof First, consider symmetric semigroups S_{2q} with even *edim*. The total number of syzygies degrees (including those which are related in couples), appeared in (53), is given by $2\zeta_{2q} - 1$. Replacing by this number the total Betti number B_m in (32), we get $\widetilde{g}_{2q} = 2\zeta_{2q} - 1 - (2q - 1) = 2(\zeta_{2q} - q)$, which does not relate to equalities (51). Keeping in mind the supplementary relations (51) for genera G_k, we have to decrease the last number twice, i.e., we arrive at (56).

Next, consider symmetric semigroups S_{2q+1} with odd *edim*. By similar considerations, as in the case $m = 2q$, we arrive at $\widetilde{g}_{2q+1} = 2\zeta_{2q+1} - 1 - (2q + 1 - 1) = 2(\zeta_{2q+1} - q) - 1$, which does not related to equalities (51). Keeping in mind the supplementary relations (51) for genera G_k, we have to decrease the number $\widetilde{g}_{2q+1}$ as follows: $g_{2q+1} = (1 + \widetilde{g}_{2q+1})/2$, i.e., we arrive at (57). □

Combining Theorem 3 with (53, 55), we may specify the number g_m in more details,

$$g_{2q} = \sum_{j=1}^{q-1} \beta_j - q + 1, \qquad g_{4q+1} = 2\left(\sum_{j=1}^{q} \beta_{2j-1} - q\right), \qquad g_{4q+3} = 2\left(\sum_{j=1}^{q} \beta_{2j} - q\right) + 1. \tag{58}$$

Symmetric (Not CI) Numerical Semigroups S_4

The numerator (3) in the rational Rep of its Hilbert series $H(S_4; z)$ reads [1],

$$Q(S_4; z) = 1 - \sum_{j=1}^{5} z^{x_j} + \sum_{j=1}^{5} z^{Q_4 - x_j} - z^{Q_4}, \qquad \beta_1 = 5, \quad g_4 = 4.$$

We present polynomial equations (6, 22) for the 10 first symmetric polynomials, $X_k = \sum_{j=1}^{5} x_j^k$. Among them, equations of the first and second degrees are coincided. Together with (49) they give

$$X_1 = 2Q_4 = 4K_0. \tag{59}$$

The rest of eight equations might be decomposed in couples of the odd and even degrees,

$$
\begin{aligned}
2X_3 - 3Q_4X_2 + 2Q_4^3 &= 3\pi_4, \\
2X_3 - 3Q_4X_2 + 2Q_4^3 &= \frac{4!}{0!}\frac{K_0\pi_4}{2Q_4}, \\
2X_5 - 5Q_4X_4 + 10Q_4^2X_3 - 10Q_4^3X_2 + 6Q_4^5 &= \frac{5!}{1!}K_1\pi_4, \\
2X_5 - 5Q_4X_4 + \frac{20}{3}Q_4^2X_3 - 5Q_4^3X_2 + \frac{8}{3}Q_4^5 &= \frac{6!}{2!}\frac{K_2\pi_4}{3Q_4}, \\
2X_7 - 7Q_4X_6 + 21Q_4^2X_5 - 35Q_4^3X_4 + 35Q_4^4X_3 - 21Q_4^5X_2 + 10Q_4^7 &= \frac{7!}{3!}K_3\pi_4, \\
2X_7 - 7Q_4X_6 + 14Q_4^2X_5 - \frac{35}{2}Q_4^3X_4 + 14Q_4^4X_3 - 7Q_4^5X_2 + 3Q_4^7 &= \frac{8!}{4!}\frac{K_4\pi_4}{4Q_4},
\end{aligned}
$$

$$
\begin{aligned}
2X_9 - 9Q_4X_8 + 36Q_4^2X_7 - 84Q_4^3X_6 + 126Q_4^4X_5 - 126Q_4^5X_4 + 84Q_4^6X_3 - 36Q_4^7X_2 + 14Q_4^9 &= \frac{9!}{5!}K_5\pi_4, \\
2X_9 - 9Q_4X_8 + 24Q_4^2X_7 - 42Q_4^3X_6 + \frac{252}{5}Q_4^4X_5 - 42Q_4^5X_4 + 24Q_4^6X_3 - 9Q_4^7X_2 + \frac{16}{5}Q_4^9 &= \frac{10!}{6!}\frac{K_6\pi_4}{5Q_4}
\end{aligned}
$$

Their successive solution gives,

$$
\begin{aligned}
&K_2 = 2\left(K_1 - \frac{1}{3}K_0^2\right)K_0, \qquad K_4 = 4\left(K_3 - 2K_0^2K_1 + \frac{4}{5}K_0^4\right)K_0, \\
&K_6 = 6\left(K_5 - \frac{20}{3}K_0^2K_3 + 16K_0^4K_1 - \frac{136}{21}K_0^6\right)K_0,
\end{aligned} \tag{60}
$$

where K_0, K_1, K_3, K_5 are four independent coefficients, in accordance with (58). Formulas (60) and (23, 52) are strongly related. Namely, the former may be obtained by a straightforward substitution of (52) into (23).

The list of formulas (60) may be continued if we consider Eqs. (22) for higher degrees,

$$
K_{2r} = 2r\left(K_{2r-1} - \rho_{2r}^1K_0^2K_{2r-3} + \cdots - (-1)^r\rho_{2r}^{r-1}K_0^{2(r-1)}K_1 + (-1)^r\rho_{2r}^rK_0^{2r}\right)K_0, \quad \rho_{2r}^j \in \mathbb{Q},
$$

where K_{2j+1}, $j \geq 3$, are algebraic (not polynomial) functions of K_0, K_1, K_3, K_5.

Supplementary Relations for K_r and G_r in Symmetric CI Semigroups

This kind of numerical semigroups is described by a simple Hilbert series (2) with a numerator $Q(S_m; z)$

$$
Q(S_m; z) = \left(1 - z^{e_1}\right)\left(1 - z^{e_2}\right)\cdots\left(1 - z^{e_{m-1}}\right),
$$

built on $m-1$ degrees e_j. The alternating power sum $\mathbb{C}_k(S_m)$ reads,

$$\mathbb{C}_n(S_m) = \sum_{j=1}^{m-1} e_j^n - \sum_{j>r=1}^{m-1} (e_j + e_r)^n + \ldots - (-1)^{m-1} \left(\sum_{j=1}^{m-1} e_j \right)^n . \tag{61}$$

Then, according to the Rep (24, 25), the expression in (61) may be written as follows,

$$\mathbb{C}_n(S_m) = \frac{(-1)^m \, n!}{(n-m+1)!} \varepsilon_{m-1} T_{n-m+1}(E), \qquad \mathbb{C}_{m-1}(S_m) = (-1)^m (m-1)\,!\, \varepsilon_{m-1}, \tag{62}$$

where polynomials $T_r(E) = T_r(E_1, \ldots, E_r)$ are built by replacing $X_r \to E_r = \sum_{j=1}^{m-1} e_j^r$ in polynomials $T_r(X_1, \ldots, X_r)$, defined in (25), and $\varepsilon_{m-1} = \prod_{j=1}^{m-1} e_j$.

Combining formulas (22) and (62), we obtain,

$$\pi_m = \varepsilon_{m-1}, \qquad K_p = \frac{T_{p+1}(E)}{p+1}, \qquad p \geq 0, \tag{63}$$

where the first equality was established earlier (see formula (5.4) in [7]) while the second relation leads to an infinite number of equalities. An universality of (63) disappears if we consider symmetric (not CI) semigroups, see e.g., formula (59) for K_0. The number g_m of independent genera G_r is given by the number of degrees of syzygies, bearing in mind the first relation in (63),

$$g_m = m - 2 \tag{64}$$

In symmetric semigroups S_m, there holds a strict inequality $\mu > m$ (see [5]) that bounds a genus from below, $G_0 \geq m+1$, and leads, in combination with (64), to another inequality $g_m < G_0$.

Below we present three examples with symmetric CI semigroups S_m, where $m = 2, 3, 4$.

Example 1 CI semigroup S_2, $g_2 = 0$, $e = \pi_2$, $\qquad E_r = \pi_2^r$,

By formula (14) from [11] and (63), we obtain

$$T_r(e) = \frac{\pi_2^r}{r+1}, \qquad K_r = \frac{\pi_2^{r+1}}{(r+1)(r+2)}. \tag{65}$$

Assuming that Conjecture 1 is true, substitute the last into (27),

$$(p+1) \sum_{r=0}^{p} \binom{p}{r} T_{p-r}(\sigma) G_r = \frac{\pi_2^{p+1}}{p+2} - 2^{p+1} T_{p+1}(\delta),$$

which may be resolved with respect to G_r if we make use of the inverse matrix $\left(\binom{p}{r} T_{p-r}(\sigma)\right)^{-1}$. We give explicit expressions for the five first genera G_r,

$$G_0 = \frac{\pi_2 - \sigma_1 + 1}{2}, \qquad G_1 = \frac{\pi_2 - \sigma_1 + 1}{2}\,\frac{2\pi_2 - \sigma_1 - 1}{6}, \tag{66}$$
$$G_2 = \frac{\pi_2 - \sigma_1 + 1}{2}\,\frac{\pi_2(\pi_2 - \sigma_1)}{6},$$
$$G_3 = \frac{\pi_2 - \sigma_1 + 1}{2}\,\frac{6\pi_2^3 + \pi_2^2(4 - 9\sigma_1) + \pi_2(\sigma_1^2 - 2\sigma_1 - 1) + (\sigma_1 + 1)(\sigma_1^2 + 1)}{60},$$
$$G_4 = \frac{\pi_2 - \sigma_1 + 1}{2}\,\frac{\pi_2(\pi_2 - \sigma_1)}{6}\,\frac{2\pi_2^2 - \pi_2(2\sigma_1 - 3) - \sigma_1^2}{5}.$$

Formulas (66) coincide with expressions for genera, derived [16] in terms of generators d_1, d_2, e.g.,

$$G_0 = \frac{(d_1 - 1)(d_2 - 1)}{2}, \quad G_1 = \frac{(d_1 - 1)(d_2 - 1)(2d_1d_2 - d_1 - d_2 - 1)}{12}.$$

Example 2 CI semigroup S_3, $g_3 = 1$, $\qquad E_1 = 2K_0$.

There exists one independent power sum E_1, while the sums E_{2r} may be expressed as follows,

$$E_2 = E_1^2 - 2\pi_3, \qquad E_4 = E_1^4 - 4\pi_3E_1^2 + 2\pi_3^2, \qquad E_6 = E_1^6 - 6\pi_3E_1^4 + 9\pi_3^2E_1^2 - 2\pi_3^3.$$

Substituting E_{2k} into expressions for $T_r(E)$ in Appendix 6 and subsequently into (63) we obtain

$$K_1 = \frac{1}{3}\left(2K_0^2 - \frac{\pi_3}{4}\right), \tag{67}$$
$$K_2 = \frac{1}{3}\left(2K_0^2 - \frac{\pi_3}{2}\right)K_0,$$
$$K_3 = \frac{1}{5}\left(4K_0^4 - \frac{3\pi_3}{2}K_0^2 + \frac{\pi_3^2}{12}\right),$$
$$K_4 = \frac{4}{15}\left(4K_0^4 - 2\pi_3K_0^2 + \frac{\pi_3^2}{4}\right)K_0,$$
$$K_5 = \frac{1}{21}\left(32K_0^6 - 20\pi_3K_0^4 + 4\pi_3^2K_0^2 - \frac{\pi_3^3}{8}\right),$$
$$K_6 = \frac{1}{7}\left(16K_0^6 - 12\pi_3K_0^4 + \frac{10\pi_3^2}{3}K_0^2 - \frac{\pi_3^3}{4}\right)K_0.$$

Combining (67) with formulas (23) and (52), we obtain for genera G_r the polynomial expressions in G_0. We present here only the four first formulas; expressions for G_5, G_6 are extremely lengthy.

$$
\begin{aligned}
G_1 &= \frac{2}{3}G_0^2 + \left(\frac{\delta_1}{3} - \frac{1}{2}\right) G_0 + \frac{\gamma}{6}, \qquad \gamma = \delta_1^2 - \delta_2 - \frac{\pi_3}{2}, \qquad (68)\\
G_2 &= \frac{2}{3}G_0^3 + \frac{2}{3}(\delta_1 - 1) G_0^2 + \frac{1}{3}\left(\gamma - \delta_1 + \frac{1}{2}\right) G_0 - \frac{\gamma}{6},\\
G_3 &= \frac{4}{5}G_0^4 + \left(\frac{6\delta_1}{5} - 1\right) G_0^3 + \left(\frac{2\gamma}{3} + \frac{2\delta_1^2}{15} + \frac{\pi_3}{30} - \delta_1 + \frac{1}{3}\right) G_0^2 +\\
&\left[\gamma\left(\frac{\delta_1}{5} - \frac{1}{2}\right) - \frac{\delta_1}{3}\left(\frac{2\delta_2}{5} - \frac{1}{2}\right)\right] G_0 + \frac{\delta_1^2(\delta_1^2 - \pi_3)}{20} - \frac{\delta_2^2 + 2\gamma\delta_2 - \gamma}{12} + \frac{\delta_4}{30} + \frac{\pi_3^2}{60}\\
G_4 &= \frac{16}{15}G_0^5 + \frac{8}{5}\left(\frac{4\delta_1}{3} - 1\right) G_0^4 + 2\left(\frac{1}{3} - \frac{6\delta_1}{5} + \frac{4\delta_1^2 + \pi_3}{15} + \frac{2\gamma}{3}\right) G_0^3 +\\
&\left(\frac{2\delta_1}{3} + 4\gamma\left(\frac{\delta_1}{5} - \frac{1}{3}\right) - \frac{4\delta_1^2 + 8\delta_1\delta_2 + \pi_3}{15}\right) G_0^2 + \left(\frac{3\delta_1^4 + 4\delta_1\delta_2 - 5\delta_2^2 + 2\delta_4}{15} -\right.\\
&\left.\gamma\frac{6\delta_1 + 10\delta_2 - 5}{15} + \frac{\pi_3^2}{15} - \frac{1}{30}\right) G_0 + \frac{5\delta_2^2 - 3\delta_1^4 - 2\delta_4}{15} + \frac{\gamma\delta_2}{3} + \frac{\pi_3}{10}\left(\delta_1^2 - \frac{\pi_3}{3}\right).
\end{aligned}
$$

In [4], formulas for G_r were derived in terms of three diagonal elements of matrix of minimal relations for generators d_j, which makes them less convenient from computational point of view than formulas (67, 68).

Example 3 CI semigroup S_4, $g_4 = 2$, $\qquad E_1 = 2K_0$, $\qquad E_2 = 12(2K_1 - K_0^2)$.

There exist two independent power sums E_1, E_2, while the rest E_k may be expressed as follows,

$$
\begin{aligned}
2E_3 &= 3E_1E_2 - E_1^3 + 6\pi_4,\\
2E_4 &= E_2^2 + 2E_1^2E_2 - E_1^4 + 8\pi_4E_1,\\
4E_5 &= 5E_1E_2^2 - E_1^5 + 10\pi_4(E_1^2 + E_2),\\
4E_6 &= 6E_1^2E_2^2 + E_2^3 - 3E_1^4E_2 + 24\pi_4E_1E_2 + 12\pi_4^2.
\end{aligned}
$$

Substituting E_k into expressions for $T_r(E)$ in Appendix 6 and subsequently into (63) we obtain

$$
\begin{aligned}
K_2 &= 2\left(K_1 - \frac{1}{3}K_0^2\right) K_0, \qquad (69)\\
K_3 &= \frac{1}{5}\left(12K_1^2 + 2K_0^2K_1 - \frac{8}{3}K_0^4 - \frac{\pi_4}{12}K_0\right),\\
K_4 &= \frac{1}{5}\left(48K_1^2 - 32K_0^2K_1 + \frac{16}{3}K_0^4 - \frac{\pi_4}{3}K_0\right) K_0,\\
K_5 &= \frac{1}{7}\left(72K_1^3 + 48K_1^2K_0^2 - 80K_1K_0^4 + \frac{64}{3}K_0^6 - \pi_4K_1K_0 - \frac{2\pi_4}{3}K_0^3 + \frac{\pi_4^2}{72}\right),\\
K_6 &= \frac{1}{7}\left(432K_1^3 - 384K_1^2K_0^2 + 80K_1K_0^4 + \frac{16}{3}K_0^6 - 6\pi_4K_1K_0 + \frac{2\pi_4}{3}K_0^3 + \frac{\pi_4^2}{12}\right) K_0.
\end{aligned}
$$

By comparison (60) and (69), formulas for K_4 and K_6 in (69) may be obtained if we substitute K_3 and K_5 in (69) into K_4 and K_6 in (60). Combining (69) with formulas (23), we obtain for genera G_r the polynomial expressions in G_0, G_1, e.g.,

$$G_2 = \frac{2G_0 - 1}{3}\left(3G_1 - G_0^2 + G_0\right), \tag{70}$$

$$G_3 = \frac{12G_1^2}{5} + \frac{G_1}{20}\left(10 + 8G_0^2 + 2\sigma_1 - \sigma_1^2 - 4G_0(2+\sigma_1) - \sigma_2\right) - \frac{8G_0^4}{15} + \frac{2G_0^3}{15} + \frac{G_0^2}{60}\left(\sigma_2 + 3\sigma_1^2 - 12\sigma_1 - 46\right) - \frac{G_0}{120}\left(\sigma_1(\sigma_2 - \sigma_1^2 + 6\sigma_1 - 10) + 2\pi_4 + 2\sigma_2 - 28\right) + \frac{1}{240}\left(\sigma_1(\sigma_2 - \sigma_1^2 + 2\sigma_1 - 2) + 2\pi_4\right).$$

6 Concluding Remarks

We study polynomial identities of arbitrary degree n for syzygies degrees of numerical semigroups S_m and show in (22) that for $n \geq m$ they contain higher genera $G_r = \sum_{s\in\mathbb{Z}_>\setminus S_m} s^r$ of S_m,

$$\sum_{j=1}^{\beta_1} C_{1,j}^n - \sum_{j=1}^{\beta_2} C_{2,j}^n + \ldots - (-1)^{m-1}\sum_{j=1}^{\beta_{m-1}} C_{m-1,j}^n = \frac{(-1)^m n\,!}{(n-m)\,!}\pi_m K_{n-m}(G_0, \ldots, G_p),$$

where a coefficient $K_p(G_0, \ldots, G_p)$ is a linear combination of genera. We calculate explicitly several first expressions (23) for K_p, $0 \leq p \leq 6$, and put forward Conjecture 1 related K_p and $G_0, \ldots, G_p$ for any integer $p \geq 0$. In symbolic calculus [17], this relationship (27) reads

$$K_p = (T(\sigma) + G)^p + \frac{2^{p+1}}{p+1} T_{p+1}(\delta),$$

where after binomial expansion the symbolic powers $T^{p-r}(\sigma)G^r$ are converted into $T_{p-r}(\sigma)G_r$. Symmetric polynomials $T_r(\sigma) = T_r(\sigma_1, \ldots, \sigma_r)$ and $T_r(\delta) = T_r(\delta_1, \ldots, \delta_r)$ are arisen in the theory of *symmetric CI* semigroups [7, 11]

$$\sum_{j=1}^{m-1} e_j^n - \sum_{j>r=1}^{m-1} (e_j + e_r)^n + \ldots - (-1)^{m-1}\left(\sum_{j=1}^{m-1} e_j\right)^n = \frac{(-1)^m\, n!}{(n-m+1)!}\varepsilon_{m-1}T_{n-m+1}(E),$$

where polynomials $T_r(E) = T_r(E_1, \ldots, E_r)$ are related (see [11], formula (16)) to the polynomial part of the partition function, which gives a number of partitions of $s \geq 0$ into $m-1$ positive integers. Thus, in the relationship (23, 27), there coexist genera G_r of *non-symmetric* semigroup S_m and characteristic polynomials $T_r(\sigma)$ associated with semigroup generators $d_1, \ldots, d_m$.

Based on a finite number of syzygies degrees and homogeneity of the $m-2$ first polynomial identities (6), we find a number g_m of algebraically independent coefficients K_p for different kinds of semigroups. Due to the relationship (27), this leads to g_m algebraically independent genera G_p. However, the polynomial equations, related K_p, $p \geq g_m$, with independent coefficients K_j, $0 \leq j < g_m$, read much shorter than their countpartners, related G_p, $p \geq g_m$, with independent genera G_j. It can be seen for non-symmetric and symmetric (not CI) semigroups, (see Sect. 5), but, in particular, for symmetric CI semigroups comparing relations (65), (67) and (69) with (66), (68) and (70). These observations make us to suppose that K_p has deeper algebraic meaning than a simple combination of G_p.

Appendix 1: Stirling Numbers of the 1st Kind

Making use of recurrence Eq. (12), we calculate the first formulas for $\left[{n \atop n-k}\right]$ up to $k = 9$.

$$\begin{aligned}
&\left[{n \atop n}\right] = 1, \quad \left[{n \atop n-1}\right] = \binom{n}{2}, \quad \left[{n \atop n-2}\right] = \binom{n}{3}\left(\frac{3n}{4} - \frac{1}{4}\right), \quad \left[{n \atop n-3}\right] = \binom{n}{4}\binom{n}{2},\\
&\left[{n \atop n-4}\right] = \binom{n}{5}\left(\frac{5n^3}{16} - \frac{5n^2}{8} + \frac{5n}{48} + \frac{1}{24}\right),\\
&\left[{n \atop n-5}\right] = \binom{n}{6}\left(\frac{3n^3}{16} - \frac{5n^2}{8} + \frac{5n}{16} + \frac{1}{8}\right)n, \qquad \text{(A1)}\\
&\left[{n \atop n-6}\right] = \binom{n}{7}\left(\frac{7n^5}{64} - \frac{35n^4}{64} + \frac{35n^3}{64} + \frac{91n^2}{576} - \frac{7n}{96} - \frac{1}{36}\right),\\
&\left[{n \atop n-7}\right] = \binom{n}{8}\left(\frac{n^5}{16} - \frac{7n^4}{16} + \frac{35n^3}{48} + \frac{7n^2}{144} - \frac{7n}{24} - \frac{1}{9}\right)n.\\
&\left[{n \atop n-8}\right] = \binom{n}{9}\left(\frac{9n^7}{256} - \frac{21n^6}{64} + \frac{105n^5}{128} - \frac{7n^4}{32} - \frac{469n^3}{768} - \frac{9n^2}{64} + \frac{101n}{960} + \frac{3}{80}\right)\\
&\left[{n \atop n-9}\right] = \binom{n}{10}\left(\frac{5n^7}{256} - \frac{15n^6}{64} + \frac{105n^5}{128} - \frac{7n^4}{12} - \frac{665n^3}{768} + \frac{25n^2}{192} + \frac{101n}{192} + \frac{3}{16}\right)n.
\end{aligned}$$

Appendix 2: Derivatives $\Psi_{z=1}^{(r)}$

Find expressions for ratio of derivatives $\Psi_{z=1}^{(r)}/\Psi_{z=1}^{(0)}$, $r \le 4$,

$$\frac{\Psi_{z=1}^{(1)}}{\Psi_{z=1}^{(0)}} = \sum_{j=1}^{m} \frac{\Psi_{j,z=1}^{(1)}}{\Psi_{j,z=1}}, \qquad \frac{\Psi_{z=1}^{(2)}}{\Psi_{z=1}^{(0)}} = \left(\sum_{j=1}^{m} \frac{\Psi_{j,z=1}^{(1)}}{\Psi_{j,z=1}}\right)^2 + \sum_{j=1}^{m} \frac{\Psi_{j,z=1}^{(2)}}{\Psi_{j,z=1}} - \sum_{j=1}^{m} \left(\frac{\Psi_{j,z=1}^{(1)}}{\Psi_{j,z=1}}\right)^2,$$

$$\begin{aligned}\frac{\Psi_{z=1}^{(3)}}{\Psi_{z=1}^{(0)}} = {} & \left(\sum_{j=1}^{m} \frac{\Psi_{j,z=1}^{(1)}}{\Psi_{j,z=1}}\right)^3 + \sum_{j=1}^{m} \frac{\Psi_{j,z=1}^{(3)}}{\Psi_{j,z=1}} + 2\sum_{j=1}^{m} \left(\frac{\Psi_{j,z=1}^{(1)}}{\Psi_{j,z=1}}\right)^3 + 3\sum_{j=1}^{m} \frac{\Psi_{j,z=1}^{(1)}}{\Psi_{j,z=1}} \sum_{j=1}^{m} \frac{\Psi_{j,z=1}^{(2)}}{\Psi_{j,z=1}} \\ & - 3\sum_{j=1}^{m} \frac{\Psi_{j,z=1}^{(1)}}{\Psi_{j,z=1}} \sum_{j=1}^{m} \left(\frac{\Psi_{j,z=1}^{(1)}}{\Psi_{j,z=1}}\right)^2 - 3\sum_{j=1}^{m} \frac{\Psi_{j,z=1}^{(1)}\Psi_{j,z=1}^{(2)}}{(\Psi_{j,z=1})^2},\end{aligned} \tag{B1}$$

$$\begin{aligned}\frac{\Psi_{z=1}^{(4)}}{\Psi_{z=1}^{(0)}} = {} & \left(\sum_{j=1}^{m} \frac{\Psi_{j,z=1}^{(1)}}{\Psi_{j,z=1}}\right)^4 + \sum_{j=1}^{m} \frac{\Psi_{j,z=1}^{(4)}}{\Psi_{j,z=1}} - 6\sum_{j=1}^{m} \left(\frac{\Psi_{j,z=1}^{(1)}}{\Psi_{j,z=1}}\right)^4 + 3\left(\sum_{j=1}^{m} \left(\frac{\Psi_{j,z=1}^{(1)}}{\Psi_{j,z=1}}\right)^2\right)^2 \\ & - 6\sum_{j=1}^{m} \left(\frac{\Psi_{j,z=1}^{(1)}}{\Psi_{j,z=1}}\right)^2 \left(\sum_{j=1}^{m} \frac{\Psi_{j,z=1}^{(1)}}{\Psi_{j,z=1}}\right)^2 + 8\sum_{j=1}^{m} \frac{\Psi_{j,z=1}^{(1)}}{\Psi_{j,z=1}} \sum_{j=1}^{m} \left(\frac{\Psi_{j,z=1}^{(1)}}{\Psi_{j,z=1}}\right)^3 \\ & + 6\left(\sum_{j=1}^{m} \frac{\Psi_{j,z=1}^{(1)}}{\Psi_{j,z=1}}\right)^2 \sum_{j=1}^{m} \frac{\Psi_{j,z=1}^{(2)}}{\Psi_{j,z=1}} + 12\sum_{j=1}^{m} \left(\frac{\Psi_{j,z=1}^{(1)}}{\Psi_{j,z=1}}\right)^2 \frac{\Psi_{j,z=1}^{(2)}}{\Psi_{j,z=1}} + 3\left(\sum_{j=1}^{m} \frac{\Psi_{j,z=1}^{(2)}}{\Psi_{j,z=1}}\right)^2 \\ & - 6\sum_{j=1}^{m} \left(\frac{\Psi_{j,z=1}^{(1)}}{\Psi_{j,z=1}}\right)^2 \sum_{j=1}^{m} \frac{\Psi_{j,z=1}^{(2)}}{\Psi_{j,z=1}} - 3\sum_{j=1}^{m} \left(\frac{\Psi_{j,z=1}^{(2)}}{\Psi_{j,z=1}}\right)^2 - 4\sum_{j=1}^{m} \frac{\Psi_{j,z=1}^{(1)}\Psi_{j,z=1}^{(3)}}{(\Psi_{j,z=1})^2} \\ & - 12\sum_{j=1}^{m} \frac{\Psi_{j,z=1}^{(1)}}{\Psi_{j,z=1}} \sum_{j=1}^{m} \frac{\Psi_{j,z=1}^{(1)}\Psi_{j,z=1}^{(2)}}{(\Psi_{j,z=1})^2} + 4\sum_{j=1}^{m} \frac{\Psi_{j,z=1}^{(1)}}{\Psi_{j,z=1}} \sum_{j=1}^{m} \frac{\Psi_{j,z=1}^{(3)}}{\Psi_{j,z=1}}.\end{aligned}$$

Using a summation rule in a finite calculus (see formula (2.50) in [3]), we find a ratio $\Psi_{j,z=1}^{(r)}/\Psi_{j,z=1}$,

$$\sum_{l=0}^{d-1} (l)_r = \frac{(d)_{r+1}}{r+1} \quad \longrightarrow \quad \frac{\Psi_{j,z=1}^{(r)}}{\Psi_{j,z=1}} = \frac{(d_j - 1)_r}{r+1}. \tag{B2}$$

Substituting (B2) into (B1), we arrive at formulas (17).

Appendix 3: Symmetric Polynomials $T_k(X_1, \ldots, X_m)$

We present formulas for the first symmetric polynomials $T_k(X_1, \ldots, X_m)$ up to $k = 7$.

$$\begin{aligned}
T_0 &= 1, \\
T_1 &= \frac{1}{2}X_1, \\
T_2 &= \frac{1}{3}\frac{3X_1^2 + X_2}{4}, \\
T_3 &= \frac{1}{4}\frac{X_1^2 + X_2}{2}X_1, \\
T_4 &= \frac{1}{5}\frac{15X_1^4 + 30X_1^2X_2 + 5X_2^2 - 2X_4}{48}, \\
T_5 &= \frac{1}{6}\frac{3X_1^4 + 10X_1^2X_2 + 5X_2^2 - 2X_4}{16}X_1, \\
T_6 &= \frac{1}{7}\frac{63X_1^6 + 315X_1^4X_2 + 315X_1^2X_2^2 - 126X_1^2X_4 + 35X_2^3 - 42X_2X_4 + 16X_6}{576}, \\
T_7 &= \frac{1}{8}\frac{9X_1^6 + 63X_1^4X_2 + 105X_1^2X_2^2 - 42X_1^2X_4 + 35X_2^3 - 42X_2X_4 + 16X_6}{144}X_1.
\end{aligned} \tag{C1}$$

References

1. Bresinsky H.: *Symmetric semigroups of integers generated by four elements*, Manuscripta Math., **17**, 205–219 (1975)
2. Gelfand I.M., Kapranov M.M., Zelevinski A.V.: *Discriminants, Resultants and Multidimensional Determinants*, BirkhWauser, Boston (1994)
3. Graham R.L., D.E. Knuth D.E., Patashnik O.: *Concrete Mathematics: a foundation for computer science*, Addison-Wesley, NY, 2nd ed. (1994)
4. Fel L.G., Rubinstein B.Y.: *Power sums related to semigroups $S(d_1, d_2, d_3)$*, Semigroup Forum, **74**, 93–98 (2007)
5. Fel L.G.: *Duality relation for the Hilbert series of almost symmetric numerical semigroups*, Israel J. Math. **185**, 413–444 (2011)
6. Fel L.G.: *On Frobenius numbers for symmetric (not complete intersection) semigroups generated by four elements* Semigroup Forum, **93**, 423–426 (2016)
7. Fel L.G.: *Restricted partition functions and identities for degrees of syzygies in numerical semigroups*, Ramanujan J., **43**, 465–491 (2017)
8. Fel L.G.: *Symmetric (not complete intersection) semigroups generated by five elements*, Integers: The Electronic J. of Comb. Number Theory, **18**, # A44 (2018)
9. Fel L.G.: *Symmetric (not complete intersection) semigroups generated by six elements*, In *Numerical Semigroups*, Springer INdAM Series **40**, 93–109 (2020)
10. Fel L.G.: *Symmetric (not complete intersection) semigroups generated by 4,5,6 elements*, International Meeting Numerical Semigroups, 2018, Cortona, Italy,
11. Fel L.G.: *Symmetric polynomials associated with numerical semigroups*, Discrete Math. Lett., **5**, 56–62 (2021)

12. Hejmej B.: *A note about irreducibility of a resultant*, Bulletin dela Société des Sciences et des Letters de Lódź, **68**, 27–32 (2018)
13. Herzog J., Kühl M.: *On the Betti numbers of finite pure and linear resolutions*, Comm. Algebra, **12**, 1627–1646 (1984)
14. Macdonald I.G.: *Symmetric functions and Hall polynomials*, Oxford: Clarendon Press (1995)
15. Macaulay F.S.: *Some formulæ in elimination*, Proc. London Math. Soc., **35**, 3–27 (1902)
16. Rödseth Ö.J.: *A note on Brown and Shiue's paper on a remark related to the Frobenius problem*, Fibonacci Quarterly, **32**, 407–408 (1994)
17. S.M. Roman S.M., Rota G.-C.: *The umbral calculus*, Adv. Math. **27**, 95–188 (1978)
18. Rosales J.C., García-Sánchez P.A.: *Numerical semigroups* , Developments in Mathematics **20**, Springer, New York (2009)

L^p Estimates for Bilinear Generalized Radon Transforms in the Plane

A. Greenleaf, A. Iosevich, B. Krause, and A. Liu

1 Introduction

A classical result due to Littman [22] and Strichartz [31] (see also [25]) says that, for $d \geq 2$, the spherical averaging (or F. John [20]) operator,

$$Af(x) = \int_{S^{d-1}} f(x-y)\, d\sigma(y) = \sigma * f(x), \tag{1}$$

where σ is the surface measure on S^{d-1}, is bounded from $L^p(\mathbb{R}^d)$ to $L^q(\mathbb{R}^d)$ iff $\left(\frac{1}{p}, \frac{1}{q}\right)$ is in the closed triangle with the vertices

$$(0,0),\ (1,1),\ \text{and}\ \left(\frac{d}{d+1}, \frac{1}{d+1}\right). \tag{2}$$

The work of the first listed author was partially supported by NSF Grants DMS-1362271 and -1906186, the second by NSA Grant H98230-15-1-0319, and the third by an NSF Postdoctoral Fellowship.

A. Greenleaf · A. Iosevich (✉)
Department of Mathematics, University of Rochester, Rochester, NY 14627, USA
e-mail: iosevich@math.rochester.edu

A. Greenleaf
e-mail: allan@math.rochester.edu

B. Krause
Department of Mathematics, King's College, London, UK
e-mail: ben.krause@kcl.ac.uk

A. Liu
Department of Mathematics, Massachusetts Institute of Technology, Cambridge, MA 02139, USA
e-mail: cliu568@MIT.EDU

M. B. Nathanson (ed.), *Combinatorial and Additive Number Theory V*,
Springer Proceedings in Mathematics & Statistics 395,
https://doi.org/10.1007/978-3-031-10796-2_9

The operator A is, along with the classical Radon transform, a model for the generalized Radon transforms studied by Guillemin and Sternberg [18] and Phong and Stein (see, e.g., [26, 29], and the references contained therein). These are linear operators of the form

$$\mathcal{R}f(x) = \int_{\phi(x,y)=t} f(y)\psi(x,y)\,d\sigma_{x,t}(y), \tag{3}$$

where $t \in \mathbb{R}$, $\phi \in C^\infty(\mathbb{R}^d \times \mathbb{R}^d)$ is a defining function, i.e., $d_{x,y}\phi \neq (0,0)$ on

$$Z_t := \left\{(x,y) \in \mathbb{R}^d \times \mathbb{R}^d : \phi(x,y) = t\right\},$$

ψ is a smooth cut-off, $\sigma_{x,t}$ is surface measure on Z_t, and ϕ satisfies the following condition [26]:

Definition 1 A defining function $\phi : \mathbb{R}^d \times \mathbb{R}^d \to \mathbb{R}$ satisfies the *Phong-Stein rotational curvature condition* at t if, for all $(x,y) \in Z_t$,

$$det \begin{pmatrix} 0 & \nabla_x\phi \\ -(\nabla_y\phi)^T & \frac{\partial^2\phi}{\partial x_i \partial y_j} \end{pmatrix} \neq 0. \tag{4}$$

Under the rotational curvature assumption, $\mathcal{R} : L^p(\mathbb{R}^d) \to L^q(\mathbb{R}^d)$ for $\left(\frac{1}{p}, \frac{1}{q}\right)$ as in (2) above. This is a folk theorem (as far as we know) and follows by substituting the $L^2 \to L^2_{\frac{d-1}{2}}$ boundedness of Fourier integral operators associated with canonical graphs into Strichartz's proof [31] in the case of the spherical averaging operator. Note that if $\phi(x,y) = |x-y|$, the Euclidean distance, and $t = 1$, we recover the spherical averaging operator A of (1).

The purpose of this paper is to study natural bilinear variants of the linear generalized Radon transforms, with the considerations limited to two dimensions. A family of model operators, arising from combinatorial geometry, is given by

$$B_\theta(f,g)(x) = \int f(x-y)g(x-\Theta y)\,d\sigma(y), \tag{5}$$

where σ is the arc-length measure on S^1 and Θ denotes the counter-clockwise rotation by an angle $\theta \neq 0$. We exclude the degenerate case $\theta = 0$, since $B_0(f,g) = A(f \cdot g)$, with the linear circular mean operator A as in (1).

Before stating the main theorem, we describe some motivating applications. Consider n points in a point set $P \subset \mathbb{R}^2$ and the problem of counting equilateral triangles of side-length 1 among them (see Fig. 1 above for a particularly triangle-rich configuration). We have

Fig. 1 An equilateral triangular grid (mathforum.org)

$$\#\{(x, y, z) \in P^3 : |x - y| = |x - z| = |y - z| = 1\}$$

$$= \sum_{x,y,z} 1_C(x - y)1_C(x - z)1_C(y - z)1_P(x)1_P(y)1_P(z), \qquad (6)$$

where C is the circle of radius 1 centered at the origin. This expression equals

$$\sum_{x,u,v} K(u, v)1_P(x - u)1_P(x - v)1_P(x),$$

where K is the indicator function of the set

$$\{(u, v) \in C \times C : |u - v| = 1\} = \{(u, \Theta u) : u \in C\} \cup \{(u, \Theta^{-1}u) : u \in C\},$$

where Θ is the rotation by $\frac{\pi}{3}$.

It follows that the trilinear form in (6) equals

$$\sum_x \mathcal{B}_{\frac{\pi}{3}}(1_P, 1_P)(x)1_P(x) + \sum_x \mathcal{B}_{-\frac{\pi}{3}}(1_P(x), 1_P(x))1_P(x),$$

where $\mathcal{B}_\theta$ is the discrete version of the bilinear operator defined in (5) above. Different values of θ are similarly associated with counting triangles of different congruence types. See, for example, [5, 12, 16] where operators of this types are studied, in one form or another, in the context of point configuration problems in geometric measure theory.

The spherical averaging operator can be similarly interpreted as the continuous analogue of an operator counting pairs of distances. Indeed, arguing as above, let P be a finite point set, let C be the unit circle, and consider

$$\#\{(x,y)\in P\times P : |x-y|=1\} = \sum_{x,y} 1_C(x-y)1_P(x)1_P(y)$$

$$= \sum_x \left\{ \sum_y 1_P(x-y)1_C(y) \right\} 1_P(x) = \sum_x (\mathcal{A}1_P)(x)1_P(x),$$

where $\mathcal{A}$ is the discrete analogue of the spherical averaging operator $Af(x)$.

A way to understand the operators $Af(x)$ and $B_\theta(f,g)(x)$ in terms of a coherent geometric paradigm is the following. Let E be a compact subset of $\mathbb{R}^d$. Define a graph by designating the vertices to be the points of E, and connecting two vertices x and y by an edge iff $|x-y|=1$. Then the spherical averaging operator $Af(x)$ may be viewed as the edge operator on this graph. Now define a hyper-graph on E by connecting a triple x, y, z by a hyper-edge iff $|x-y|=|x-z|=|y-z|=1$. The hyper-edge operator is then precisely the bilinear operator $B_{\frac{\pi}{3}}(f,g)$ (or $B_{-\frac{\pi}{3}}(f,g)$). One can define similar objects by replacing the distance function $|x-y|$ with a more general function $\phi(x,y)$, as in (3).

These examples suggest a class of bilinear generalized Radon transforms in two dimensions, of the form

$$B(f,g)(x) = \lim_{\epsilon\downarrow 0} \epsilon^{-3} \int\int_{\{|\phi_1(x,y)-t_1|<\epsilon;|\phi_2(x,z)-t_2|<\epsilon;|\phi_3(y,z)-t_3|<\epsilon\}} f(y)g(z)\psi(x,y,z)dydz, \tag{7}$$

where ψ is a smooth cut-off function and ϕ_j's are suitably regular functions. In the case when $\phi_j(x,y)\equiv|x-y|$, we recover the operator B_θ, $\theta=\frac{\pi}{3}$ defined in (5).

There is a considerable literature on bilinear and multilinear singular integral and pseudodifferential operators, e.g., [8, 9, 13, 15, 23, 24], among many others. There has also been extensive but less complete work on bilinear and multilinear Radon transforms, e.g., Bennett-Carbery-Wright [4], Bejenaru-Herr-Tataru [1], Bennett-Bez [2], Stovall [30], Bennett-Bez-Gutiérrez [3] and Koch-Steinerberger [21]. In particular, J. Bennett, A. Carbery, and J. Wright, as an application of their nonlinear Loomis-Whitney inequality [4], obtained unrestricted $L^2\times L^2\to L^2$ estimates for a class of bilinear generalized Radon transforms in the plane.

For the model operators B_θ, $0<\theta<\pi$, defined in (5), the restricted strong type L^2 estimates we prove are in fact covered by the unrestricted results of [4]. (We thank Mike Christ and Betsy Stovall for bringing to our attention, after the completion of the first version of this paper, the relevance of [4, 30] and related works.) However, our method does apply to some operators not satisfying a basic structural assumption of [4, 30], namely that the projections from the incidence relation be submersions. We believe the proof of the $L^{2,1}\times L^{2,1}\to L^2$ boundedness presented here, although in the spirit of [4], is simpler and worth recording since, it extends to the general operators of the form (7) that we consider, which include geometries not covered by those results. See Sect. 10 for further discussion.

Our result for the model operators B_θ is the following; an extension to a more general class of bilinear generalized Radon transforms is in Sect. 2 below.

Theorem 2 *For $\theta \in (0, 2\pi)$, let B_θ be defined as in (5) above.*

(i) Suppose that $\theta \neq \pi$. Then the type set $Q(B_\theta)$ of B_θ, i.e., those $\left(\frac{1}{p}, \frac{1}{q}, \frac{1}{r}\right) \in [0, 1]^3$ such that

$$B_\theta : L^p(\mathbb{R}^2) \times L^q(\mathbb{R}^2) \to L^r(\mathbb{R}^2),$$

is the closed polyhedron with vertices $(0, 0, 0)$, $(\frac{2}{3}, \frac{2}{3}, 1)$, $(0, \frac{2}{3}, \frac{1}{3})$, $(\frac{2}{3}, 0, \frac{1}{3})$, $(1, 0, 1)$, $(0, 1, 1)$ and $(\frac{1}{2}, \frac{1}{2}, \frac{1}{2})$, except for $(\frac{1}{2}, \frac{1}{2}, \frac{1}{2})$, where a restricted strong type bound holds, i.e., $B_\theta : L^{2,1} \times L^{2,1} \to L^2$. (See Fig. 2.)[1]

(ii) If $\theta = \pi$, the operator is bounded if $\left(\frac{1}{p}, \frac{1}{q}, \frac{1}{r}\right) \in Q(B_\pi)$, the closed polyhedron with vertices $(0, 0, 0)$, $(\frac{2}{3}, \frac{2}{3}, 1)$, $(0, \frac{2}{3}, \frac{1}{3})$, $(\frac{2}{3}, 0, \frac{1}{3})$, $(0, 1, 1)$, and $(1, 0, 1)$. (See Fig. 3.)

(iii) Moreover, in the Banach cube $p, q, r \geq 1$, the exponents in both (i) and (ii) are best possible, except for the question of whether in case (i) there is a strong type estimate for $(\frac{1}{2}, \frac{1}{2}, \frac{1}{2})$, which is unknown at this time.

Remark 3 By reflection about the horizontal axis, it suffices to consider $0 < \theta \leq \pi$.

Remark 4 Examination of the proof shows that all the non-trivial endpoints in the degenerate case, $\theta = \pi$, follow, in one way or another, from the $L^p(\mathbb{R}^2) \to L^q(\mathbb{R}^2)$ bounds (2) for the circular averaging operator $f \to Af$ defined in (1) above. In the non-degenerate case, there is restricted strong type boundedness at an additional vertex, $\left(\frac{1}{2}, \frac{1}{2}, \frac{1}{2}\right)$, and the estimates which follow from that by interpolating with the estimates valid in the degenerate case.

Remark 5 As observed in [30, Sect. 2], the form $S_\theta(f, g, h) := \langle B_\theta(f, g), h\rangle$ can not be bounded outside of the the Banach cube $p, q, r \geq 1$: Suppose that an estimate

$$|S_\theta(f, g, h)| \lesssim ||f||_p ||g||_q ||h||_{r'} \tag{8}$$

holds. Then take f, g and h to be the characteristic functions of disks of radius δ, 1 and 1, resp. The LHS of (8) is $\simeq \delta^2$, while the RHS is $\simeq \delta^{2/p}$; letting $\delta \to 0$, (8) implies that $2 \geq 2/p$ and hence $p \geq 1$. Permuting, one sees that $q, r \geq 1$, as well. The same argument applies to the general class of B considered in the next section.

[1] In fact, unrestricted strong type at $(\frac{1}{2}, \frac{1}{2}, \frac{1}{2})$ holds by the results of Bennett, Carbery and Wright [4].

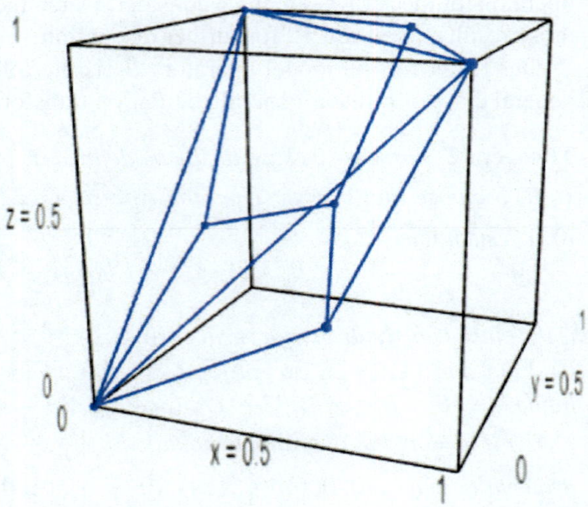

Fig. 2 The typeset $Q(B_\theta)$ for the non-degenerate cases, $0 < \theta < \pi$

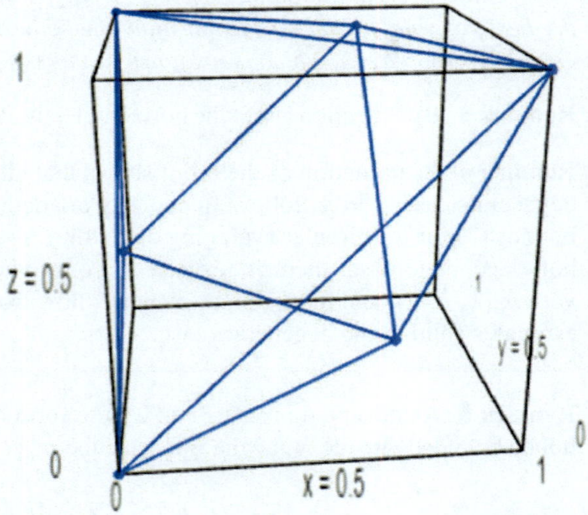

Fig. 3 The typeset $Q(B_\pi)$ in the degenerate case, $\theta = \pi$

2 A General Class of Bilinear Generalized Radon Transforms

We now extend Theorem 2 to a class of bilinear operators modeled on the linear generalized Radon transforms of [18, 26]. Bilinear generalized Radon transforms (and more singular variants) have arisen in inverse problems (e.g., [11, 17]), although in those problems the relevant estimates are in terms of Sobolev spaces.

Let (ϕ_1, ϕ_2, ϕ_3) be a trio of smooth, real-valued functions on $\mathbb{R}^2 \times \mathbb{R}^2$, and for $(t_1, t_2, t_3) \in \mathbb{R}^3$, and let K be the product delta distribution on $\mathbb{R}^2 \times \mathbb{R}^2 \times \mathbb{R}^2$,

$$K(x, y, z) = \delta(\phi_1(x, y) - t_1) \cdot \delta(\phi_2(x, z) - t_2) \cdot \delta(\phi_3(y, z) - t_3) \cdot \psi(x, y, z), \tag{9}$$

where $\psi \in C_0^\infty(\mathbb{R}^6)$, $\psi \geq 0$. Define a bilinear generalized Radon transform by

$$B(f, g)(x) = \int \int K(x, y, z)\, f(y)g(z)dydz, \tag{10}$$

defined weakly by

$$\langle B(f, g), h\rangle = \langle K, h(x)f(y)g(z)\rangle. \tag{11}$$

If one defines

$$Z_1 = \{(x, y) \in \mathbb{R}^2 \times \mathbb{R}^2 : \phi_1(x, y) = t_1\}, \quad Z_2 = \{(x, z) \in \mathbb{R}^2 \times \mathbb{R}^2 : \phi_2(x, z) = t_2\},$$
$$\text{and } Z_3 = \{(y, z) \in \mathbb{R}^2 \times \mathbb{R}^2 : \phi_3(y, z) = t_3\}, \tag{12}$$

then Z_1, Z_2 and Z_3 are smooth submanifolds in $\mathbb{R}^2 \times \mathbb{R}^2$ under the assumption that the ϕ_j are *defining functions* in the sense that

$$d\phi_j \neq (0, 0) \text{ on } Z_j,\ j = 1, 2, 3. \tag{13}$$

We now slightly strengthen this assumption: Setting

$$Z = \{(x, y, z) \in \mathbb{R}^2 \times \mathbb{R}^2 \times \mathbb{R}^2 : (x, y) \in Z_1, (x, z) \in Z_2, (y, z) \in Z_3\},$$

which is the support of $K(x, y, z)$, it will follow that $Z \subset \mathbb{R}^6$ is a smooth, codimension 3 submanifold, and the product in (9) is well-defined, if $(\phi_1(x, y), \phi_2(x, z), \phi_3(y, z))$ is a set of defining functions for Z, i.e., one has

$$rank \begin{bmatrix} d_x\phi_1 & d_y\phi_1 & 0 \\ d_x\phi_2 & 0 & d_z\phi_2 \\ 0 & d_y\phi_3 & d_z\phi_3 \end{bmatrix} = 3, \quad \forall (x, y, z) \in Z. \tag{14}$$

K is then a conormal distribution with respect to Z of order 0, denoted $K \in I^0(Z)$, since it has a representation

$$K(x, y, z) = \int_{\mathbb{R}^3} e^{i[\phi_1(x,y)\tau_1 + \phi_2(x,z)\tau_2 + \phi_3(y,z)\tau_3]} \psi(x, y, z) \cdot 1(\tau)\, d\tau, \tag{15}$$

interpreted as an oscillatory integral in the sense of Hörmander [19]. Equivalently, K is a Fourier integral distribution associated with the Lagrangian manifold $N^*Z \subset T^*\mathbb{R}^6 \setminus 0$, $K \in I^0(N^*Z)$, where N^*Z denotes the conormal bundle of Z,

$$N^*Z = \Big\{(x, y, z, \xi, \eta, \zeta) : (x, y, z) \in Z,\ \xi = \tau_1 d_x\phi_1 + \tau_2 d_x\phi_2,\ \eta = \tau_1 d_y\phi_1 + \tau_3 d_y\phi_3,\ \zeta = \tau_2 d_z\phi_2 + \tau_3 d_z\phi_3,\quad (\tau_1, \tau_2, \tau_3) \in \mathbb{R}^3 \setminus 0\Big\}. \tag{16}$$

The order of K as a conormal distribution just equals the order, 0, of its amplitude $\psi \cdot 1(\tau)$, while its order as a Fourier integral distribution is determined by $0 + \frac{3}{2} - \frac{6}{4} = 0$.

The following is the extension of Theorem 2 to this more general class of bilinear generalized Radon transforms. The second-order rotational curvature condition was discussed in the Introduction, while the first order conditions on the defining functions will be discussed in Sect. 9 below.

Theorem 6 *A bilinear generalized Radon transform B defined by (10) with kernel (9) conormal for a smooth $Z \subset \mathbb{R}^2 \times \mathbb{R}^2 \times \mathbb{R}^2$ with defining functions ϕ_1, ϕ_2, ϕ_3 satisfying (14) has the same type set as the non-degenerate model operators B_θ in Theorem 2 (i) if*

(a) at all $(x_0, y_0, z_0) \in supp(K)$,

$$\begin{vmatrix} d_y\phi_1(x_0, y_0) \\ d_y\phi_3(y_0, z_0) \end{vmatrix} \neq 0, \qquad \begin{vmatrix} d_z\phi_2(x_0, z_0) \\ d_z\phi_3(y_0, z_0) \end{vmatrix} \neq 0, \tag{17}$$

and

(b) ϕ_3 satisfies the rotational curvature condition (4).

3 Sharpness Examples

In this section we show that the ranges of exponents in Theorem 2 are the best possible.

General Rotation θ:

Let $f(x) = \chi_{B_\delta}(x)$ and $g(x) = \chi_{A_\delta}(x)$, where A_δ is the annulus of radius $\sqrt{2}(1 - \cos(\theta))^{\frac{1}{2}}$ and width δ. Then $B_\theta(f, g)(x) \approx \delta$ on the annulus of radius 1 and width δ. It follows that

$$||B_\theta(f, g)||_r \approx \delta^{1+\frac{1}{r}},$$

which leads to the relation

$$\delta^{1+\frac{1}{r}} \lesssim \delta^{\frac{2}{p}} \cdot \delta^{\frac{1}{q}},$$

or, equivalently

$$\frac{2}{p} + \frac{1}{q} \leq 1 + \frac{1}{r}. \tag{18}$$

By symmetry we also obtain

$$\frac{2}{q} + \frac{1}{p} \leq 1 + \frac{1}{r}. \tag{19}$$

Taking $p = q$ yields

$$\frac{3}{p} \leq 1 + \frac{1}{r}.$$

Taking $r = 1$, this yields the condition $p \geq \frac{3}{2}$. When $r = 2$, we get $p \geq 2$.

Let $E = F = B_R$, the ball of radius R centered at the origin. It follows that $B_\theta(\chi_E, \chi_F)(x) \approx 1$ on a set of measure $\approx R^2$. It follows that

$$R^{\frac{2}{r}} \lesssim R^{\frac{2}{p}} \cdot R^{\frac{2}{q}},$$

which leads to the restriction

$$\frac{1}{r} \leq \frac{1}{p} + \frac{1}{q}. \tag{20}$$

The Case $\theta = \pi$:

Let χ_E be the indicator function of the ϵ by ϵ^2 rectangle tangent to S^1 at the north pole and let χ_F be of the same form, but located at the south pole. Then $B(\chi_E, \chi_F)(x) \approx \epsilon$ on a set of area $\approx \epsilon^3$. It follows that

$$\epsilon \cdot \epsilon^{\frac{3}{r}} \leq C\epsilon^{\frac{3}{p}+\frac{3}{q}},$$

which forces

$$\frac{3}{p} + \frac{3}{q} \leq 1 + \frac{3}{r}, \tag{21}$$

which is a stricter condition than (18) or (19).

The Case $\theta \neq \pi$

Let χ_E be the indicator function of the ϵ by ϵ^2 rectangle tangent to S^1 at the north pole and let χ_F be of the same form, but tangent to S^1 at the angle $\theta \neq \pi$. Then $B(\chi_E, \chi_F)(x) \approx \epsilon$ on a set of area $\approx \epsilon^4$. It follows that

$$\epsilon \cdot \epsilon^{\frac{4}{r}} \leq C\epsilon^{\frac{3}{p}+\frac{3}{q}},$$

which forces

$$\frac{3}{p} + \frac{3}{q} \leq 1 + \frac{4}{r}, \tag{22}$$

which is, once again a stricter condition than (18) and (19).

Sharpness Examples from Duality

Fixing a function g, define $T_1 f(x) = B_\theta(f, g)(x)$. Then

$$< T_1 f, h >= \int\int f(x-y)g(x-\Theta y)h(x)d\sigma(y)dx.$$

Let $x' = x - y$, one obtains

$$\int\int f(x')g(x'+y-\theta y)h(x'+y)dx'd\sigma(y) =< h, T_1^* f >,$$

where

$$T_1^* h(x') = \int h(x'+y)g(x'+y-\Theta y)d\sigma(y) =: B_\theta^1(h, g)(x').$$

Similarly, with f fixed, let $T_2 g(x) = B_\theta(f, g)(x)$. Then

$$T_2^* h(x') = \int h(x'+\Theta y)f(x'+\Theta y - y)d\sigma(y) =: B_\theta^2(f, h)(x').$$

It is not difficult to see that T_j^*, $j = 1, 2$, satisfies the same bounds T does. This is because T_j^* has essentially the same form with respect to another curve with strictly positive curvature. In other words, if $Q(B_\theta)$ denotes the type set of triples $\left(\frac{1}{p}, \frac{1}{q}, \frac{1}{r}\right)$ such that $B_\theta : L^p(\mathbb{R}^2) \times L^q(\mathbb{R}^2) \to L^r(\mathbb{R}^2)$, then, for both $j = 1, 2$, $\left(\frac{1}{p}, \frac{1}{q}, \frac{1}{r}\right) \in Q(B_\theta)$ if and only if $\left(\frac{1}{p}, \frac{1}{q}, \frac{1}{r}\right) \in Q(B_\theta^j)$.

Applying this idea to (18) and (19), we obtain the constraint

$$\frac{1}{p} + \frac{1}{q} \leq \frac{2}{r}. \tag{23}$$

On the other hand, applying duality to (22) yields constraints

$$\frac{4}{p} + \frac{3}{q} \leq 2 + \frac{3}{r} \tag{24}$$

and

$$\frac{3}{p} + \frac{4}{q} \leq 2 + \frac{3}{r}. \tag{25}$$

4 Summary of Sharpness Conditions and Vertices of $Q(B_\theta)$

The following is the list of necessary conditions on the exponents.

- $0 \leq \frac{1}{p}, \frac{1}{q}, \frac{1}{r} \leq 1$, (Banach cube).
- $\frac{2}{p} + \frac{1}{q} \leq 1 + \frac{1}{r}$, $\frac{1}{p} + \frac{2}{q} \leq 1 + \frac{1}{r}$, (Small ball and annulus).
- $\frac{1}{p} + \frac{1}{q} \leq \frac{2}{r}$, (Dual of small ball and annulus).
- $\frac{3}{p} + \frac{3}{q} \leq 1 + \frac{4}{r}$, (Tangent rectangles $\theta \neq \pi$).
- $\frac{3}{p} + \frac{3}{q} \leq 1 + \frac{3}{r}$, (Tangent rectangles $\theta = \pi$).
- $\frac{4}{p} + \frac{3}{q} \leq 2 + \frac{3}{r}$, $\frac{3}{p} + \frac{4}{q} \leq 2 + \frac{3}{r}$, (Dual of tangent rectangles $\theta \neq \pi$).
- $\frac{1}{r} \leq \frac{1}{p} + \frac{1}{q}$, (Large ball).

Remark 7 Boxes with dimension $\delta \times \delta \times \cdots \times \delta \times \delta^2$, tangent to the unit sphere, are often referred to as C. Fefferman boxes [10] in the harmonic analysis literature.

Vertices of $Q(B_\theta)$

Using SageMath [27], one can compute the vertices of the polyhedron determined by the inequalities above. For $0 < \theta < \pi$, the vertices are

(i) $(0, 0, 0)$
(ii) $(0, 1, 1)$
(iii) $(1, 0, 1)$
(iv) $\left(\frac{2}{3}, 0, \frac{1}{3}\right)$
(v) $\left(0, \frac{2}{3}, \frac{1}{3}\right)$
(vi) $\left(\frac{2}{3}, \frac{2}{3}, 1\right)$
(vii) $\left(\frac{1}{2}, \frac{1}{2}, \frac{1}{2}\right)$

In the degenerate case $\theta = \pi$, there is one fewer vertex:

(i) $(0, 0, 0)$
(ii) $(0, 1, 1)$
(iii) $(1, 0, 1)$
(iv) $\left(\frac{2}{3}, 0, \frac{1}{3}\right)$
(v) $\left(0, \frac{2}{3}, \frac{1}{3}\right)$
(vi) $\left(\frac{2}{3}, \frac{2}{3}, 1\right)$

Remark 8 We refer to $\left(\frac{2}{3}, \frac{2}{3}, 1\right)$ as the *universal* (non-trivial) vertex because it arises for every $\theta \in (0, \pi]$. We refer to $\left(\frac{1}{2}, \frac{1}{2}, \frac{1}{2}\right)$ as the *non-degenerate* vertex because it is only present for θ non-degenerate, i.e., $\theta \neq \pi$. Note that the remaining vertices are the same for both the non-degenerate and degenerate cases.

5 Trivial Bounds

We will establish the boundedness of B_θ at the vertices described above. The full range of exponents is then recovered using multilinear interpolation; see, e.g., [14] and the references contained therein. One may assume throughout that $f, g \geq 0$, since the general case can be recovered by writing f and g in terms of their real and imaginary parts, and then these as differences between their positive and negative parts. We have the pointwise estimate

$$|B_\theta(f, g)(x)| = \int f(x - y) g(x - \Theta y) d\sigma(y) \leq ||f||_{L^\infty} \cdot ||g||_{L^\infty},$$

hence $(0, 0, 0)$, the vertex (i) in Sect. 4, is in the simplex of exponents where B_θ is bounded.

Similarly,

$$|B_\theta(f, g)(x)| \leq \left(\int f(x - y) d\sigma(y) \right) \cdot ||g||_{L^\infty}$$

and

$$|B_\theta(f, g)(x)| \leq ||f||_{L^\infty} \cdot \left(\int g(x - \Theta y) d\sigma(y) \right),$$

so that

$$B_\theta : L^\infty(\mathbb{R}^d) \times L^q(\mathbb{R}^2) \to L^r(\mathbb{R}^d) \text{ and } B_\theta : L^p(\mathbb{R}^2) \times L^\infty(\mathbb{R}^2) \to L^r(\mathbb{R}^2)$$

for $(\frac{1}{q}, \frac{1}{r})$ and $(\frac{1}{p}, \frac{1}{r})$ in the typeset for the circular averaging operator, A, which is the triangle with vertices $(0, 0)$, $(1, 1)$ and $(\frac{2}{3}, \frac{1}{3})$ (i.e., (2) for $d = 2$). This proves boundedness of B_θ at the vertices (ii)–(v) of Sect. 4 for all $\theta \in (0, \pi]$.

6 The $L^{\frac{3}{2}} \times L^{\frac{3}{2}} \to L^1$ Estimate for the Model Operators

For $f, g \geq 0$, writing $\int B_\theta(f,g)(x)\,dx$ as

$$\int\int f(x-y)g(x-\Theta y)d\sigma(y)dx = \int f(x)\left\{\int g(x+y-\Theta y)d\sigma(y)\right\}dx =: \int f(x)\cdot(A_\theta g)(x)dx,$$

and applying Hölder, we see that

$$||B_\theta(f,g)||_{L^1(\mathbb{R}^2)} \leq ||f||_{L^{\frac{3}{2}}(\mathbb{R}^2)} \cdot ||A_\theta g(x)||_{L^3(\mathbb{R}^2)}. \tag{26}$$

Observe that

$$|y-\Theta y|^2 = 2(1- < y, \Theta y >),$$

which for $|y| = 1$ is a constant depending only on θ, non-zero provided that $\Theta \neq I$, the identity map. Hence, A_θ is a rescaled version of the circular averaging operator A from (1) and satisfies the same estimates (2), in particular the $L^{\frac{3}{2}} \to L^3$ bound. Therefore, if $\Theta \neq I$,

$$||B_\theta(f,g)||_{L^1(\mathbb{R}^2)} \lesssim ||f||_{L^{\frac{3}{2}}(\mathbb{R}^2)} ||g||_{L^{\frac{3}{2}}(\mathbb{R}^2)}$$

by the classical result of Strichartz [31] and Littman [22]. This establishes boundedness of B_θ at the vertex (vi) in Sect. 4 for all $\theta \in (0, \pi]$.

7 The $L^{2,1} \times L^{2,1} \to L^2$ Estimate for the Model Operators, $0 < \theta < \pi$

We want to show that B_θ is of restricted strong type, that is, $B_\theta : L^{2,1}(\mathbb{R}^2) \times L^{2,1}(\mathbb{R}^2) \to L^2(\mathbb{R}^2)$, for $\theta \neq \pi$. Thus, we need to show that if $E, F \subset \mathbb{R}^2$, then $||B_\theta(\chi_E, \chi_F)||_{L^2} \lesssim |E|^{\frac{1}{2}}|F|^{\frac{1}{2}}$. Assuming without loss of generality that $|E| \leq |F|$, write

$$
\begin{aligned}
||B_\theta(\chi_E, \chi_F)||^2_{L^2(\mathbb{R}^2)} &= \int\int\int \chi_E(x-y)\chi_F(x-\Theta y)\chi_E(x-y')\chi_F(x-\Theta y')\, d\sigma(y)d\sigma(y')dx \qquad (27)\\
&= \int_{|\alpha-\alpha'|<\frac{\theta}{2}}\int\int \chi_E(x-y)\chi_F(x-\Theta y)\chi_E(x-y')\chi_F(x-\Theta y')\, d\sigma(y)d\sigma(y')dx\\
&+ \int_{|\alpha-\alpha'|\geq\frac{\theta}{2}}\int\int \chi_E(x-y)\chi_F(x-\Theta y)\chi_E(x-y')\chi_F(x-\Theta y')\, d\sigma(y)d\sigma(y')dx\\
&\leq \int_{|\alpha-\alpha'|<\frac{\theta}{2}}\int\int \chi_E(x-y)\chi_F(x-\Theta y')\, d\sigma(y)d\sigma(y')dx\\
&+ \int_{|\alpha-\alpha'|\geq\frac{\theta}{2}}\int\int \chi_E(x-y)\chi_E(x-y')\, d\sigma(y)d\sigma(y')dx,
\end{aligned}
$$

where $y = (\cos(\alpha), \sin(\alpha))$, $y' = (\cos(\alpha'), \sin(\alpha'))$.

To make a change of variables for the first integral in the last expression, set

$$u_1 = x_1 - \cos(\alpha),\ u_2 = x_2 - \sin(\alpha),\ v_1 = x_1 - \cos(\alpha' + \theta), v_2 = x_2 - \sin(\alpha' + \theta); \qquad (28)$$

for the second integral, make the change of variables

$$u_1 = x_1 - \cos(\alpha),\ u_2 = x_2 - \sin(\alpha),\ v_1 = x_1 - \cos(\alpha'), v_2 = x_2 - \sin(\alpha'). \qquad (29)$$

Then the Jacobian for the first is

$$\sin(\alpha - \alpha' - \theta),$$

while the Jacobian of the second is

$$\sin(\alpha - \alpha').$$

Both of these quantities are bounded away from 0 because of the constraints on the angle between y and y'. Note that this argument fails when $\theta = \pi$ since if $|\alpha - \alpha'| \geq \frac{\theta}{2}$, the Jacobian goes to 0 regardless of which terms we keep.

As long as $0 < \theta < \pi$, though, we have that the Jacobian in both cases is bounded from below by $\frac{1}{2}\sin(\frac{\theta}{2})$. It follows that

$$||B_\theta(\chi_E, \chi_F)||_{L^2(\mathbb{R}^2)} \leq C_\theta(|E|^2 + |E||F|)^{\frac{1}{2}} \leq 2C|E|^{\frac{1}{2}}|F|^{\frac{1}{2}}.$$

for some constant C depending only on θ. Moreover, it is not difficult to see from the argument above that

$$C_\theta \leq C'\frac{1}{\min\{\theta, \pi - \theta\}},$$

where C' is a uniform constant independent of θ. This establishes the boundedness of B_θ at the vertex (vii) for non-degenerate θ, i.e., $\theta \in (0, \pi)$.

8 The $L^{\frac{3}{2}} \times L^{\frac{3}{2}} \to L^1$ Estimate for General Operators

We now turn to the general class of bilinear Radon transforms in Theorem 6. In this section, we show that if for all $(x, y, z) \in Z$, (i) $\phi_1(x, y), \phi_2(x, z), \phi_3(y, z)$ satisfy (14), and (ii) $\phi_3(y, z)$ satisfies the Phong-Stein condition (4), then $B : L^{\frac{3}{2}}(\mathbb{R}^2) \times L^{\frac{3}{2}}(\mathbb{R}^2) \to L^1(\mathbb{R}^2)$. Recall that the bilinear kernel is given by (9). As for the model operators, since $K \geq 0$, one can assume that $f, g \geq 0$, and write

$$||B(f, g)||_{L^1} = \int B(f, g)(x)dx = \int f(y) \left[\int \left[\int K(x, y, z)\, dx\right] g(z)\, dz\right] dy.$$

The expression inside the outermost of the square brackets is a linear operator, $Tg(y)$. It hence suffices to show that $T : L^{\frac{3}{2}}(\mathbb{R}^2) \to L^3(\mathbb{R}^2)$; by the Phong-Stein condition (4) for ϕ_3, this will follow if we show that T is a generalized Radon transform of associated to Z_3, where Z_3 is as in (12). The kernel of T is

$$L(y, z) = \int K(x, y, z)\, dx = \pi_*(K)(y, z),$$

where π_* denotes pushforward of distributions under the projection $\pi(x, y, z) = (y, z)$. The operator $\pi_* : \mathcal{E}'(\mathbb{R}^6) \to \mathcal{E}'(\mathbb{R}^4)$ is itself an FIO, $\pi_* \in I^{-\frac{1}{2}}(C_\pi)$, associated to the canonical relation

$$C_\pi = \left\{(y, \eta, z, \zeta; x, y, z, 0, \eta, \zeta) : (x, y, z) \in \mathbb{R}^6,\ (\eta, \zeta) \in \mathbb{R}^4 \setminus 0\right\};$$

see Guillemin and Sternberg [18]. C_π is a non-degenerate canonical relation in $T^*\mathbb{R}^4 \times T^*\mathbb{R}^{10}$, and thus its application to N^*Z is covered by the transverse intersection calculus. A direct calculation shows that $C_\pi \circ N^*Z = N^*Z_3$. and thus $L \in I^{-\frac{1}{2}}(N^*Z_3)$. Hence, T is a linear generalized Radon transform on $\mathbb{R}^2$ satisfying the Phong-Stein condition, and has the same mapping properties as any such operator; in particular, $T : L^{\frac{3}{2}}(\mathbb{R}^2) \to L^3(\mathbb{R}^2)$.

9 The $L^{2,1} \times L^{2,1} \to L^2$ Estimate for General Operators

Now, to prove the restricted strong type $L^{2,1}(\mathbb{R}^2) \times L^{2,1}(\mathbb{R}^2) \to L^2(\mathbb{R}^2)$ result for B, consider as for the model operators treated in Sect. 7 the L^2 norm squared of the operator applied to indicator functions of measurable sets E and F:

$$\int\int\int\int\int \chi_E(y)\chi_E(y')\chi_F(z)\chi_F(z')K(x,y,z)K(x,y',z')\,dydy'dzdz'dx. \tag{30}$$

Modifying somewhat the argument for the model operators in Sect. 7, one can show that if any one of the following four bounded properties holds, one obtains $||B(\chi_E,\chi_F)||_{L^2} \lesssim |E|^{\frac{1}{2}}|F|^{\frac{1}{2}}$:

$$K_{yz'} := \int\int\int K(x,y,z)K(x,y',z')\,dy'dzdx \in L^\infty(\mathbb{R}^6_{x,y,z'}), \tag{31}$$

$$K_{zy'} := \int\int\int K(x,y,z)K(x,y',z')\,dydz'dx \in L^\infty(\mathbb{R}^6_{x,z,y'}), \tag{32}$$

$$K_{zz'} := \int\int\int K(x,y,z)K(x,y',z')\,dydy'dx \in L^\infty(\mathbb{R}^6_{x,z,z'}), \tag{33}$$

and

$$K_{yy'} := \int\int\int K(x,y,z)K(x,y',z')\,dzdz'dx \in L^\infty(\mathbb{R}^6_{x,y,y'}). \tag{34}$$

If (31) holds, we eliminate $z,\ y'$ in (30) by noting that $\chi_E(y')\chi_F(z) \le 1$ and obtain an upper bound $C|E||F|$ for (30). If (32) holds, we proceed in a similar fashion, using $\chi_E(y)\chi_F(z') \le 1$. If (33) holds, we employ $\chi_E(y)\chi_E(y') \le 1$ and bound the whole expression in (30) by $C|F|^2$, which, if $|F| \le |E|$, is bounded by $C|E||F|$. On the other hand, if $|E| \le |F|$, we may use the boundedness of the expression in (34), together with $\chi_F(z)\chi_F(z') \le 1$, to bound (30) by $C|E|^2$, which is $\le C|E||F|$. Thus, regardless of whether $|E| \le |F|$ or $|F| \le |E|$, we obtain $||B(\chi_E,\chi_F)||_{L^2} \lesssim |E|^{\frac{1}{2}}|F|^{\frac{1}{2}}$.

This argument holds more generally if any (x,y,z,y',z') has a neighborhood in $\mathbb{R}^{10}$ on which one of

$$(31) \text{ or } (32) \text{ or } [(33) \text{ and } (34)] \text{ holds}, \tag{35}$$

with the conjunction in the last term to cover both of the cases $|E| \le |F|$ and $|F| \le |E|$. Taking a subordinate partition of unity of the domain of integration in (30), we can then apply the above arguments to still obtain $||B(\chi_E,\chi_F)||_{L^2} \lesssim |E|^{\frac{1}{2}}|F|^{\frac{1}{2}}$.

In the framework of the product conormal kernels above, we can formulate first order conditions on the ϕ_j, and hence on Z and K, which imply that one of (31)–(34) holds locally. To start, let

$$K\hat{\otimes}K(x,y,z,y',z') = K(x,y,z)\cdot K(x,y',z') \in \mathcal{E}'(\mathbb{R}^{10})$$

and

$$Z\hat{\times}Z = \{(x,y,z,y',z') : (x,y,z) \in Z,\ (x,y',z') \in Z\} \subset \mathbb{R}^{10}.$$

Analogous to (14), we assume

$$rank \begin{bmatrix} d_x\phi_1(x,y) & d_y\phi_1 & 0 & 0 & 0 \\ d_x\phi_2(x,z) & 0 & d_z\phi_2 & 0 & 0 \\ 0 & d_y\phi_3(y,z) & d_z\phi_3 & 0 & 0 \\ d_x\phi_1(x,y') & 0 & 0 & d_{y'}\phi_1 & 0 \\ d_x\phi_2(x,z') & 0 & 0 & 0 & d_{z'}\phi_2 \\ 0 & 0 & 0 & d_{y'}\phi_3(y',z') & d_{z'}\phi_3 \end{bmatrix} = 6, \tag{36}$$

for all $(x, y, z, y', z') \in Z \hat{\times} Z$. Then $Z \hat{\times} Z$ is smooth and 4-dimensional, and $K \hat{\otimes} K$ is a smooth density on it. The kernel $K_{yz'}$ in (31) above is just $\pi_*^{yz'}(K \hat{\otimes} K)$ $(y, z') \in \mathcal{D}'(\mathbb{R}^4)$, the pushforward by $\pi^{yz'}(x, y, z, y', z') = (y, z')$. This will be a smooth density on $\mathbb{R}^4$, hence with a smooth (thus locally bounded) Radon-Nikodym derivative, if $Z \hat{\times} Z$ is a graph over the y, z' variables, and by the implicit function theorem, this holds when the 6×4 submatrix consisting of the y and z' columns of the 6×10 matrix in (36) has rank 4. Deleting the two rows of zeros, this holds at a point $(x_0, y_0, z_0, y_0', z_0') \in Z \hat{\times} Z$ iff

$$det \begin{bmatrix} d_y\phi_1(x_0,y_0) & 0 \\ d_y\phi_3(y_0,z_0) & 0 \\ 0 & d_{z'}\phi_3(x_0,z_0') \\ 0 & d_{z'}\phi_3(y_0',z_0') \end{bmatrix} \neq 0. \tag{37}$$

Similarly, considering the pushforward maps $\pi_*^{zy'}$, $\pi_*^{zz'}$ and $\pi_*^{yy'}$, we see that the local boundedness of (32), (33) and (34) follow from

$$det \begin{bmatrix} d_z\phi_2(x_0,z_0) & 0 \\ d_z\phi_3(y_0,z_0) & 0 \\ 0 & d_{y'}\phi_1(x_0,y_0') \\ 0 & d_{y'}\phi_3(y_0',z_0') \end{bmatrix} \neq 0, \tag{38}$$

$$det \begin{bmatrix} d_z\phi_2(x_0,z_0) & 0 \\ d_z\phi_3(y_0,z_0) & 0 \\ 0 & d_{z'}\phi_2(x_0,y_0') \\ 0 & d_{z'}\phi_3(y_0',z_0') \end{bmatrix} \neq 0, \tag{39}$$

and

$$det \begin{bmatrix} d_y\phi_1(x_0,y_0) & 0 \\ d_y\phi_3(y_0,z_0) & 0 \\ 0 & d_{y'}\phi_1(x_0,y_0') \\ 0 & d_{y'}\phi_3(y_0',z_0') \end{bmatrix} \neq 0, \quad \text{resp.} \tag{40}$$

Conditions (37)–(40) are of course open conditions, so if any one holds at a point $(x_0, y_0, z_0, y_0', z_0') \in Z \hat{\times} Z$, then it holds on a neighborhood. Thus, if at every $(x_0, y_0, z_0, y_0', z_0') \in Z \hat{\times} Z$,

$$\text{at least one of (37) or (38) or [(39) and (40)] holds,} \tag{41}$$

then (35) holds. Combined with the discussion in Sect. 8, valid if ϕ_3 satisfies the rotational curvature condition, this finishes the proof of Theorem 6.

Now, we may simplify conditions (36) and (41) as follows. By a partition of unity on $supp(K)$, one may localize and work near a basepoint (x_0, y_0, z_0); thus, in the considerations above, open conditions which hold at $(x_0, y_0, z_0, y_0, z_0) \in Z \hat{\times} Z$ will still be valid at nearby points of $(x_0, y_0, z_0, y_0', z_0') \in Z \hat{\times} Z$. Then (36) simply reduces back to (14), while in (37)–(40), it suffices to have the determinants non-zero at $y_0' = y_0$ and $z_0' = z_0$, and (41) is equivalent with

$$\begin{vmatrix} d_y\phi_1(x_0, y_0) \\ d_y\phi_3(y_0, z_0) \end{vmatrix} \neq 0, \qquad \begin{vmatrix} d_z\phi_2(x_0, z_0) \\ d_z\phi_3(y_0, z_0) \end{vmatrix} \neq 0, \tag{42}$$

at all $(x_0, y_0, z_0) \in supp(K)$, which is (17) of Theorem 6.

10 Comparison with Loomis-Whitney Conditions

The nonlinear Loomis-Whitney inequality of [4] (in its nonquantitative version) says the following in our setting: Suppose that μ is a smooth density on $Z \subset \mathbb{R}^2_x \times \mathbb{R}^2_y \times \mathbb{R}^2_z$ is smooth, with $dim(Z) = 3$ and such that each of the projections π_x, π_y and $\pi_z : Z \to \mathbb{R}^2$ are submersions on $supp(\mu)$. Further, suppose that at each $(x_0, y_0, z_0) \in supp(\mu)$,

$$span\left\{ker(d\pi_x),\ ker(d\pi_y),\ ker(d\pi_z)\right\} = TZ. \tag{43}$$

Then

$$\left| \int_Z f_1(x) f_2(y) f_3(z) d\mu(x, y, z) \right| \le \prod_{j=1}^{3} ||f_j||_{L^2(\mathbb{R}^2)}. \tag{44}$$

I.e., the bilinear generalized Radon transform,

$$B(f_2, f_3)(x) = \int\int K(x, y, z) f_2(y) f_3(z)\, d\mu(\cdot, y, z)$$

is bounded $L^2(\mathbb{R}^2) \times L^2(\mathbb{R}^2) \to L^2(\mathbb{R}^2)$.

For the Z_θ corresponding to the non-degenerate model operators B_θ, $0 < \theta < \pi$, one can check that these conditions are satisfied, so that unrestricted strong type boundedness in fact holds at the non-degenerate vertex $(\frac{1}{2}, \frac{1}{2}, \frac{1}{2})$. Indeed, if we parametrize Z_θ as $\left\{(x, x-\omega, x-\theta\omega) : x \in \mathbb{R}^2, \omega \in S^1\right\}$, then one sees that the projections, $\pi_1(x,\omega) = x$, $\pi_2(x,\omega) = x-\omega$, $\pi_3(x,\omega) = x-\theta\omega$ are submersions, and their kernels,

$$Ker(D\pi_1) = \mathbb{R}\cdot\partial_\omega,\ Ker(D\pi_2) = \mathbb{R}\cdot(\omega\cdot\nabla_x + \partial_\omega),\ Ker(D\pi_3) = \mathbb{R}\cdot((\theta\omega)\cdot\nabla_x + \partial_\omega),$$

span the tangent space at each point.

However, there are operators covered by our Theorem 6 which are not covered by the Loomis-Whitney result. A simple example of one such is given by

$$Z_{\mathbf{t}} = \{(x,y,z) : \phi_1(x,y) = t_1,\ \phi_2(x,z) = t_2,\ \phi_3(y,z) = t_3\},$$

with

$$\phi_1(x,y) = x_1 - y_1,\ \phi_2(x,z) = x_1 - z_1,\ \phi_3(y,z) = |y-z|^2 - t_3, \tag{45}$$

and with the cut-off $\psi(x,y,z)$ supported where $(y-z) \not\parallel e_1$. One can check that $Z_{\mathbf{t}} \neq \emptyset$ if $(t_1-t_2)^2 < t_3$, and the conditions in (14) and (42) are satisfied, so that Theorem 6 applies. However, taking $x,\ y_2$ as coordinates on $Z_{\mathbf{t}}$, we see that

$$ker(d\pi_y) = ker(d\pi_z) = \mathbb{R}\cdot\frac{\partial}{\partial x_2},$$

violating the transversality condition (43) everywhere.

It would be interesting to unify the results presented here with those in [4] and [30], or even later works such as [2, 3], at the possible price of obtaining only restricted strong type estimates at the non-degenerate vertex.

References

1. I. Bejenaru, S. Herr and D. Tataru, *A convolution estimate for two-dimensional hypersurfaces*, Rev. Mat. Iberoam. **26** (2010), 707–728.
2. J. Bennett and N. Bez, *Some nonlinear Brascamp-Lieb inequalities and applications to harmonic analysis*, J. Funct. Analysis **259**, 2520–2556 (2010).
3. J. Bennett, N. Bez and S. Gutiérrez, *Transversal multilinear Radon-like transforms: local and global estimates*, Rev. Math. Iberoam. **29** (2013), 765–788.
4. J. Bennett, A. Carbery and J. Wright, *A non-linear generalisation of the Loomis-Whitney inequality and applications*, Math. Res. Lett. **12** (2005), 443–457.
5. J. Bourgain, *A Szemeredi type theorem for sets of positive density*, Israel J. Math. **54** (1986), no. 3, 307–331.
6. P. Brass, W. Moser and J. Pach, *Research Problems in Discrete Geometry.* Springer, New York, (2005).
7. M. Christ, *Convolution, curvature and combinatorics: a case study*, Internat. Math. Res. Not. **19** (1998), 1033-1048.
8. R. Coifman and Y. Meyer, *Wavelets. Calderón-Zygmund and multilinear operators.* Translated from the 1990 and 1991 French originals by David Salinger. Cambridge Studies in Advanced Mathematics, **48**. Cambridge University Press, Cambridge, (1997).

9. C. Demeter, T. Tao and C. Thiele, *Maximal multilinear operators*, Trans. Amer. Math. Soc. 360 (2008), no. 9, 4989–5042.
10. C. Fefferman, *A note on spherical summation multipliers*, Israel J. Math. **15** (1973), 44-52.
11. R. Felea, R. Gaburro, A. Greenleaf and C. Nolan, *Bilinear operators and Fréchet differentiability in seismic inversion*, in preparation.
12. H. Furstenberg, Y. Katznelson, and B. Weiss, *Ergodic theory and configurations in sets of positive density*, Mathematics of Ramsey Theory, 184–198, Algorithms Combin., **5**, Springer, Berlin, 1990.
13. L. Grafakos and X. Li, *Uniform bounds for the bilinear Hilbert transforms I*, Ann. of Math. (2) **159** (2004), no. 3, 889–933.
14. L. Grafakos and T. Tao, *Multilinear interpolation between adjoint operators*, J. Funct. Analysis **199** (2003), no. 2, 379–385.
15. L. Grafakos and R. Torres, *Multilinear Calderón-Zygmund theory*, Adv. Math. **165** (2002), no. 1, 124–164.
16. A. Greenleaf and A. Iosevich, *Three point configuration, a bilinear operator and applications discrete geometry*, Analysis and PDE , **5**, no. 2 (2012), 397–409.
17. A. Greenleaf, M. Lassas, M. Santacesaria, S. Siltanen and G. Uhlmann, *Propagation and recovery of singularities in the inverse conductivity problem*, Analysis & PDE, **11-8** (2018), 1901–1943.
18. V. Guillemin and S. Sternberg, *Geometric Asymptotics.* Mathematical Surveys, **14**. Amer. Math. Soc., Providence, R.I., 1977.
19. L. Hörmander, *The Analysis of Linear Partial Differential Operators, IV*, Springer, New York, 1985.
20. F. John, *Plane Waves and Spherical Means Applied to Partial Differential Equations.* Interscience, New York,1955.
21. H. Koch and S. Steinerberger, *Convolution estimates for singular measures and some global nonlinear Brascamp-Lieb inequalities*, Proc. Roy. Soc. Edinb. Sect. A **145** (2015), 1223–1237.
22. W. Littman, *Lp-Lq-estimates for singular integral operators arising from hyperbolic equations*, Partial differential equations (Proc. Sympos. Pure Math., Vol. XXIII, Univ. California, Berkeley, Calif., 1971), 479–481. Amer. Math. Soc., Providence, R.I., (1973).
23. M. Lacey and C. Thiele, *L^p estimates on the bilinear Hilbert transform for $2<p<\infty$*, Ann. of Math. **146** (1997), 693–724.
24. C. Muscalu, T. Tao and C. Thiele, *Multi-linear operators given by singular multipliers*, J. Amer. Math. Soc. **15** (2002), no. 2, 469–496.
25. D. Oberlin, *Two estimates for curves in the plane*, Proc. Amer. Math. Soc. **132** (2004), no. 11, 3195–3201.
26. D. Phong and E. Stein, *Radon transforms and torsion*, Internat. Math. Res. Notices (1991), no. 4, 49–60.
27. SageMath, an open source mathematical software system. https://www.sagemath.org.
28. C. D. Sogge, *Fourier integral in classical analysis*, Cambridge University Press, 1993.
29. E. M. Stein, *Harmonic Analysis: Real Variable Methods.* Princeton University Press, Princeton, 1993.
30. B. Stovall, *L^p improving multilinear Radon-like transforms*, Rev. Mat. Iberoam. **27** (2011), 1059–1085.
31. R. Strichartz, *Convolutions with kernels having singularities on a sphere*, Trans. Amer. Math. Soc. **148** (1970), 461–471.

Expansion, Divisibility and Parity: An Explanation

Harald Andrés Helfgott

1 Introduction

This paper is meant as an informal exposition of [2]. The main result is a statement on how a linear operator defined in terms of divisibility by primes has small norm. In this exposition, we will choose to start from one of its main current applications, namely,

$$\frac{1}{\log x}\sum_{n\leq x}\frac{\lambda(n)\lambda(n+1)}{n}=O\left(\frac{1}{\sqrt{\log\log x}}\right), \tag{1}$$

which strengthens results by Tao [6] and Tao-Teräväinen [7]. (Here $\lambda(n)$ is the Liouville function, viz., the completely multiplicative function such that $\lambda(p)=-1$ for every prime p.) There are other corollaries, some of them subsuming the above statement. It is also true that above statement is an improvement on a bound, whereas the main result is a result that is new also in a qualitative sense. One may thus ask oneself whether it is right to center the exposition on (1).

All the same, (1) is a concrete statement that is obviously interesting, being a step towards Chowla's conjecture ("logarithmic Chowla in degree 2"), and so it is a convenient initial goal.

First, some meta comments. We may contrast two possible ways of writing a paper—what may be called the *incremental* and the *retrospective* approaches.

H. A. Helfgott (✉)
IMJ-PRG, UMR 7586, 58 avenue de France, Bâtiment S. Germain, case 7012, 75013 Paris, CEDEX 13, France

Mathematisches Institut, Georg-August Universität Göttingen, Bunsenstraße 3-5, 37073 Göttingen, Germany
e-mail: harald.helfgott@gmail.com

M. B. Nathanson (ed.), *Combinatorial and Additive Number Theory V*,
Springer Proceedings in Mathematics & Statistics 395,
https://doi.org/10.1007/978-3-031-10796-2_10

- In the *incremental* approach, we write a paper while we solve a problem, letting complications and detours accreted. There is much that can be said against this approach: it is hard to distinguish it from simply lazy writing; the end product may be unclear; just how one got to the solution of the problem may be inessential or even misleading.
- The *retrospective* approach consists in writing the paper once the proof is done, from the perspective that one has reached by the time one has solved the problem.

These are of course two extremes. Few people write nothing down while solving a problem, and the way one followed to reach the solution generally has some influence on the finished paper. In the case of [2], what we followed was mainly the retrospective approach, with some incremental elements, mainly to deal with technical complications that we had to deal with after we had an outline of a proof. It is tempting to say that is still too much incrementality, but, in fact, some of the feedback we have received suggest a drawback of the retrospective approach that I had not thought of before.

When faced with a result with a lengthy proof, readers tend to come up with their own "natural strategy". So far, so good: active reading is surely a good thing. What then happens, though, is that readers may see necessary divergences from their "natural strategy" as technical complications. They may often be correct; however, they may miss why the "natural strategy" may not work, or how it leads to the main, essential difficulty—the heart of the problem, which they may then miss for following the complications.

What I will do in this write-up is follow, not an incremental approach, but rather an idealized view of what the path towards the solution was or could have been like; a recreated incrementality with the benefit of hindsight, then, starting from a "natural strategy", with an emphasis on what turns out to be essential.

Notation. We will use notation that is usual within analytic number theory. In particular, given two functions f, g on $\mathbb{R}^+$ or $\mathbb{Z}^+$, $f(x) = O(g(x))$ means that there exists a constant $C > 0$ such that $|f(x)| \leq Cg(x)$ for all large enough x, and $f(x) = o(g(x))$ means that $\lim_{x\to\infty} f(x)/g(x) = 0$ (and $g(x) > 0$ for x large enough). By $O^*(B)$, we will mean "a quantity whose absolute value is no larger than B"; it is a useful bit of notation for error terms. We define $\omega(n)$ to be the number of prime divisors of an integer n.

Initial Setup

Let us set out to reprove Tao's "logarithmic Chowla" statement, that is,

$$\frac{1}{\log x} \sum_{n\leq x} \frac{\lambda(n)\lambda(n+1)}{n} \to 0$$

as $x \to \infty$. Now, Tao's method gives a bound of $O(1/(\log\log\log\log x)^{\alpha})$ on the left side (as explained in [3], with $\alpha = 1/5$), while Tao-Teräväinen should yield a bound of $O(1/(\log\log\log x)^{\alpha})$ for some $\alpha > 0$. Their work is based on depleting entropy, or, more precisely, on depleting mutual information. Our method gives stronger bounds (namely, $O(1/\sqrt{\log\log x})$) and is also "stronger" in ways that will later become apparent. Let us focus, however, simply on giving a different proof, and welcome whatever might come from it.

The first step will consist of a little manipulation as in Tao, based on the fact that λ is multiplicative. Let $W = \sum_{n\le x} \lambda(n)\lambda(n+1)/n$. For any prime (or integer!) p,

$$\begin{aligned}\frac{1}{p}W &= \sum_{n\le x} \frac{\lambda(pn)\lambda(pn+p)}{pn}\\ &= \sum_{n\le px:\ p|n} \frac{\lambda(n)\lambda(n+p)}{n} = \sum_{n\le x:\ p|n} \frac{\lambda(n)\lambda(n+p)}{n} + O\left(\frac{\log p}{p}\right).\end{aligned}$$

Hence, for any set of primes $\mathbf{P}$,

$$\sum_{p\in\mathbf{P}} \sum_{n\le x:\ p|n} \frac{\lambda(n)\lambda(n+p)}{n} = W\mathscr{L} + O\left(\sum_{p\in\mathbf{P}} \frac{\log p}{p}\right),$$

where $\mathscr{L} = \sum_{p\in\mathbf{P}} 1/p$. If H is such that $p \le H$ for all $p \in \mathbf{P}$, then, by the prime number theorem, $\sum_{p\in\mathbf{P}} (\log p)/p \ll \log H$. Thus

$$W = \frac{1}{\mathscr{L}} \sum_{n\le x} \sum_{p\in\mathbf{P}:p|n} \frac{\lambda(n)\lambda(n+p)}{n} + O\left(\frac{\log H}{\mathscr{L}}\right).$$

Assuming $H = x^{o(1)}$ (so that $\log H = o(\log x)$) and $\mathscr{L} \ge 1$, and using a little partial summation, we see that, to prove that $W = o(\log x)$, it is enough to show that $S_0 = o(N\mathscr{L})$, where

$$S_0 = \sum_{N<n\le 2N} \sum_{p\in\mathbf{P}:p|n} \lambda(n)\lambda(n+p).$$

Let us make this sum a little more symmetric. Let $\mathbf{N} = \{n \in \mathbb{Z} : N < n \le 2N\}$, and define

$$S = \sum_{n\in\mathbf{N}} \sum_{\sigma=\pm 1} \sum_{\substack{p\in\mathbf{P}:p|n\\ n+\sigma p\in\mathbf{N}}} \lambda(n)\lambda(n+\sigma p).$$

Then $S = 2S_0 + O(\sum_{p\in\mathbf{P}} 1) = 2S_0 + O(H)$, and thus it is enough to prove that

$$S = o(N\mathscr{L}).$$

Objectives. Tao showed that there *exists* a set $\mathbf{P}$ of primes (very small compared to N) such that $S = o(N\mathscr{L})$. It is our aim to prove that $S = o(N\mathscr{L})$ for *every* set $\mathbf{P}$ of primes satisfying some simple conditions. (As we said, we assume $p \leq H$, and it is not hard to see that we have to assume $\mathscr{L} \to \infty$; it will also be helpful to assume that no $p \in \mathbf{P}$ is tiny compared to H.) We will in fact be able to show that $S = O(N\sqrt{\mathscr{L}})$, which is essentially optimal.

2 A First Attempt

We now set out on our own.

Old Habits Die Hard. A Reduction

It is a deep-seated instinct for an analytic number theorist to apply Cauchy-Schwarz:

$$\begin{aligned} S^2 &\leq \left(\sum_{n\in\mathbf{N}} \sum_{\sigma=\pm1} \sum_{\substack{p\in\mathbf{P}:p|n\\ n+\sigma p\in\mathbf{N}}} \lambda(n)\lambda(n+\sigma p) \right)^2 \\ &\leq N \sum_{n\in\mathbf{N}} \sum_{\sigma_1,\sigma_2=\pm1} \sum_{\substack{p_1,p_2\in\mathbf{P}\\ p_i|n,\ n+\sigma_i p_i\in\mathbf{N}}} \lambda(n+\sigma_1 p_1)\lambda(n+\sigma_2 p_2) \\ &\leq \sum_{n\in\mathbf{N}} \sum_{\sigma_1,\sigma_2=\pm1} \sum_{\substack{p_1,p_2\in\mathbf{P}\\ p_1|n,\ p_2|n+\sigma_1 p_1\\ n+\sigma_1 p_1,\, n+\sigma_1 p_1+\sigma_2 p_2\in\mathbf{N}}} \lambda(n)\lambda(n+\sigma_1 p_1+\sigma_2 p_2), \end{aligned}$$

where we are changing variables in the last step.

We iterate, applying Cauchy-Schwarz ℓ times:

$$S^{2^\ell} \leq N^{2^\ell-1} \sum_{n\in\mathbf{N}} \sum_{\substack{\sigma_i=\pm1,\ p_i\in\mathbf{P}\\ \forall 1\leq i\leq 2^\ell: p_i|n+\sigma_1 p_1+\ldots+\sigma_{i-1}p_{i-1}}} \lambda(n)\lambda(n+\sigma_1 p_1+\ldots+\sigma_{2^\ell} p_{2^\ell}).$$

We can see $n+\sigma_1 p_1+\ldots+\sigma_{2^\ell} p_{2^\ell}$ as the outcome of a "walk" of length 2^ℓ.

Suppose for a moment that, for $k = 2^\ell$ large, the number of walks of length k from n to m is generally about $\psi(m)$, where ψ is a nice continuous function. Then S^{2^ℓ} would tend to

$$N^{2^\ell - 1} \sum_{n\in\mathbf{N}} \sum_m \lambda(n)\lambda(n+m)\psi(m).$$

The main result (Theorem 1) in Matomäki-Radziwiłł [4] would then give us a bound on that double sum. Let us write that bound in the form

$$\left| \sum_{n\in\mathbf{N}} \sum_m \lambda(n)\lambda(n+m)\psi(m) \right| \leq \mathrm{err}_2 \cdot N\mathscr{L}^k,$$

since $|\psi|_1$ should be about $\mathscr{L}^k$, and write our statement on convergence to ψ in the form

$$\sum_{n\in\mathbf{N}} \sum_m \left| \psi(m) - \sum_{\substack{\sigma_i=\pm 1,\ p_i\in\mathbf{P} \\ \forall 1\leq i\leq 2^\ell : p_i | n+\sigma_1 p_1+\ldots+\sigma_{i-1}p_{i-1} \\ \sigma_1 p_1+\ldots+\sigma_k p_k = m}} 1 \right| = \mathrm{err}_1 \cdot N\mathscr{L}^k. \tag{2}$$

Then

$$S \leq (\mathrm{err}_1^{1/k} + \mathrm{err}_2^{1/k}) \cdot N\mathscr{L}.$$

Here already we would seem to have a problem. The "width" M of the distribution ψ (meaning its scale) should be $\ll \sqrt{k} \cdot \mathbb{E}(p : p \in \mathbb{P}) \leq \sqrt{k}H$; the distribution could be something like a Gaussian at that scale, say. Now, the bound from [4] is roughly of the quality $\mathrm{err}_2 \leq 1/\log M$. One can use intermediate results in the same paper to obtain a bound on err_2 roughly of the form $1/M^\delta$, $\delta > 0$, if we remove some integers from $\mathbf{N}$. At any rate, it seems clear that we would need k to be $o(\log H)$.

As it turns out, all of that is a non-issue, in that there is a way to avoid taking the kth root of err_2 altogether. Let us make a mental note, however.

Walks of Different Kinds

The question now is how large ℓ has to be for the number of walks of length $k = 2^\ell$ from n to $n+m$ to approach a continuous distribution $\psi(m)$. Consider first the walks $n, n+\sigma_1 p_1, \ldots, n+\sigma_1 p_1 + \cdots + \sigma_k p_k$ such that no prime p_i is repeated. Fix σ_i, p_i and let n vary. By the Chinese Remainder Theorem, the number of $n \in \mathbf{N}$ such that

$$p_1 | n,\ \ p_2 | n+\sigma_1 p_1,\ \ \ldots,\ \ p_k | n+\sigma_1 p_1 + \ldots + \sigma_{k-1} p_{k-1}$$

is almost exactly $N/p_1 p_2 \cdots p_k$. In other words, the probability of that walk being allowed is almost exactly $1/p_1 \ldots p_k$. We may thus guess that ψ has the same shape (scaled up by a factor of $\mathscr{L}^k$) as the distribution of the endpoint of a random walk where each edge of length p is taken with probability $1/p_i$ (divided by $\mathscr{L}$, so that the probabilities add up to 1). That distribution should indeed tend to a continuous distribution—namely, a Gaussian—fairly quickly. Of course, here, we are just talking about the contribution of walks with distinct edges p_i to

$$\sum_{n\in\mathbf{N}} \sum_{m} \left(\psi(m) - \sum_{\substack{\sigma_i=\pm 1,\ p_i\in\mathbf{P} \\ \forall 1\leq i\leq 2':\, p_i | n+\sigma_1 p_1+\ldots+\sigma_{i-1} p_{i-1} \\ \sigma_1 p_1+\ldots+\sigma_k p_k = m}} 1 \right),$$

without absolute values, and we do need to take absolute values as in (2). However, we can get essentially what we want by looking at the variance

$$\sum_{n\in\mathbf{N}} \sum_{m} \left(\psi(m) - \sum_{\substack{\sigma_i=\pm 1,\ p_i\in\mathbf{P} \\ \forall 1\leq i\leq k:\, p_i | n+\sigma_1 p_1+\ldots+\sigma_{i-1} p_{i-1} \\ \sigma_1 p_1+\ldots+\sigma_k p_k = m}} 1 \right)^2,$$

and considering the contribution to this variance made by closed walks

$$\begin{aligned} n, &n + \sigma_1 p_1, \ldots, n + \sigma_1 p_1 + \cdots + \sigma_k p_k = m, \\ &n + \sigma_1 p_1 + \cdots + \sigma_k p_k - \sigma_{k+1} p_{k+1}, \ldots, m - (\sigma_{k+1} p_{k+1} + \ldots + \sigma_{2k} p_{2k}) = n \end{aligned}$$

with $p_1, p_2, \ldots, p_{2k}$ distinct:

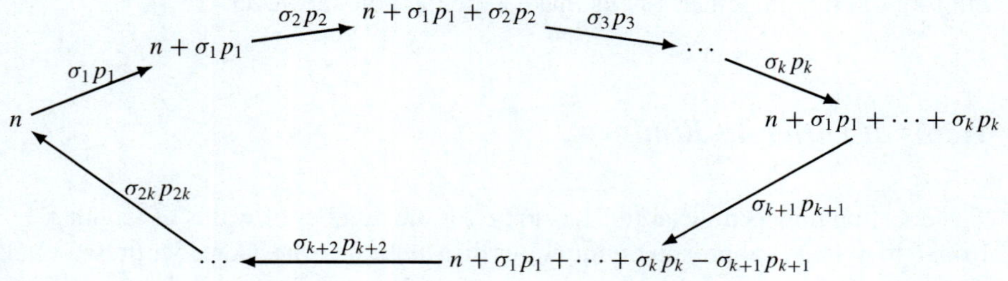

The contribution of these closed walks is almost exactly what we would obtain from the naïve model we were implicitly considering, viz., a random walk where each edge p_i is taken with probability $1/(\mathscr{L} p_i)$, and so we should have the same limiting distribution as in that model.

What about walks where some primes p_i do repeat? At least some of them may make a large contribution that is not there in our naïve model. For instance, consider walks of length $2k$ that retrace their steps, so that the $(n+1)$th step is the nth step backwards, the $(n+2)$th step is the $(n-1)$th step backwards, etc.:

$$n, n+\sigma_1 p_1, \ldots, n+\sigma_1 p_1+\cdots+\sigma_k p_k,$$
$$n+\sigma_1 p_1+\cdots+\sigma_{k-1} p_{k-1}, \ldots, n+\sigma_1 p_1, n,$$

with

$$p_1|n,\ p_2|n+\sigma_1 p_1,\ \ldots,\ p_k|n+\sigma_1 p_1+\ldots+\sigma_{k-1} p_{k-1},$$

$$p_k|n+\sigma_1 p_1+\ldots+\sigma_{k-1} p_{k-1}+\sigma_k p_k,\ \ldots,\ p_2|n+\sigma_1 p_1+\sigma_2 p_2,\ p_1|n+\sigma_1 p_1.$$

The second row of divisibility conditions here is obviously implied by the first row. Hence, again by the Chinese Remainder Theorem, the walk is valid for almost exactly $N/p_1 p_2 \cdots p_k$ elements $n \in \mathbf{N}$, rather than for $N/(p_1 p_2 \cdots p_k)^2$ elements. The contribution of such walks to

$$\sum_{n\in\mathbf{N}} \sum_{\substack{\forall 1\leq i\leq 2k: \sigma_i=\pm 1,\ p_i\in\mathbf{P} \\ \forall 1\leq i\leq 2k: p_i|n+\sigma_1 p_1+\ldots+\sigma_{i-1} p_{i-1} \\ \sigma_1 p_1+\ldots+\sigma_{2k} p_{2k}=0}} 1$$

(which is the interesting part of the variance we wrote down before) is clearly $N\mathscr{L}^k$. In order for it not to be of greater order than what one expects from the limiting distribution, we should have $N\mathscr{L}^k \ll N\mathscr{L}^{2k}/M$, where M, the width of the distribution, is, as we saw before, very roughly $\sqrt{k}H$. Thus, we need $k \gg (\log H)/(\log \mathscr{L})$.

There are of course other walks that make similar contributions; take, for instance,

$$n, n+p_1, n, n-p_3, n-p_3+p_4, n-p_3, n-p_3+p_6, n-p_3, n$$

for $k=3$. These are what we may call *trivial walks*, in the sense that a word is *trivial* when it reduces to the identity. It is tempting to say that their number is $2^k C_k$, where $C_k \leq 2^{2k}$ is the kth Catalan number (which, among other things, count the number of expressions containing k pairs of parentheses correctly matched: for example, ()(()) would correspond to the trivial walk above). In fact, the matter becomes more subtle because some primes may reappear without taking us one step further back to the origin of the walk; for instance, in the above, we might have $p_4 = p_1$, and that is a possibility that is not recorded by a simple pattern of correctly matched parentheses— yet it must be considered separately. Here again, we make a mental note.

It is, incidentally, no coincidence that, when we try to draw the trivial walk above, we produce a tree:

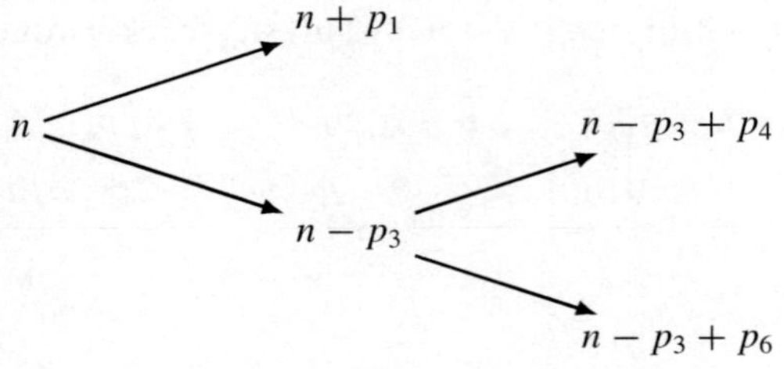

Any trivial walk gives us a tree (or rather a tree traversal) when drawn.

Now let us look at walks that fall into neither of the two classes just discussed; that is, walks where we do have some repeated primes $p_i = p_{i'}$ even after we reduce the walk. (When we say we *reduce* a walk, we mean an analogous procedure to that of reducing a word.) Then, far from being independent, the condition

$$p_i | n + \sigma_1 p_1 + \ldots + \sigma_{i-1} p_{i-1}$$

either implies or contradicts the condition

$$p_i = p_{i'} | n + \sigma_1 p_1 + \ldots + \sigma_{i'-1} p_{i'-1}$$

for given $\{(\sigma_i, p_i)\}_i$, depending on whether

$$p_i | \sigma_i p_i + \sigma_{i+1} p_{i+1} + \ldots + \sigma_{i'-1} p_{i'-1}.$$

We may draw another graph, emphasizing the two edges with the same label $\pm p_i$:

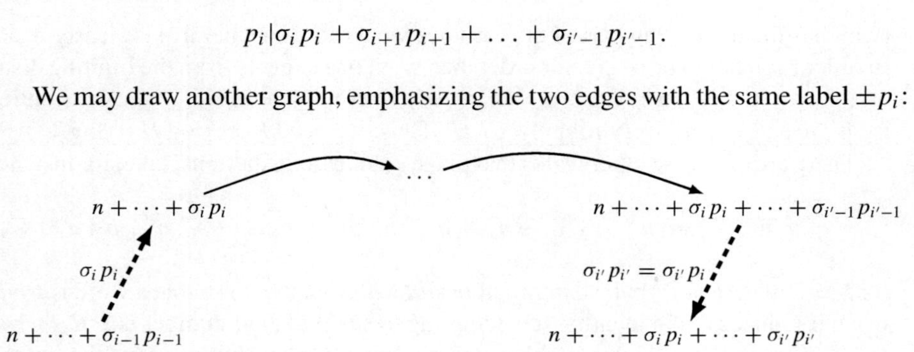

At this point, it becomes convenient to introduce the assumption that $p \geq H_0$ for all $p \in \mathbf{P}$. Then it is clear that, if $i' - i > 1$ and $p_j \neq p_i$ for all $i < j < i'$, the divisibility condition $p_i | \sigma_{i+1} p_{i+1} + \ldots + \sigma_{i'-1} p_{i'-1}$ may hold only for a proportion $\ll 1/H_0$ of all tuples $(p_{i+1}, \ldots, p_{i'-1})$.

So far, so good, except that it is not enough to save one factor of H_0, and indeed we should save a factor of at least M, which is roughly in the scale of H, not H_0. Obviously, for $\mathscr{L} \to \infty$ to hold, we need $H_0 = H^{o(1)}$, and so we need to save more than any constant number of factors of H_0.

We have seen three rather different cases. In general, we would like to have a division of all walks into three classes:

1. walks containing enough non-repeated primes p_i that their contribution is one would expect from the hoped-for limiting distribution;
2. rare walks, such as, for example, trivial walks;
3. walks for which there are many independent conditions of the form $p_i|n + \sigma_{i+1}p_{i+1} + \ldots + \sigma_{i'-1}p_{i'-1}$ as above.

Some initial thoughts on the third case. We should think a little about what we mean or should mean by "independent". It is clear that, if we have several conditions $p|L_j(p_1, \ldots, p_{2k})$, where the L_j are linear forms spanning a space of dimension D, then, in effect, we have only D distinct conditions. It is also clear that, while having several primes p_i divide the same quantity $L(p_1, \ldots, p_{2k})$ ought to give us more information than just knowing one prime divides it, that is true only up to a point: if $L(p_1, \ldots, p_{2k}) = 0$ (something that we expect to happen about $1/\sqrt{k}H$ of the time), then every condition of the form $p_i|L(p_1, \ldots, p_{2k})$ holds trivially.

It is also the case that we should be careful about which primes do the dividing. Say two indices i, i' are equivalent if $p_i = p_{i'}$. Choose your equivalence relation $\sim$, and paint the indices i in some equivalence classes blue, while painting the indices i in the other equivalence classes red. It is not hard to show, using a little geometry of numbers, that, if $p_{i_j}|L_j(p_1, \ldots, p_{2k})$ for some blue indices i_j and linear forms L_j, $j \in J$, and the space spanned by the forms L_j *considered as formal linear combinations on the variables x_i for i red* is D, we can gain a factor of at least H_0^D or so: the primes p_i for i red have to lie in a lattice of codimension D and index $\geq H_0^D$. A priori, however, it is not clear which primes we should color blue and which ones red.

We have, at any rate, arrived at what may be called the core of the problem—how to classify our walks in three classes as above, and how to estimate their contribution accordingly.

3 Graphs, Operators and Eigenvalues

It is now time to step back and take a fresh look at the problem. Matters will become clearer and simpler, but, as we will see, the core of the problem will remain.

We have been talking about walks. Now, walks are taken in a graph. Thinking about it for a moment, we see that we have been considering walks in the graph $V = \mathbf{N}$ as its set of vertices and $E = \{\{n, n+p\} : n, n+p \in \mathbf{N}, p \in \mathbf{P}, p|n\}$ as its set of edges. (In other words, we draw an edge between n and $n+p$ if and only if p divides n.) We also considered random walks in what we called the "naïve model"; those are walks in the weighted graph Γ' having $\mathbf{N}$ as its set of vertices and an edge of weight $1/p$ between any $n, n+p \in \mathbf{N}$ with $p \in \mathbf{P}$, regardless of whether $p|n$.

Adjacency, Eigenvalues and Expansion

Questions about walks in a graph Γ are closely tied to the *adjacency operator* Ad_Γ. This is a linear operator on functions $f : V \to \mathbb{C}$ taking f to a function $\mathrm{Ad}_\Gamma f : V \to \mathbb{C}$ defined as follows: for $v \in V$,

$$(\mathrm{Ad}_\Gamma f)(v) = \sum_{w:\{v,w\}\in E} f(w).$$

In other words, Ad_Γ replaces the value of f at a vertex v by the sum of its values $f(w)$ at the neighbors w of v. The connection with walks is not hard to see: for instance, it is very easy to show that, if $1_v : V \to \mathbb{C}$ is the function taking the value 1 at v and 0 elsewhere, then, for any $w \in V$ and any $k \geq 0$, $((\mathrm{Ad}_\Gamma)^k 1_v)(w)$ is the number of walks of length k from v to w.

The connection between Ad_Γ and our problem is very direct, in that it can be stated without reference to random walks. We want to show that

$$\sum_{n\in\mathbf{N}} \sum_{\sigma=\pm 1} \sum_{\substack{p\in\mathbf{P}:p|n\\ n+\sigma p\in\mathbf{N}}} \lambda(n)\lambda(n+\sigma p) = o(N\mathscr{L}).$$

That is exactly the same as showing that

$$\langle \lambda, \mathrm{Ad}_\Gamma \lambda\rangle = o(\mathscr{L}),$$

where $\langle\cdot,\cdot\rangle$ is the inner product defined by

$$\langle f, g\rangle = \frac{1}{N}\sum_{n\in\mathbf{N}} f(n)\overline{g(n)}$$

for $f, g : V \to \mathbb{C}$.

The behavior of random walks on a graph—in particular, the limit distribution of their endpoints—is closely related to the notion of *expansion*. A regular graph Γ (that is, a graph where every vertex has the same degree d) is said to be an *expander graph* with parameter $\epsilon > 0$ if, for every eigenvalue γ of Ad_Γ corresponding to an eigenfunction orthogonal to constant functions,

$$|\gamma| \leq (1-\epsilon)d.$$

(A few basic remarks may be in order. Since Γ is regular of degree d, a constant function on V is automatically an eigenfunction with eigenvalue d. Now, Ad_Γ is a symmetric operator, and thus it has full real spectrum: the space of all functions $V \to \mathbb{C}$ is spanned by a set of eigenfunctions of Ad_Γ, all orthogonal to each other; the corresponding eigenvalues are all real, and it is easy to see that all of them are at most d in absolute value.)

It is clear that we need something both stronger and weaker than expansion. (We cannot use the definition of expansion above "as is" anyhow, in that our graph Γ is not regular; its *average* degree is $\mathscr{L}$.) We need a stronger bound than what expansion provides: we want to show, not just that $|\langle\lambda, \mathrm{Ad}_\Gamma\lambda\rangle|$ is $\leq (1-\epsilon)\mathscr{L}$, but that it is $= o(\mathscr{L})$. There is nothing unrealistically strong here—in the strongest kind of expander graph (*Ramanujan graphs*), the absolute value of every eigenvalue is at most $2\sqrt{d-1}$.

At the same time, we cannot ask for $\langle f, \mathrm{Ad}_\Gamma f\rangle/|f|_2^2 = o(\mathscr{L})$ to hold for every f orthogonal to constant functions. Take $f = 1_{I_1} - 1_{I_2}$, where I_1, I_2 are two disjoint intervals of the same length $\geq 100\,H$, say. Then f is orthogonal to constant functions, but $(\mathrm{Ad}_\Gamma f)(n)$ is equal to $\omega(n)f(n)$, except possibly for those n that lie at a distance $\leq H$ of the edges of I_1 and I_2. Hence, $\langle f, \mathrm{Ad}_\Gamma f\rangle/|f|_2^2$ will be close to $\mathscr{L}$. It follows that Ad_Γ will have at least one eigenfunction orthogonal to constant functions and with eigenvalue close to $\mathscr{L}$; in fact, it will have many.

(This observation is related to the fact that endpoint of a short random walk on Γ *cannot* be approximately equidistributed, as it is in an expander graph: the edges of Γ are too short for that. The most we could hope for is what we were aiming for, namely, that the distribution of the endpoint converges to a nice distribution, centered at the starting point.)

We could aim to show that $\langle f, \mathrm{Ad}_\Gamma f\rangle/|f|_2^2$ is small whenever f is approximately orthogonal to approximately locally constant functions, say. Since the main result in [4] can be interpreted as the statement that λ is approximately orthogonal to such functions, we would then obtain what we wanted to prove for $f = \lambda$.

We will find it cleaner to proceed slightly differently. Recall our weighted graph Γ', which was meant as a naïve model for Γ. It has an adjacency operator $\mathrm{Ad}_{\Gamma'}$ as well, defined as before. (Since Γ' has weights $1/p$ on its edges, $(\mathrm{Ad}_{\Gamma'} f)(n) = \sum_{p\in\mathbf{P}}(f(n+p)+f(n-p))/p$.) It is not hard to show, using the techniques in [4], that

$$\langle\lambda, \mathrm{Ad}_{\Gamma'}\lambda\rangle = o(\mathscr{L}).$$

(In fact, what amounts to this statement has already been shown, in [6, Lemma 3.4–3.5]; the main ingredient is [5, Theorem 1.3], which applies and generalizes the main theorem in [4]. Their bound is a fair deal smaller than $o(\mathscr{L})$.) We define the operator

$$A = \mathrm{Ad}_\Gamma - \mathrm{Ad}_{\Gamma'}.$$

It will then be enough to show that

$$\langle\lambda, A\lambda\rangle = o(\mathscr{L}),$$

as then it will obviously follow that

$$\langle\lambda, \mathrm{Ad}_\Gamma\lambda\rangle = \langle\lambda, A\lambda\rangle + \langle\lambda, \mathrm{Ad}_{\Gamma'}\lambda\rangle = o(\mathscr{L}).$$

It would be natural to guess, and try to prove, that $\langle f, Af\rangle = o(\mathscr{L})$ for *all* $f : V \to \mathbb{C}$ with $|f|_2 = 1$, i.e., that all eigenvalues of A are $o(\mathscr{L})$.

We cannot hope for quite that much. The reason is simple. For any vertex n, $\langle A1_n, A1_n\rangle$ equals the sum of the squares of the weights of the edges $\{n, n'\}$ containing n. That sum equals

$$\sum_{\substack{p\in\mathbf{P}\\ p|n}} \left(1 - \frac{1}{p}\right)^2 + \sum_{\substack{p\in\mathbf{P}\\ p\nmid n}} \frac{1}{p^2},$$

which in turn is greater than $1/4$ times the number $\omega_{\mathbf{P}}(n)$ of divisors of n in $\mathbf{P}$. Thus, A has at least one eigenvalue greater than $\sqrt{\omega_{\mathbf{P}}(n)}/2$. Now, typically, n has about $\mathscr{L}$ divisors in $\mathbf{P}$, but some integers n have many more; for some rare n, in fact, $\omega_{\mathbf{P}}(n)$ will be greater than $\mathscr{L}^2$, and so there have to be eigenvalues of A greater than $\mathscr{L}$.

It is thus clear that we will have to exclude some integers, i.e., we will define our vertex set to be some subset $\mathscr{X} \subset \mathbf{N}$ with small complement. We will set ourselves the goal of proving that all of the eigenvalues of the operator $A|_{\mathscr{X}}$ defined by

$$(A|_{\mathscr{X}})(f) = (A(f|_{\mathscr{X}}))|_{\mathscr{X}}$$

are $o(\mathscr{L})$. (Here $f|_{\mathscr{X}}$ is just the function taking the value $f(n)$ for $n \in \mathscr{X}$ and 0 for $n \notin \mathscr{X}$.) Then, for $f = \lambda$, or for any other f with $|f|_\infty \leq 1$,

$$\langle f, Af\rangle = \langle f, (A|_{\mathscr{X}})f\rangle + O\left(\sum_{n\in\mathbf{N}\setminus\mathscr{X}} 2\left(\omega_{\mathbf{P}}(n) + \mathscr{L}\right)\right),$$

where, if $\mathbf{N} \setminus \mathscr{X}$ is small enough (as it will be), it will not be hard to show that the sum within $O(\cdot)$ is quite small. We will then be done: obviously $\langle f, (A|_{\mathscr{X}})f\rangle$ is bounded by the largest eigenvalue of $A|_{\mathscr{X}}$ times $|f|_2$ (which is $\leq |f|_\infty \leq 1$), and so we will indeed have $\langle f, Af\rangle = o(\mathscr{L})$.

We will in fact be able to prove something stronger: there is a subset $\mathscr{X} \subset \mathbf{N}$ with small complement such that all eigenvalues of $A|_{\mathscr{X}}$ are

$$O(\sqrt{\mathscr{L}}).$$

(This bound is optimal up to a constant factor.) This is our main theorem.

We hence obtain that

$$\langle \lambda, A\lambda\rangle = O(\sqrt{\mathscr{L}}). \tag{3}$$

From (3), we deduce the bound

$$\frac{1}{\log x} \sum_{n\leq x} \frac{\lambda(n)\lambda(n+1)}{n} = O\left(\frac{1}{\sqrt{\log\log x}}\right) \tag{4}$$

we stated at the beginning.

More generally, we get $\langle f, Af\rangle = O(\sqrt{\mathscr{L}})$ for any f with $|f|_\infty \leq 1$, or for that matter by any f with $|f|_4 \leq e^{100\mathscr{L}}$ and $|f|_2 \leq 1$. We obtain plenty of consequences besides (4).

Powers, Eigenvalues and Closed Walks

Now that we know what we want to prove, let us come up with a strategy.

There is a completely standard route towards bounds on eigenvalues of operators such as A (or $A|_{\mathscr{X}}$), relying on the fact that the trace is invariant under conjugation. Because of this invariance, the trace of a power A^{2k} is the same whether A is written taking a full family of orthogonal eigenvectors as a basis, or just taking the characteristic functions 1_n as our basis. Looking at matters the first way, we see that

$$\mathrm{Tr}(A|_{\mathscr{X}})^{2k} = \sum_{i=1}^{N} \lambda_i^{2k},$$

where $\lambda_1, \lambda_2, \ldots, \lambda_N$ are the eigenvalues corresponding to the basis made out of eigenvectors. Looking at matters the second way, we see that $\mathrm{Tr}(A|_{\mathscr{X}})^{2k} = N_{2k}$, where N_{2k} is the sum over all closed walks of length $2k$ of the products of the weights of the edges in each walk:

$$N_{2k} = \sum_{n\in\mathscr{X}} \sum_{\substack{p_1,\ldots,p_{2k}\in\mathbf{P} \\ \sigma_1,\ldots,\sigma_{2k}\in\{-1,1\} \\ \forall 1\leq i\leq 2k: n+\sigma_1 p_1+\ldots+\sigma_i p_i\in\mathscr{X} \\ \sigma_1 p_1+\ldots+\sigma_{2k} p_{2k}=0}} \prod_{i=1}^{2k} \left(1_{p_i | n+\sigma_1 p_1+\ldots+\sigma_{i-1} p_{i-1}} - \frac{1}{p_i}\right)$$

where we adopt the convention $1_{\text{true}} = 1$, $1_{\text{false}} = 0$.

Since all eigenvalues are real, it is clear that

$$\lambda_i^{2k} \leq N_{2k}$$

for every eigenvalue λ_i. Often, and also now, that inequality is not enough in itself for a good bound on λ_i. What is then often done is to show that every eigenvalue must have multiplicity $\geq M$, where M is some large quantity. Then it follows that, for every eigenvalue γ,

$$M\gamma^{2k} \leq N_{2k},$$

and so $|\gamma| \leq (N_{2k}/M)^{1/2k}$.

We do not quite have high multiplicity here (why would we?) but we have something that is almost as good: if there is one large eigenvalue, then there are many mutually orthogonal functions g_i of norm 1 with $\langle g_i, Ag_i\rangle$ large. Then we can bound

$\mathrm{Tr}A^{2k}$ from below, using these functions g_i (and some arbitrary functions orthogonal to them) as our basis, and, since $\mathrm{Tr}A^{2k}$ also equals N_{2k}, we can hope to obtain a contradiction with an upper bound on N_{2k}.

For simplicity, let us start by sketching a proof that, if $|\langle f, Af\rangle|$ is large ($\geq \rho\mathscr{L}$, say) for some f with $|f|_\infty \leq 1$, then there are many orthogonal functions g_i of norm 1 and $\langle g_i, Ag_i\rangle$ large (with "large" meaning $\geq \rho\mathscr{L}/2$, say). This weaker statement suffices for our original goal, since we may set f equal to the Liouville function λ.

Let $I_1, I_2, \ldots \subset \mathbb{N}$ be disjoint intervals of length $\geq 10\, H/\rho$ (say) covering $\mathbb{N}$. Edges starting at a vertex v in I_i end at another vertex in I_i, unless they are close to the edge. Hence, $\sum_i |\langle f|_{I_i}, A\left(f|_{I_i}\right)\rangle|$ is not much smaller than $|\langle f, Af\rangle|$, and then it follows easily that $\langle f|_{I_i}, A\left(f|_{I_i}\right)\rangle/|f|_{I_i}|_2^2$ must be large for many i. Thus, setting $g_i = f|_{I_i}/|f|_{I_i}|$ for these i, we obtain the desired statement.

To prove truly that A has no large eigenvalues, we should proceed as we just did, but assuming only that $|f|_2 \leq 1$, not that $|f|_\infty \leq 1$. The basic idea is the same, except that (a) pigeonholing is a little more delicate, (b) if f is almost entirely concentrated in a small subset of $\mathbf{N}$, then we can extract only a few mutually orthogonal functions g_i from it. Recall that we are anyhow restricting to a set $\mathscr{X} \subset \mathbf{N}$. A brief argument suffices to show that we can avoid the problem posed by (b) simply by making $\mathscr{X}$ a little smaller (essentially: deleting the support of such g_i, and then running through the entire procedure again), while keeping its complement $\mathbf{N} \setminus \mathscr{X}$ very small.

In any event: we obtain that, if, for some $X \subset \mathbf{N}$, $\mathrm{Tr}(A|_X)^{2k}$ is not too large (smaller than $(\rho\mathscr{L}/2)^{2k} N/H$ or so) then there is a subset $\mathscr{X} \subset X$ with $X \setminus \mathscr{X}$ small such that every eigenvalue of $A|_{\mathscr{X}}$ is small ($\leq \rho\mathscr{L}$). It thus remains to prove that $\mathrm{Tr}(A|_X)^{2k}$ is small for some $X \subset \mathbf{N}$ with small complement $\mathbf{N} \setminus X$.

Recall that $\mathrm{Tr}(A|_X)^{2k} = N_{2k}$ (with N_{2k} defined as above, except with X instead of $\mathscr{X}$) and that X should not include integers n with many more prime divisors in $\mathbf{P}$ than average. Our task is to bound N_{2k}.

A Brief Look Back

We have come full circle, or rather we have arrived twice at the same place. We started with a somewhat naïve approach that lead us to random walks. Then we took a step back and analyzed the situation in a way that turned out to be cleaner; for instance, the problem involving $\mathrm{err}_2^{1/k}$ vanished. As it happens, that cleaner approach took us to random walks again. Surely this is a good sign.

It is also encouraging to see signs that other people have thought in the same direction. The paper by Matomäki-Radziwiłł-Tao on sign patterns of λ and μ is based on the examination of a graph equivalent to Γ; what they show is, in essence, that Γ is almost everywhere locally connected. Being connected may be a much weaker property than expansion, but it is a step in the same direction. As for expansion itself, (Sect. 4) comments that "some sort of expander graph property" may hold for that graph (equivalent to Γ) "or [for] some closely related graph". He goes on to say:

> Unfortunately, we were unable to establish such an expansion property, as the edges in the graph [...] do not seem to be either random enough or structured enough for standard methods of establishing expansion to work.

And so we will set about to establish expansion by our methods (standard or not).

In any event, our initial discussion of random walks is still pertinent. Recall the plan with which we concluded, namely, to divide walks into three kinds: walks with few non-repeated primes, walks imposing many independent divisibility conditions, and rare walks. This plan will shape our approach to bounding N_{2k} in the next section.

4 Main Part of the Proof: Counting Closed Walks

Let us recapitulate. Let $\mathbf{N} = \{n \in \mathbb{Z} : N < n \leq 2N\}$. We have defined a linear operator A on functions $f : \mathbf{N} \to \mathbb{C}$ as the difference of the adjacency operators of two graphs Γ, Γ':

$$A = \mathrm{Ad}_\Gamma - \mathrm{Ad}_{\Gamma'}.$$

We would like to show that there is a subset $X \subset \mathbf{N}$ with small complement $\mathbf{N} \setminus X$ such that, for some k that is not too small, the trace

$$\mathrm{Tr}(A|_X)^{2k}$$

is substantially smaller than $\mathscr{L}^{2k} N$. Indeed, we will prove that there is a constant C such that

$$\mathrm{Tr}(A|_X)^{2k} \leq (C\mathscr{L})^k N,$$

where $\mathscr{L} = \sum_{p \in \mathbf{P}} 1/p$.

Incidentally, when we say "k not too small", we mean "k is larger than $\log H$ or so"; we already saw that we stand to lose a factor of $H^{1/k}$ when going from (a) a trace bound as above to (b) a bound on eigenvalues, which is our ultimate goal. If $k \gg \log H$, then $H^{1/k}$ is just a constant.

For comparison: if, as will be the case, we define X so that every $n \in X$ has at most $K\mathscr{L}$ prime factors, the trivial bound is

$$\mathrm{Tr}(A|_X)^{2k} \leq ((K+1)\mathscr{L})^{2k} N.$$

We also saw that $\mathrm{Tr}(A|_X)^{2k}$ can be expressed as a sum over closed walks, i.e., walks that end where they start:

$$\mathrm{Tr}(A|_X)^{2k} = \sum_{n \in X} \sum_{\substack{p_1, \ldots, p_{2k} \in \mathbf{P} \\ \sigma_1, \ldots, \sigma_{2k} \in \{-1,1\} \\ \forall 1 \leq i \leq 2k:\ n + \sigma_1 p_1 + \ldots + \sigma_i p_i \in X \\ \sigma_1 p_1 + \ldots + \sigma_{2k} p_{2k} = 0}} \prod_{i=1}^{2k} \left(1_{p_i | n + \sigma_1 p_1 + \ldots + \sigma_{i-1} p_{i-1}} - \frac{1}{p_i} \right).$$

Here the double sum just goes over closed walks of length $2k$ in the weighted graph $\Gamma - \Gamma'$, which has X as its set of vertices and an edge between any two vertices n, n' whose difference $n' - n$ is a prime p in our set of primes $\mathbf{P}$; the weight of the edge is then $1 - 1/p$ if $p|n$, and $-1/p$ otherwise. The contribution of a walk equals the product of the weights of its edges.

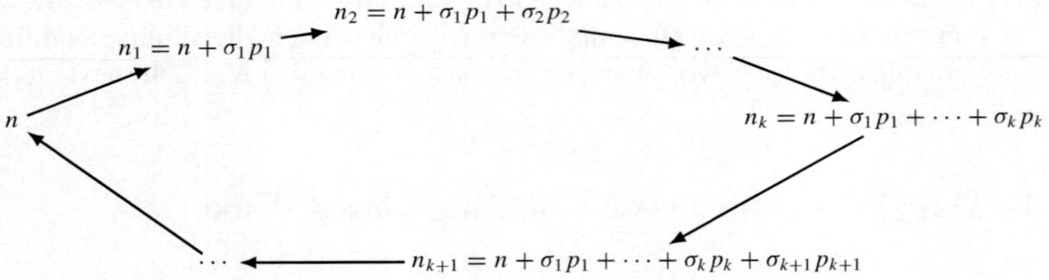

Cancellation

It might be nicer to work with an expression with yet simpler weights. First, though, let us see what gains we can get from cancellation. Let $p_1, \ldots, p_{2k} \in \mathbf{P}$ and $\sigma_1, \ldots, \sigma_{2k} \in \{-1, 1\}$ be given, and consider the total contribution of the paths they describe as n varies in X. Say there is a p_i that appears only once, i.e., $p_j \neq p_i$ for all $j \neq i$. The weight of the edge from $n_{i-1} = n + \sigma_1 p_1 + \ldots + \sigma_{i-1} p_{i-1}$ to $n_i = n + \sigma_1 p_1 + \ldots + \sigma_i p_i$ is $1 - 1/p$ if $p|n_{i-1}$ and $1/p$ otherwise. The weights of all the other edges depend on the congruence classes $n \bmod p_j$ for all $j \neq i$.

Suppose for a moment that $X = \mathbf{N}$. Then, for $\vec{p}, \vec{\sigma}$ fixed, and n in a given congruence class $n \bmod p_j$ for every $j \neq i$ (that is, n in a given congruence class $a + P\mathbb{Z}$ for $P = \prod_{p \in \{p_1, \ldots, p_{i-1}, p_{i+1}, \ldots, p_{2k}\}} p$, by the Chinese remainder theorem), the probability that p_i divides n_{i-1} is almost exactly $1/p_i$: the number of n in $\mathbf{N}$ in our congruence class mod P is $N/P + O^*(1)$ (that is, no less than $N/P - 1$ and no more than $N/P + 1$), and, for such n, again by the Chinese remainder theorem, $p|n_{i-1}$ if and only if n lies in a certain congruence class modulo $p_i \cdot P$; the number of n in $\mathbf{N}$ in that congruence class is $N/(p_i P) + O^*(1)$.

Hence, among all n in $\mathbf{N} \cap (a + P\mathbb{Z})$, a proportion almost exactly $1/p$ have a weight $1 - 1/p$ on the edge from n_{i-1} to n_i, and a proportion almost exactly $1 - 1/p$ have a weight $-1/p$ there instead. Since all other weights are fixed, we obtain practically total cancellation:

$$\frac{1}{p}\left(1 - \frac{1}{p}\right) - \left(1 - \frac{1}{p}\right)\frac{1}{p} = 0.$$

In other words, the contribution of paths where at least one p_i appears only once is practically nil. Hence, we can assume that, in our paths, every p_i appears at least twice among $p_1, p_2, \ldots, p_{2k}$.

Of course we do not actually want to set $X = \mathbf{N}$, and in fact we cannot, as we have already seen. If X is well distributed in arithmetic progressions, then we should still get cancellation but it will not be total—there will be an error term. Much of the pain here comes from the fact that we have to exclude numbers with too many prime factors (meaning: $> K\mathscr{L}$ prime factors). Suppose for simplicity that X is the set of all numbers in $\mathbf{N}$ with $\leq K\mathscr{L}$. Recall that all vertices n, $n_1 = n + \sigma_1 p_1$, $n_2 = n + \sigma_1 p_1 + \sigma_2 p_2 + \ldots$ have to be in X; in particular, $n_{i-1} \in X$. As a consequence, the likelihood that $p|n_{i-1}$ is slightly lower than $1/p$: if $n_{i-1} = pm$, then m is constrained to have $\leq K\mathscr{L} - 1$ prime factors, and it is slightly more difficult for m to satisfy that constraint than it is for an $n \in \mathbf{N}$ to have $\leq K\mathscr{L}$ prime factors. We do have cancellation, but it is not total, as it is for $X = \mathbf{N}$. The techniques involved in estimating how much cancellation we do have are standard within analytic number theory.

Later, we will also exclude some other integers from X, besides those having $> K\mathscr{L}$ prime factors. We will then need to show that the effect on cancellation is minor. Doing so will require some arguably new techniques; we will cross that bridge when we come to it.

To cut a long story short, the effect of cancellation will be, not that every p_i appears at least twice among $p_1, p_2, \ldots, p_{2k}$, but that the number of "singletons" (primes that appear only once) is small. More precisely, a path with m singletons will have to pay a penalty of a factor of $\mathscr{L}^{-m/2}$.

Shapes. Geometry of Numbers and Ranks

Let us see what we have. Write $\mathbf{k} = \{1, 2, \ldots, 2k\}$. Let $\mathbf{l}$ range among all subsets of $\mathbf{k}$. Here $\mathbf{l}$ will be our set of "lit" indices, corresponding to the set of indices i such that $p_i | n + \sigma_1 p_1 + \ldots + \sigma_{i-1} p_{i-1}$ in the above. Every "unlit" index i gives us a weight of $1/p_i$. We define an equivalence relation $\sim$ on $\mathbf{k}$ by letting $i \sim j$ if and only if $p_i = p_j$. Given an equivalence class $[i]$, we define $p_{[i]}$ to equal p_i for any (and hence every) $i \in [i]$. If an equivalence class $[i]$ is not completely unlit (that is, if $[i] \cap \mathbf{l} \neq \emptyset$), then it gives us a weight of $1/p_{[i]}$ (coming from $p_i | n + \sigma_1 p_1 + \ldots + \sigma_{i-1} p_{i-1}$ for some lit index $i \in [i]$). It is also the case that, when two indices $i \sim j$ are both lit, they impose the condition

$$p_{[i]} | \sigma_{i+1} p_{i+1} + \ldots + \sigma_j p_j,$$

coming from $p_i | n + \sigma_1 p_1 + \ldots + \sigma_i p_i$ and $p_i = p_j | n + \sigma_1 p_1 + \ldots + \sigma_j p_j$. Let us write β_i as shorthand for $\sigma_1 p_1 + \ldots + \sigma_i p_i$; then our condition becomes

$$p_{[i]} | \beta_j - \beta_i.$$

Given a walk $n, n + \sigma_1 p_1, n + \sigma_1 p_1 + \sigma_2 p_2, \ldots$, we define its *shape* to be $(\sim, \vec{\sigma})$, where $\sim$ is the equivalence relation it induces (as above). In fact, let us start with shapes, meaning pairs $(\sim, \vec{\sigma})$, where $\sim$ is an equivalence class on $\{1, 2, \ldots, k\}$ and $\vec{\sigma} \in \{-1, 1\}^{2k}$. For any given shape, we will bound the contribution of all walks of that shape. There will be some shapes for which we will not be successful; we will later treat walks of those shapes and show that their contribution is small in some other way.

To rephrase what we said just before: given $\mathbf{l} \subset \mathbf{k}$, the contribution of a shape $(\sim, \vec{\sigma})$ will be at most

$$\mathscr{L}^{-\frac{|\mathcal{S}(\sim)|}{2}} \sum_{\substack{\{p_{[i]}\}_{[i]\in\Pi},\, p_{[i]}\in\mathbf{P} \\ i_1\sim i_2 \wedge (i_1, i_2 \in \mathbf{l}) \Rightarrow p_{i_1} | \beta_{i_2} - \beta_{i_1}}} \prod_{i\notin\mathbf{l}} \frac{1}{p_{[i]}} \prod_{\substack{[i]\in\Pi \\ [i]\not\subset \mathbf{k}\setminus\mathbf{l}}} \frac{1}{p_{[i]}}, \tag{5}$$

where Π is the set of equivalence classes of $\sim$ and $\mathcal{S}(\sim)$ is the set of singletons of $\sim$, where a "singleton" is an equivalence class with exactly one element. (Here and later, we write $|S|$ for the number of elements of a set S.) It is not hard to see then that the contribution of a shape must be $O(\mathscr{L}^k)$. That bound will not do in general, however, as there are by far too many shapes; we should aim for a better bound than $O(\mathscr{L}^k)$ for most shapes.

What we have to do is, in essence, bound the number of solutions $(p_{[i]})_{[i]\in\Pi}$ to a system of divisibility conditions

$$p_{[i]} | \sigma_{i+1} p_{[i+1]} + \ldots + \sigma_j p_{[j]}. \tag{6}$$

It would be convenient if the divisors $p_{[i]}$ were all distinct from the primes in the sums being divided. Then we could apply directly the following Lemma, which is really grade-school-level geometry of numbers.

Lemma 1 *Let $\mathbf{M} = (b_{i,j})_{1\le i,j\le m}$ be a non-singular m-by-m matrix with integer entries. Assume $|b_{i,j}| \le C$ for all $1 \le i, j \le m$. Let $\vec{c} \in \mathbb{Z}^m$, and let $d_1, \ldots, d_m \ge D$, where $D \ge 1$. Let $N_1, \ldots, N_m$ be real numbers $\ge D$. Then the number of solutions $\vec{n} \in \mathbb{Z}^m$ to*

$$d_i | (\mathbf{M}\vec{n} + \vec{c})_i \quad \forall 1 \le i \le m$$

with $N_i \le n_i \le 2N_i$ is at most

$$\left(\frac{2\,Cm}{D}\right)^m \prod_{i=1}^m N_i.$$

The trivial bound is clearly $\prod_{i=1}^m (N_i + 1)$.

Proof Divide the box $\prod_{i=1}^m [N_i, 2N_i]$ into $\le \prod_{i=1}^m \left(\frac{N_i}{D} + 1\right) \le \left(\frac{2}{D}\right)^m \prod_{i=1}^m N_i$ m-dimensional boxes of side D. The image of such a box under the map $\vec{n} \mapsto \mathbf{M}\vec{n} + \vec{c}$ is contained in a box whose edges are open or half-open intervals of length CmD. Since $d_i \ge D$, that box contains at most $(Cm)^m$ solutions $\vec{m}$ to the equations $d_i | m_i$.

Of course, we *can* make our set of divisors and our set of variables disjoint: we can choose to color some equivalence classes $[i]$ blue and some other equivalence classes $[j]$ red, and consider only those divisibility relations (6) in which $[i]$ is colored blue. We fix $p_{[i]}$ for $[i]$ blue, and in fact for all non-red $[i]$, and treat $p_{[j]}$ with $[j]$ as our variables. We can then use the Lemma above to bound the number of values of $(p_{[j]})_{[j] \text{ red}}$ that satisfy our divisibility relations.

To be precise: let $x_{[j]}$ be a formal variable for each red equivalence class $[j]$. Define

$$v(i) = \sum_{j<i:\ [j] \text{ red}} \sigma_j x_{[j]}. \tag{7}$$

Let r be the dimension of the space spanned by the differences $v(i_2) - v(i_1)$ with $[i_1] = [i_2]$ blue. Then we can select r divisibility relations of the form (6) and r red equivalence classes such that the matrix consisting of a row $(\sum_{j\in[j]} \sigma_j)_{[j] \text{ red}}$ for each equivalence relation is non-singular. (We are just saying that a matrix of rank r has a non-singular r-by-r submatrix.) We can then apply our lemma.

After some book-keeping, we obtain a bound on our sum from (5), namely,

$$\sum_{\substack{\{p_{[i]}\}_{[i]\in\Pi},\ p_{[i]}\in\mathbf{P} \\ i_1\sim i_2 \wedge (i_1,i_2\in\mathbf{l}) \Rightarrow p_{i_1} | \beta_{i_2}-\beta_{i_1}}} \prod_{i\notin\mathbf{l}} \frac{1}{p_{[i]}} \prod_{\substack{[i]\in\Pi \\ [i]\not\subset\mathbf{k}\setminus\mathbf{l}}} \frac{1}{p_{[i]}} \leq \frac{1}{H_0^r} \left(\frac{4kr\log H}{\mathscr{L}\log 2}\right)^r \mathscr{L}^{|\Pi|}. \tag{8}$$

Here the important factor is $1/H_0^r$. We see that we "win" if r is at least somewhat large. The question is then how to choose which equivalence classes to color red or blue so as to make the rank r large.

Ranks and a New Graph. Sets with Large Boundary

To address this question, let us define a new graph. First, though, let us define the *reduction* of a shape $(\sim, \sigma)$. A shape clearly induces a word

$$w = x_{[1]}^{\sigma_1} x_{[2]}^{\sigma_2} \dots x_{[2k]}^{\sigma_{2k}}.$$

This word can be reduced (if it isn't already), and the resulting reduced word induces a "reduced shape" $(\sim', \sigma')$. If all representatives of an equivalence class of $(\sim, \sigma)$ disappear during the reduction, we color that class yellow. It is the non-yellow classes that we will color red or blue.

We define a graph $\mathscr{G}_{(\sim,\sigma)}$ to be an undirected graph having the non-yellow equivalence classes as its vertices, and an edge between two vertices v_1, v_2 if there are $i_1 \in v_1$, $i_2 \in v_2$ such that every equivalence class containing at least one index $j \in \{i_1 + 1, i_1 + 2, \dots, i_2 - 1\}$ is yellow.

(We define matters in this way, rather than simply reduce the word and join two vertices v_1, v_2 if there are $i_1 \in v_1$, $i_2 \in v_2$ such that $i_2 = i_1 + 1$, because reducing the word could create more singletons. At any rate, the idea is that, if there are only yellow indices between i_1 and i_2, then $v(i_1) = v(i_2)$, where $v(i)$ is defined as in (+). But we are getting ahead of ourselves.)

Let us see two examples. Let $k = 3$, and let $\sim$ have equivalence classes

$$\{\{1, 4\}, \{2, 5\}, \{3\}, \{6\}\},$$

with $\sigma \in \{-1, 1\}^{2k}$ arbitrary. Then the graph $\mathcal{G}_{(\sim,\sigma)}$ is

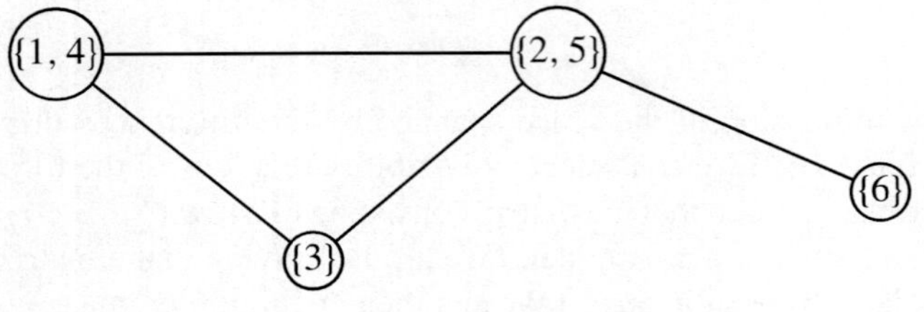

As for our second example, let $k = 5$, $\vec{\sigma} = (1, -1, 1, -1, 1, -1, 1, 1, 1, -1)$, and let $\sim$ have equivalence classes $\{\{1, 8\}, \{2, 9\}, \{3\}, \{4, 7\}, \{5, 6, 10\}\}$. Then the induced word is

$$w = x_{[1]}x_{[2]}^{-1}x_{[3]}x_{[4]}^{-1}x_{[5]}x_{[5]}^{-1}x_{[4]}x_{[1]}x_{[2]}x_{[5]}^{-1},$$

which has reduction

$$w' = x_{[1]}x_{[2]}^{-1}x_{[3]}x_{[1]}x_{[2]}x_{[5]}^{-1}.$$

Hence, the equivalence class $\{4, 7\}$ is colored yellow, and the graph $\mathcal{G}_{(\sim,\vec{\sigma})}$ is

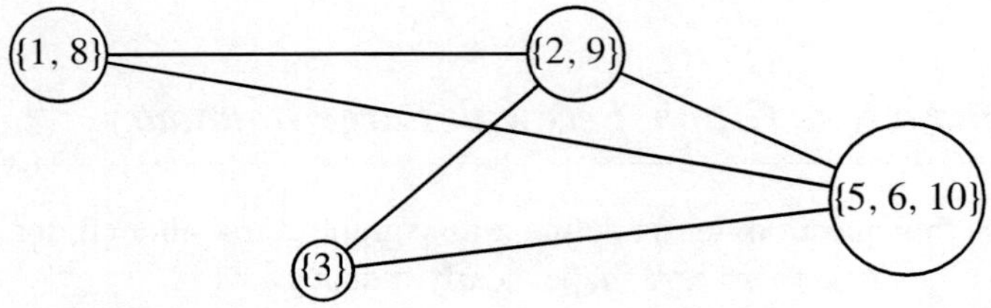

It is clear from the definition that $\mathcal{G}_{(\sim,\vec{\sigma})}$ is always connected: if $i_1 < i_2 < i_3 < \ldots$ are the indices in non-yellow equivalence classes, then there is an edge from $[i_1]$ to $[i_2]$, an edge from $[i_2]$ to $[i_3]$, etc.

Given a subset V' of the set of vertices V of a graph $\mathcal{G}$, write $\mathcal{G}|_{V'}$ for the restriction of $\mathcal{G}$ to V', i.e., the subgraph of $\mathcal{G}$ having V' as its set of vertices and set of all edges in $\mathcal{G}$ between elements of V' as its set of edges. What happens if we choose our coloring so that the restriction $\mathcal{G}|_{\mathbf{blue}}$ to the set of blue vertices (named **blue**) is connected?

Lemma 2 *Let $(\sim, \sigma)$ be a shape, and let $\mathcal{G} = \mathcal{G}_{(\sim,\vec{\sigma})}$ and $v(i)$ be as above. Color some non-yellow vertices red and some other non-yellow vertices blue, in such a way that, for* **blue** *the set of blue vertices, the restriction $\mathcal{G}|_{\mathbf{blue}}$ is connected.*

Then the space V spanned by the vectors

$$v(i_2) - v(i_1) \qquad \textit{with} \quad [i_1] = [i_2] \textbf{\textit{ blue}}$$

equals the space W spanned by all vectors

$$v(i_2) - v(i_1) \qquad \textit{with} \quad [i_1], [i_2] \textit{ both } \textbf{blue}.$$

The proof is an exercise, and its idea may be best made clear by an example. *Sketch of proof (or rather, a worked example)* Say we have three blue equivalence classes, corresponding to letters x, y, z in the induced word, and let them be disposed as follows:

$$\underbrace{\qquad\quad}\, x \,\underbrace{\qquad\qquad}\, z \,\underbrace{\quad}\, yx \,\underbrace{\quad\;}\, zy \,\underbrace{\quad}$$

Call the indices of the six letters we have written $i_1, i_2, \ldots, i_6$. Then the space V in the Lemma is the space spanned by

$$v_{i_4} - v_{i_1}, \; v_{i_6} - v_{i_3}, \; v_{i_5} - v_{i_2},$$

where the space W in the Lemma is the space spanned by all vectors

$$v_{i_r} - v_{i_s}, \qquad 1 \leq r, s \leq 6.$$

It is clear that $V \subset W$, but why is $W \subset V$? Why, say, is $v_{i_2} - v_{i_1}$ in V? Well, let us follow a path in $\mathscr{G}|_{\mathbf{blue}}$ going from x (the first blue letter, i.e., the letter at position i_1) to z (the second blue letter, i.e., the letter at position i_2): there is an edge from (the equivalence class labeled) x to (the equivalence class label-led) y, and an edge from y to z. So:

$v_{i_4} - v_{i_1}$ is in V	because i_1 and i_4 are both in the equivalence class x
v_{i_4} equals v_{i_3}	because i_4 and i_3 are adjacent (meaning there cannot be red indices between them; note that there is an edge from x to y precisely because i_4 and i_3 are adjacent)
$v_{i_6} - v_{i_3}$ is in V	because i_6 and i_3 are both in y
v_{i_5} equals v_{i_6}	because i_6 and i_5 are adjacent, as is again reflected in the fact that there is an edge from y to z,
$v_{i_2} - v_{i_6}$ is in V	because i_2 and i_6 are both in z

Hence, $v_{i_2} - v_{i_1} \in V$, as we have shown by following a path in $\mathscr{G}|_{\mathbf{blue}}$ from x to z. The same argument works in general for any two indices in blue equivalence classes.

Now we have to bound the rank of W from below. We first reduce our word; yellow letters disappear. The most optimistic expectation would be that the rank of W equal the number of gaps between blue "chunks" indicated by braces in our example from before:

$$\underbrace{\qquad\quad}\, x \,\underbrace{\qquad\qquad}\, z \,\underbrace{\quad}\, yx \,\underbrace{\quad\;}\, zy \,\underbrace{\quad}$$

(Chunks may have merged during reduction.) The number of gaps here is 4, considered cyclically (so that the first and last gap become one). The gaps correspond to $v(i_2) - v(i_1)$, $v(i_3) - v(i_2)$, $v(i_5) - v(i_4)$ and (lastly, or firstly) $v(i_1) - v(i_6)$.

Imagine for a moment that each red letter appeared only once. (From now on, all letters that are neither blue nor yellow will be colored red. In our reduced word, all letters that are not blue are red.) Then the optimistic expectation would hold: each of $v(i_2) - v(i_1)$, $v(i_5) - v(i_4)$, $\ldots$, $v(i_1) - v(i_6)$ would be a non-trivial formal linear combination of a non-zero number of symbols $x_{[j]}$, each appearing only once altogether, and so those combinations must all be linearly independent.

Of course, we cannot ensure that each red letter will appear only once, and in fact we are usually treating cases where most of them appear at least twice (i.e., singletons are rare). Let us see what we can do with a weaker assumption. What if we assume that each red letter appears at most κ times?

Let us see an easy linear-algebra lemma.

Lemma 3 *Let A be a matrix with n rows, satisfying:*

- *every row has at least one non-zero entry,*
- *no column has more than κ non-zero entries.*

Then the rank of A is $\geq n/\kappa$.

Proof We will construct a finite list S of columns, starting with the empty list. At each step, if there is a row i such that the ith entry of every column in S is 0, include at the end of S a column whose ith entry is non-zero. Stop if there is no such row.

When we stop, we must have $\kappa \cdot |S| \geq n$, as otherwise there would still be a row in which no element of S would have a non-zero entry. Since, for each column in S, there is a row in which that column has a non-zero entry and no previous column in S does, we see that the columns in S are linearly independent. Hence $\text{rank}(A) \geq |S| \geq n/\kappa$.

Now we see what to do: let A be a matrix with columns corresponding to red equivalence classes, and rows corresponding to gaps between blue chunks, with the entry $a_{i[j]}$ being the number of times the red letter corresponding to column $[j]$ appears in the gap corresponding to row i (counting appearances as $x_{[j]}^{-1}$ as negative appearances). Each column has no more than κ non-zero entries because, by assumption, each red equivalence class contains at most κ elements. We still need to show that no or few rows are full of zeros.

A row is full of zeros iff every letter x in the corresponding gap appears an equal number of times as x and as x^{-1} (i.e., every equivalence class has as many representatives i with $\sigma_i = 1$ as with $\sigma_i = -1$ within the gap). Let us call such gaps *invalid*.

There is a condition that limits how many such gaps there can be while at the same time ensuring that an equivalence class contains at most κ elements (or not quite, but something that is as good). Let $(\sim', \sigma')$ (of length $2k'$) be the reduction of the shape $(\sim, \sigma)$. Let us say that we see a *revenant* when there are indices i, i' such that (a) $i \sim i'$, and (b) there is a $j \nsim i$ with $i < j < i'$. (In other words, $x_{[i]}$ has come back after going away.) We say that there are κ *disjoint revenants* if there are

$$1 \leq i_1 < j_1 < i'_1 \leq i_2 < j_2 < i'_2 \leq \ldots \leq i_\kappa < j_\kappa < i'_\kappa \leq 2k'$$

with $i_j \sim' i'_j$ and $i_j \nsim' j_j$ for $1 \leq j \leq 2k$. Thus, for example, in

$$xxz\ldots x^{-1}yw\ldots yv\ldots y,$$

we see three disjoint revenants (with x at positions i_1 and i'_1, and y at positions i_2, $i'_2 = i_3$ and i'_3).

Let us impose the condition that there cannot be more than κ disjoint revenants in our walk. (We will be able to assume this condition by rigging the definition of X later.) Then it follows immediately that the appearances of a letter form at most κ contiguous blocks in the reduced word. Hence, a red letter cannot appear in more than κ gaps. We also see that there cannot be more than κ invalid gaps: a gap is a non-empty reduced subword, and, if a letter x appears in a reduced, non-trivial word as many times as x and as x^{-1}, either the pattern $x\ldots x^{-1}$ or the pattern $x^{-1}\ldots x$ appears in the word, with "…" standing for a non-empty subword consisting of letters that are not x. Thus, $> \kappa$ invalid gaps would give us $> \kappa$ revenants, all disjoint.

It then follows, by the easy linear-algebra lemma above, that

$$\dim(W) \geq \frac{s-\kappa}{\kappa} = \frac{s}{\kappa} - 1,$$

where s is the number of gaps.

The question is then: how do you choose which letters to color blue and which to color red so that the number s of gaps is large?

Spanning Trees and Boundaries

Let us first assume that there are no yellow letters in the non-reduced word, as that is a somewhat simpler case. Then the number of gaps equals the number of red letters x_i such that x_{i-1} is blue (or x_{2k} is blue, if $i = 1$). That number is bounded from below by

$$\frac{1}{2}|\partial\mathbf{blue}|,$$

where $\partial\mathbf{blue}$ is the set of all red equivalence classes $[i]$ such that there is a blue equivalence class $[j]$ connected to $[i]$ by an edge in $\mathscr{G} = \mathscr{G}_{\sim,\sigma}$ (meaning that $[j]$ contains an index j and $[i]$ contains an index i such that i and j are separated only by yellow letters; since there are no yellow letters, that means that $i = j+1$ or $i = j-1$ (or one of i, j is 1 and the other one is $2k$)).

The question is then how to choose the set **blue** of equivalence classes to be colored blue in such a way that $\partial\mathbf{blue}$ is large. Here **blue** can be any set of vertices such that $\mathscr{G}|_{\mathbf{blue}}$ is connected. So, in general: given a connected undirected graph $\mathscr{G}$, how do we choose a set **blue** of vertices so that $\mathscr{G}|_{\mathbf{blue}}$ is connected and ∂**blue** is large?

A *spanning tree* of a graph $\mathscr{G} = (V, E)$ is a subgraph (V, E') (where $E' \subset E$) that is a tree (i.e., has no cycles) and has the same set of vertices V as $\mathscr{G}$. Given a spanning tree of $\mathscr{G}$, we can define **blue** to be the set of internal nodes of $\mathscr{G}$, that is, the set of vertices that are not leaves. Then **blue** is connected, and ∂**blue** equals the set of leaves. The question is then: is there a spanning tree of $\mathscr{G}$ with many leaves?

Here there is a result from graph theory that we can just buy off the shelf.

Proposition 1 (Kleiman-West, 1991; see also Storer, 1981, Payan-Tchuente-Xuong, 1984, and Griggs-Kleitman-Shastri, 1989) *Let $\mathscr{G}$ be a connected graph with n vertices, all of degree ≥ 3. Then $\mathscr{G}$ has a spanning tree with $\geq n/4 + 2$ leaves.*

Using this Proposition, we prove:

Corollary 1 *Let $\mathscr{G}$ be a connected graph such that $\geq n$ of its vertices have degree ≥ 3. Then $\mathscr{G}$ has a spanning tree with $\geq n/4 + 2$ leaves.*

We omit the proof of the corollary, as it consists just of less than a page of casework and standard tricks. Alternatively, we can prove it from scratch in about a page by modifying Kleiman and West's proof.

(It is clear that some conditions on the degrees, as here, are necessary; a spanning tree of a cyclic graph (every one of whose vertices has degree 2) has no leaves.)

Before we go on to see what do we do with shapes $(\sim, \vec{\sigma})$ such that $\mathscr{G}_{(\sim,\vec{\sigma})}$ does *not* have many vertices with degree ≥ 3, let us remove the assumption that there are no yellow letters.

So, let us go back to counting gaps between blue chunks. For any two distinct non-yellow equivalence classes $[i]$, $[j]$, let us draw an arrow from $[i]$ to $[j]$ if there are representatives $i \in [i]$, $j \in [j]$ that survive in the reduced word, and such as that all letters between i and j disappear during reduction. (If $j < i$, then "between" is to be understood cyclically, i.e., the letters between i and j are those coming after i or before j.) We draw each arrow only once, that is, we do not draw multiple arrows.

For instance, in our example ($w = x_{[1]} x_{[2]}^{-1} x_{[3]} x_{[4]}^{-1} x_{[5]} x_{[5]}^{-1} x_{[4]} x_{[1]} x_{[2]} x_{[5]}^{-1}$ from before,

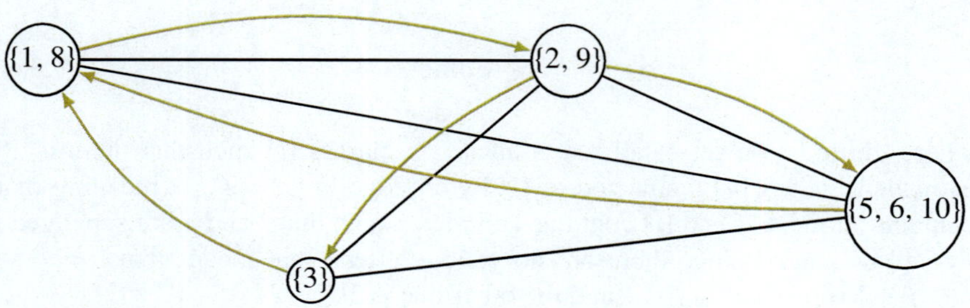

It is obvious that every vertex has an in-degree of at least 1.

For S a set of vertices, define the *out-boundary* $\vec{\partial} S$ to be the set of all vertices v not in S such that there is an arrow going from some element of S to v. Then, whether or not there are yellow letters, the number of red gaps $\underbrace{\quad}$ in the reduced word is at least $|\vec{\partial}\mathbf{blue}|$.

Lemma 4 *Let G be a directed graph such that every vertex has positive in-degree. Let S be a subset of the set vertices of G. Then there is a subset $S' \subset S$ with $|S'| \geq |S|/3$ such that, for every $v \in S'$, there is an arrow from some vertex not in S' to v.*

Proof The first step is to remove arrows until the in-degree of every vertex is exactly 1. Then G is a union of disjoint cycles. If all vertices in a cycle are contained in S, we number its vertices in order, starting at an arbitrary vertex, and include in S' the second, fourth, etc. elements. If no vertices in a cycle are in S, we ignore that cycle. If some but not all vertices in a cycle are in S, the vertices that are in S fall into disjoint subsets of the form $\{v_1, \dots v_r\}$, where there is an arrow from some v not in S to v_1, and an arrow from v_i to v_{i+1} for $1 \leq i \leq r-1$; then we include $v_1, v_3, \dots$ in S'.

We let S be the set of leaves of our spanning tree, and define **red** to be the set S' given by the Lemma; **blue** is the set of all other non-yellow equivalence classes. Then the number of gaps is $\geq (n/4+2)/3$, where n is the number of vertices of degree ≥ 3 in $\mathscr{G}_{(\sim,\vec{\sigma})}$. Hence, by our work up to now,

$$\dim(W) \geq \frac{1}{\kappa}\frac{\frac{n}{4}+2}{3} - 1 \geq \frac{n}{12\kappa} - 1,$$

and so, if n is even modestly large, we win by a large margin: we obtain a factor nearly as small as $1/H_0^{\frac{n}{12\kappa}}$ in (8).

Note. Had we been a little more careful, we would have obtained a bound of $\dim(W) \geq \frac{n}{50\log\kappa} - \frac{\kappa}{2}$ or so. This improvement—which involves drawing, and considering, multiple arrows—would affect mainly the allowable range of H_0 in the end. We will remind ourselves of the matter later.

Shapes with Low Freedom. Writer-Reader Arguments

The question now is what to do with walks of shapes $(\sim, \vec{\sigma})$ for which $\mathscr{G}_{(\sim,\vec{\sigma})}$ does not have many vertices of degree ≥ 3.

Let us first give an argument that is sufficient when the word given by our walk is already reduced; we will later supplement it with an additional argument that takes care of the reduction. Let $\mathbf{n} \subset \{1, 2, \dots, 2k\}$ be the set of indices that survive the reduction. It is enough to define an equivalence relation $\sim$ on $\mathbf{n}$ to define the graph $\mathscr{G}_\sim = \mathscr{G}_{\sim,\sigma}$ we have been considering. (We do not need to specify $\vec{\sigma}$, as its only role was to help determine which letters are yellow.) Assume that $\mathscr{G}_\sim$ has $\leq \nu$ vertices of degree ≥ 3. Let κ be, as usual, an upper bound on the number of disjoint revenants; in particular, for any equivalence class $[i]$, there are at most κ elements $i \in [i]$ such that the following element of **n** is not in $[i]$. We claim that the number of equivalence classes $\sim$ on **n** satisfying these two constraints (given by ν and κ) is

$$\leq 5^{|\mathbf{n}|}(2k)^{(\kappa-1)\nu+2}.$$

We will prove this bound by showing that we can determine an equivalence class of this kind by describing it by a string $\vec{s}$ on five letters with indices in **n**, together with some additional information at each of at most $(\kappa-1)\nu+2$ indices. The idea is that, if an index lies in an equivalence class that is a vertex of degree 1 or 2 in $\mathscr{G}_\sim$, then there are very few possibilities for the equivalence classes on which the index just thereafter may lie, namely, 1 or 2 possibilities.

We let the index i go through **n** from left to right. If $[i]$ is in an equivalence class we have not seen before, we let $s_i = *$. Assume otherwise. Let i_- be the element of **n** immediately preceding i. If $[i] = [i_-]$, let $s_i = 0$. If $[i_-]$ is a vertex of degree ≤ 2 and i is in an equivalence class that we have already seen next to $[i_-]$ (that is, just before or just after $[i_-]$ in **n**), then we let $s_i = 1$ or $s_i = 2$ depending on which one of those ≤ 2 equivalence classes we mean (the first one or the second one to appear). In all remaining cases, we let $s_i = \cdot$, and specify our equivalence class explicitly, by giving an index $j < i$ in the same equivalence class.

Let us give an example. Let $k = 84$, $\mathbf{n} = \{1, 2, \ldots, 2k\}$. Let our equivalence classes be

$$\{1, 7, 15\}, \{2, 16\}, \{3, 4, 5, 11\}, \{6, 10, 12, 14\}, \{8\}, \{9, 13\}.$$

Then $s_1 = s_2 = s_3 = s_6 = s_8 = s_9 = *$ and $s_4 = s_5 = 0$. The vertices of degree 3 are $[1]$ and $[6]$; all other vertices are of degree 2. Hence, $s_{16} = \cdot$ (since 16 follows 15, which is in $[1]$) and $s_7 = s_{11} = s_{13} = s_{15} = \cdot$ (since these indices follow 6, 10, 12, 14, which are in $[6]$). Since $3 \sim 4 \sim 5$, we let $s_4 = s_5 = 0$. It remains to consider $i = 10, 12, 14$. In the case $i = 10$, we see that $[9]$ has degree 2, but, when we come to 10, we realize that no element of $[10]$ has been seen next to an element of $[9]$ before: 8 is next to 9, but $8 \notin [10]$. Hence, we let $s_{10} = \cdot$. In the case $i = 12$, we see that $[12]$ has been seen next to $[11]$ before: $5 \in [11]$ and $6 \in [12]$. Since $[12]$ was the second equivalence class other than $[11]$ to appear next to $[11]$ (the first one was $[2]$: $2 \in [2]$, $3 \in [11]$), we write $s_{12} = 2$. The situation for $i = 14$ is analogous in that $[14]$ appeared next to $[13]$ before: $9 \in [13]$, $10 \in [14]$, and so, since $8 \notin [13], [14]$, $s_{14} = 2$.

In summary,

$$\vec{s} = ***00*\cdot**\cdot\cdot 2\cdot 2\cdot\cdot,$$

and, in addition to writing $\vec{s}$, we specify the equivalence classes of the indices i with $s_i = .$ explicitly ($[1]$ for $i = 7, 154$, $[6]$ for $i = 10$, $[3]$ for $i = 11$, $[9]$ for $i = 134$, $[2]$ for $i = 16$).

A reader can now reconstruct our equivalence classes by reading $\vec{s}$ from left to right, given that additional information. (Try it!) We should now count the number of dots $\cdot$, since that equals the number of times we have to give additional information.

For a class $[i']$ that is a vertex of degree ≤ 2, it can happen at most once (that is, for at most one element i' of $[i']$) that $s_i \neq 0, 1, 2$ for the index i in $\mathbf{n}$ right after i', unless $1 \in [i']$, in which case it can happen twice. (Someone who already has a neighbor and will end up with ≤ 2 neighbors in total can meet a new neighbor at most once.) For $[i']$ a vertex of arbitrary degree, it can happen at most κ times that $s_i \neq 0$. Hence, writing $n_{\leq 2}$ for the number of vertices of degree ≤ 2 and $n_{\geq 3}$ for the number of vertices of degree ≥ 3, we see that the total number of indices $i \in \mathbf{n}$ with $s_i \in \{*, .\}$ is at most $\kappa n_{\geq 3} + n_{\leq 2} + 1 + 1$, where the last $+1$ comes from the first index i in $\mathbf{n}$. The number of indices i with $s_i = *$ equals the number of classes, i.e., $n_{\leq 2} + n_{\geq 3}$. Hence, the number of indices i with $s_i = .$ is

$$\leq \kappa n_{\geq 3} + n_{\leq 2} + 2 - (n_{\geq 3} + n_{\leq 2}) = (\kappa - 1)n_{\geq 3} + 2 \leq (\kappa - 1)\nu + 2.$$

Each equivalence class contributes a factor of at most $\mathscr{L} = \sum_{p\in\mathbf{P}} \frac{1}{p}$ to our total in (5); singletons (equivalence classes with one element each) actually contribute $\sqrt{\mathscr{L}}$, because of the factor of $\mathscr{L}^{-\frac{|S(\sim)|}{2}}$. Recall that we are saving a factor of almost $H_0^{\frac{\nu}{12\kappa}-1}$ through (8) (let us say $H_0^{\nu/24\kappa}$, to be safe). Thus, forgetting for a moment about the yellow equivalence classes, we conclude that the contribution to (8) of the equivalence relations $\sim$ such that $G_\sim$ has ν vertices of degree ≥ 3 is

$$\ll 4^{2k} 5^{|\mathbf{n}|} (2k)^2 \left(\frac{(2k)^{\kappa-1}}{H_0^{1/24\kappa}} \right)^\nu \mathscr{L}^k,$$

where the factor of 4^{2k} is there because we also have to specify $\vec{\sigma} \in \{-1, 1\}^{\{1,\ldots,2k\}}$ and $\mathbf{l}, \mathbf{n} \subset \{1, 2, \ldots, 2k\}$. Provided that we set our parameters so that $H_0^{1/24\kappa} \geq 2(2k)^{\kappa-1}$ (and it turns out that we may do so, provided that $\log H_0$ is larger than $(\log H)^{2/3+\epsilon}$—or rather, larger than $(\log H)^{1/2+\epsilon}$, if we make the improvement through multiple arrows we mentioned a little while ago), we are done; we have a bound of size

$$\ll \mathscr{L}^k \sum_{\nu=1}^{\infty} 2^{-\nu} \ll \mathscr{L}^k,$$

which is what we wanted all along.

But wait! What about the part of the word that disappears during reduction? It is partly described by a string of matched parentheses: for example, $xx^{-1}x^{-1}yy^{-1}x$ gives us ()(()). (We also have to specify the exponents σ_i separately.) The equivalence class of the index of a closing parenthesis is the same as that of the index of the matching opening parenthesis. Thus, we need only worry about specifying the equivalence classes of the opening parentheses. There are $k - |\mathbf{n}|/2$ of them.

A naive approach would be to describe each such equivalence class $[i]$ by specifying the first index i in it each time it occurs (except for the first time). The cost of that approach could be about as large as $k^{k-|\mathbf{n}|/2}$, which is much too large. It would seem we are in a pickle. Indeed, we know we would have to be in a pickle, if we were not using the fact that we are not working in all of $\mathbf{N}$, but in a subset $X \subset \mathbf{N}$ all of whose elements have $\leq K\mathscr{L}$ divisors in $\mathbf{P}$. (If we worked in all of $\mathbf{N}$, even trivial walks, which are entirely yellow, would pose an insurmountable problem.) However, how can we use X, or the bound $\leq K\mathscr{L}$, by this point?

The point is that we need not consider all possible $(p_{[i]})$ in (5), but only those tuples that can possibly arise in a walk

$$n, n + \sigma_1 p_1, n + \sigma_1 p_1 + \sigma_2 p_2, \ldots, n + \sigma_1 p_1 + \sigma_2 p_2 + \cdots + \sigma_{2k} p_{2k} = n$$

all of whose nodes are in X. Now, if a prime p_j has appeared before as p_i (i.e., $i < j$ and $i \sim j$) and both i and j are "lit", that is $i, j \in \mathbf{l}$, then, as we know, $\sigma_i p_i + \ldots + \sigma_{j-1} p_{j-1}$ must be divisible by p_i. (Indices that are not lit do not pose a problem, due to the factors of the form $1/p$ that they contribute.) What is more: if $i \in \mathbf{l}$, $i < j$ with $p_i | \sigma_i p_i + \ldots + \sigma_{j-1} p_{j-1}$, then $n + \sigma_1 p_1 + \ldots + \sigma_{j-1} p_{j-1}$ is *forced* to be divisible by p_i (because $n + \sigma_1 p_1 + \ldots + \sigma_{i-1} p_{i-1}$ is divisible by p_i). Now, $n + \sigma_1 p_1 + \ldots + \sigma_{j-1} p_{j-1}$ has $\leq K\mathscr{L}$ divisors. Hence, given j, there are at most $K\mathscr{L}$ distinct equivalence classes $[i]$ having at least one representative $i < j$, $i \in \mathbf{l}$ such that $p_i | \sigma_i p_i + \ldots + \sigma_{j-1} p_{j-1}$. This is a property where n no longer appears.

Now, as we describe $\sim$ to our reader, when we come to an index of the one kind that remains problematic—disappearing in the reduction, corresponding to an open parenthesis, in an equivalence class that has been seen before—we need only specify an equivalence class *among those* $\leq K\mathscr{L}$ *equivalence classes that have at least one representative* $i < j$, $i \in \mathbf{l}$ *such that* $p_i | \sigma_i p_i + \ldots + \sigma_{j-1} p_{j-1}$. The reader can figure out which one those are, as that is a property given solely by $p_1, \ldots, p_{j-1}$ and $\sigma_1, \ldots, \sigma_{j-1}$. We can give them numbers 1 to $\lfloor K\mathscr{L} \rfloor$ by order of first appearance, and communicate to the reader the equivalence class we want by its number, rather than by an index. Thus we incur only in a factor of $K\mathscr{L}$, not $2k$.

In the end, we obtain a total contribution of

$$O((K\mathscr{L})^k),$$

which is what we wanted. In other words,

$$\mathrm{Tr}(A|_X)^{2k} \leq O(K\mathscr{L})^k N,$$

Q.E.D.

Incidentally, in earlier drafts of the paper, we did not have a "writer" and a "reader", but a mahout and an elephant:

They were unfortunately censored by my coauthor. As this is my exposition, here they are. The picture might be clearer now—the elephant-reader has no idea of n, or of our grand strategy, but it is an intelligent animal that can follow instructions and is endowed with a flawless memory (and the ability to test for divisibility, apparently).

5 Conclusions

Main Theorem *Let the operator A be as before, with $\mathbf{N} = \{N+1, \dots, 2N\}$ and $H_0, H, N \geq 1$ such that $H_0 \leq H$ and $\log H_0 \geq (\log H)^{1/2} (\log\log H)^2$. Let $\mathbf{P} \subset [H_0, H]$ be a set of primes such that $\mathscr{L} = \sum_{p\in\mathbf{P}} 1/p \geq e$ and $\log H \leq \sqrt{\frac{\log N}{\mathscr{L}}}$.*

Then, for any $1 \leq K \leq \frac{\log N}{\mathscr{L}(\log H)^2}$, there is a subset $\mathscr{X} \subset \mathbf{N}$ with $|\mathbf{N} \setminus \mathscr{X}| \ll N e^{-K\mathscr{L}\log K} + N/\sqrt{H_0}$ such that every eigenvalue of $A|_{\mathscr{X}}$ is

$$O\left(\sqrt{K\mathscr{L}}\right),$$

where the implied constants are absolute.

We have sketched a full proof, leaving out one, or rather two, passages—namely, the proof that we can take out from X two kinds of integers, and still keep X well-distributed enough in arithmetic progressions for cancellation to happen when we have too many lone primes. As we have said before, those two kinds of integers are: (a) integers n with $\geq K\mathscr{L}$ divisors, (b) integers n that could give rise to too many disjoint revenants. Here (b) sounds a little vague, but, if we simply take out from X the set Y_ℓ of those integers n for which there can be a "premature revenant", meaning that there exist $p \in \mathbf{P}$, $p_1, \dots, p_l \in \mathbf{P}$ with $p_i \neq p$ and $\sigma \in \{-1, 1\}^l$, $l \leq \ell$, such that

$$p|n,\ p_1|n,\ p_2|n+\sigma_1 p_1, \dots, p_l|n+\sigma_1 p_1 + \dots + \sigma_{l-1} p_{l-1},\ p|n+\sigma_1 p_1 + \dots \sigma_l p_l,$$

then we have ensured that there cannot be more than $2k/\ell$ disjoint revenants. (We have not really forgotten about the possibility that some intermediary indices may not be lit—those are taken care of by a different argument.) It is actually not hard to show that Y_ℓ is a fairly small set; what takes work is showing that it is well-distributed.

What we did was develop a new tool—a combinatorial sieve for conditions involving composite moduli. While it is somewhat technical may be interesting in that it will probably be useful for attacking other problems. Let us leave it to the appendix.

The main theorem has several immediate corollaries. First of all, we obtain what we set as our original goal.

Corollary 2 *For any $e < w \leq x$ such that $w \to \infty$ as $x \to \infty$,*

$$\frac{1}{\log w} \sum_{\frac{x}{w} \leq n \leq x} \frac{\lambda(n)\lambda(n+1)}{n} = O\left(\frac{1}{\sqrt{\log \log w}}\right).$$

We can also obtain substantially sharper results. A case in point: we can prove that $\lambda(n+1)$ averages to zero (with weight $1/n$ as above, or "at almost all scales") over integers $\leq N$ having exactly k prime factors, where k is a popular number of prime factors to have (e.g., $\lfloor \log \log N \rfloor$, or $\lfloor \log \log N \rfloor + 2021$). To see more such corollaries, look at the actual paper, or derive your own!

Subset of Acknowledgments. Bonus Track

I am grateful to many people—please read the full acknowledgments in the paper. Here I would like to thank two subsets in particular—(a) postdocs and students in Göttingen who patiently attended my online lectures during the first year of the COVID pandemic, as the proof was finally gelling, (b) inhabitants of MathOverflow. In (b), one can find, for example, Fedor Petrov, who pointed us towards Kleitman-West, besides answering other questions, but you can also find some users who chose to remain anonymous. Among them was user "BS.", who explained how one of my questions about ranks was related to topology. That relation has gone well under the surface in the current version, so let us discuss it here, for our own edification.

Consider a word w of a special kind—a word w where every letter $x_1, \ldots, x_k$ appears twice, once as x_i, once as x_i^{-1}. For $1 \leq i, j \leq k$, let $m_{i,j}$ equal 1 if either (a) x_i appears before x_i^{-1} , and x_j appears between them, but x_j^{-1} does not appear between them, or (b) x_i^{-1} appears before x_i, and x_j^{-1} appears between them, but x_j does not. Let $m_{i,j} = -1$ if either (a) or (b) is true with x_j and x_j^{-1} switched. Let $m_{i,j} = 0$ otherwise. Then the k-by-k matrix $M = (m_{i,j})$ is skew-symmetric. As people in MathOverflow kindly showed me (apparently my education in linear algebra left something to be desired…), if a skew-symmetric matrix M has rank r, then it has a minor with disjoint row and column index sets and rank $\geq r/2$. Since I was interested precisely in constructing such a minor with high rank (I and J giving us what we called "blue" and "red" vertices in the above), it made sense that I would want to know what the rank r of M might be. In particular, when is M non-singular?

What BS. showed me is that one can construct a surface S with handles corresponding to the word w in a natural way. (Apparently this construction is standard, but it was completely unknown to me.) For instance, for $w = x_1 x_2 x_1^{-1} x_2^{-1} x_3 x_3^{-1}$, the surface S looks as follows:

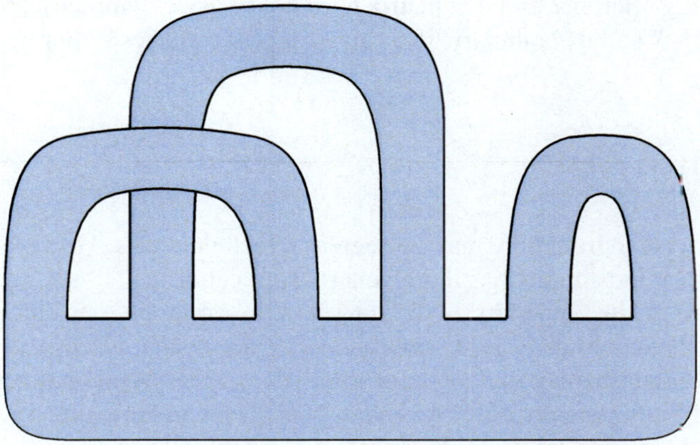

The matrix M then corresponds to the intersection form of this surface. This form is defined as an antisymmetric inner product on $H_1(S, \mathbb{Z})$, counting the number of intersections (with orientation) of two closed paths in the way you may expect. For instance, in the following, $\langle z_1, z_2 \rangle = -1$, whereas $\langle z_1, z_3 \rangle = \langle z_2, z_3 \rangle = 0$:

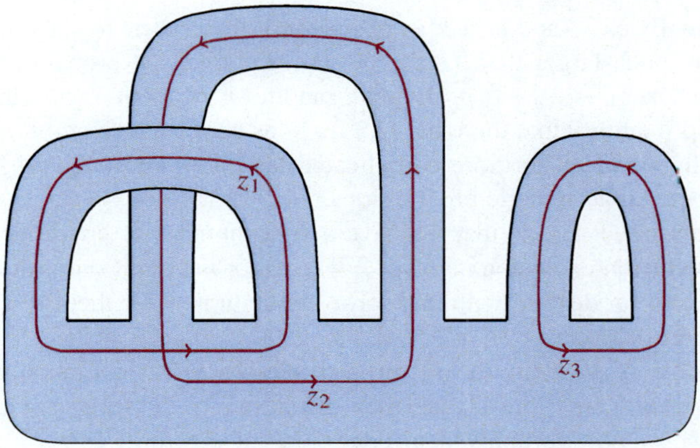

Say S has genus g and $b \geq 1$ boundary components. Then, for S_g the surface of genus g without boundary, there is an embedding $S \hookrightarrow S_g$ preserving the intersection form, with $H^1(S) \to H^1(S_g)$ having kernel of rank $b - 1$. The intersection form on $H^1(S_g)$ is non-singular. Hence, M has corank $b - 1$. In particular, M is non-singular iff $b = 1$, i.e., iff its boundary is connected.

It is an exercise to show that b equals the number of cycles in the permutation $i \mapsto \sigma(i) + 1 \bmod 2k$, where σ is the permutation of $\{1, 2, \ldots, 2k\}$ switching x_i and x_i^{-1} in w for every $1 \leq i \leq k$.

I have no idea of how to define a surface S like the above for a word w of general form—the natural generalization of M is the matrix corresponding to the system (6) of divisibility relations, and that matrix need not be skew-symmetric, or even square. However, in S and its boundary, you can already see shades of our graph $\mathscr{G}_{(\sigma,\sim)}$.

Appendix 1: Sieves

What we must address now may be seen as a technical task. However, the way we will address it most likely has more general applicability.

Our task in this appendix is to show how to exclude from our set $X \subset \mathbf{N}$ all integers n that could give rise to *premature revenants*, that is, edge lengths $p_i = p'_i$ with $i' - i$ small such that $p_j \neq p_i$ for some $i < j < i'$. (Without this last condition, we would be counting not only "revenants" but also mere repetitions.) As we already commented, it is enough to exclude the set Y_ℓ of all integers n such that there exist $p \in \mathbf{P}$, $p_1, \ldots, p_l \in \mathbf{P}$ with $p_i \neq p$ and $\sigma \in \{-1, 1\}^l$, $l \leq \ell$, for which

$$\begin{aligned} &p|n,\ p_1|n,\ p_2|n+\sigma_1 p_1, \ldots, \\ &\qquad p_l|n+\sigma_1 p_1+\ldots+\sigma_{l-1}p_{l-1},\ p|n+\sigma_1 p_1+\ldots\sigma_l p_l. \end{aligned} \tag{9}$$

It is actually easy (and in fact an exercise for the reader) to show that Y_ℓ is quite small—not much larger than $O(\mathscr{L})^\ell N/H_0$. (Outline: the probability that a given divisor p of $\sigma_1 p_1 + \ldots + \sigma_l p_l$ divide a random n is about $1/p$, which is at most $1/H_0$; the probability that there be $p_1, \ldots, p_l$ as above with $\sigma_1 p_1 + \ldots + \sigma_l p_l = 0$ is also quite small.) The more complicated task is to show that Y_ℓ is reasonably equidistributed in arithmetic progressions.

The main issue here is that we have a great number of conditions as in (9) to exclude. Inclusion-exclusion involves 2^m terms for m conditions—that is too many. There is a tool for dealing with that sort of issue in number theory, at least in some specific contexts: sieves.

We shall first show how to set up a general, abstract combinatorial sieve, for arbitrary logical conditions (rather than conditions of the form $n \equiv a \bmod p$). We will then show how to apply it to conditions of the form $n \equiv a \bmod m$, that is, congruence conditions where the moduli may be composite (as opposed to being prime, as is common in sieve theory). The matter is tricky—one has to prevent combinatorial explosion again. Rota's cross-cut theorem will be our friend.

Lastly, we will show how to apply the sieve in our context (with moduli $pp_1 \cdots p_l$ coming from (9) and sketch how to estimate the main and error terms. We will introduce *sieve graphs*.

For readers who have had some passing contact with sieve theory: while, in introductory texts on sieve theory, the emphasis is often on *counting* a set of elements S not obeying any of a set of conditions (e.g., the set S of primes n such that $n + 2$ is also a prime; its elements n do not fulfill the conditions $n \equiv 0 \bmod p$ or $n \equiv -2 \bmod p$

for any small prime p), the emphasis on much recent work, and also here, lies more generally on providing an approximation to the characteristic function 1_S of S by a function that is easier to deal with, or, if you wish, has a "simpler description" (in some precise sense). One label that has become attached to this use of sieves is "enveloping sieve", though that really describes one kind of approximation (a majorant of 1_S) and at any rate should really be called an enveloping *use* of a sieve (many sieves can be used as enveloping sieves). At any rate, that is all more or less orthogonal to the main issue here, which is that we have to develop a genuinely more general sieve.

An Abstract Combinatorial Sieve

Let $\mathbf{Q}$ be a set of conditions that an element x of a set Z may or may not fulfill. (For us, later, Z will be the set of integers, but that is of no importance at this point.) Denote by $\mathbf{Q}(x) \in 2^{\mathbf{Q}}$ the set $\{Q \in \mathbf{Q} : Q(x) \text{ is true}\}$, i.e., the set of conditions in $\mathbf{Q}$ fulfilled by x. Define $1_\emptyset(S)$ to be 1 if the set S is empty, and 0 otherwise. Then $1_\emptyset(\mathbf{Q}(x))$ equals 1 when x satisfies none of the conditions in $\mathbf{Q}$, and 0 otherwise.

We are interested in approximations to $1_\emptyset(\mathbf{Q}(x))$, i.e., the function that takes the value 1 when x satisfies none of the conditions in $\mathbf{Q}$, and 0 otherwise. This may seem to be a silly question, though it falls within the general framework we were discussing before. Let us put matters a little differently. A standard way to express $1_\emptyset(\mathbf{Q}(x))$ would be as

$$1_\emptyset(\mathbf{Q}(x)) = \sum_{\mathbf{T}\subset\mathbf{Q}(x)} (-1)^{|\mathbf{T}|},$$

and that might suit us, except that the number of subsets $\mathbf{T} \subset \mathbf{Q}(x)$ is very large. Can we obtain a reasonable approximation by means of a sum of the form

$$\sum_{\mathbf{T}\subset\mathbf{Q}(x)} g(\mathbf{T})(-1)^{|\mathbf{T}|},$$

where $g : 2^{\mathbf{Q}} \to \{0, 1\}$ is a function—preferably one whose support is much smaller than $\mathbf{Q}(x)$? (Here, as is usual, $2^{\mathbf{Q}}$ denotes the set of all subsets of $\mathbf{Q}$.)

It turns out to be possible to bound the error term in an approximation of this form in full generality. To be precise: the error term will be bounded in terms of the boundary of the support of g. Here we say that an $\mathbf{S} \subset \mathbf{Q}$ is in the boundary of a collection $B \subset 2^{\mathbf{Q}}$ if there is an element s of $\mathbf{S}$ such that exactly one of the two sets $\mathbf{S}, \mathbf{S} \setminus \{s\}$ is in B.

Lemma 5 *Let* $g : 2^{\mathbf{Q}} \to \{0, 1\}$. *Assume* $g(\emptyset) = 1$. *Choose a linear ordering for* $\mathbf{Q}$. *Then*

$$1_\emptyset(\mathbf{Q}(x)) = \sum_{\mathbf{T}\subset\mathbf{Q}(x)} g(\mathbf{T})(-1)^{|\mathbf{T}|} + \sum_{\substack{\emptyset\neq\mathbf{S}\subset\mathbf{Q}(x)\\ \min(\mathbf{Q}(x))\in\mathbf{S}}} (-1)^{|\mathbf{S}|}(g(\mathbf{S}\setminus\{\min(\mathbf{Q}(x))\}) - g(\mathbf{S})).$$

The proof is short and basically trivial (for $\mathbf{Q}(x)$ non-empty, the second sum is just a reordering of the first sum, with opposite sign). It is inspired by a passage in the proof of Brun's combinatorial sieve (see, e.g., [1, Sect. 6.2, pp. 87–89]. We do not need the linear ordering for $\mathbf{Q}$ to be in any sense natural.

Sieving by Composite Moduli

Let $\mathscr{Q}$ be a finite collection of arithmetic progressions. To each progression $P \in \mathscr{Q}$, we can associate the condition $n \in P$, for $n \in \mathbb{Z}$. Thus, we obtain a set $\mathbf{Q}$ of conditions corresponding to $\mathscr{Q}$, and apply the framework above.

We are interested in approximating $1_{n\notin P\forall P\in\mathscr{Q}}(n)$—that is, the characteristic function of the set of all n lying in no progression $P \in \mathscr{Q}$—by a sum

$$F_{\mathfrak{D}}(n) = \sum_{\substack{\mathscr{S}\subset\mathscr{Q}\\ \bigcap\mathscr{S}\in\mathfrak{D}}} (-1)^{|\mathscr{S}|} 1_{n\in\bigcap\mathscr{S}},$$

where $\mathfrak{D} \subset \mathscr{Q}^\cap$ is some set of progressions.

We will denote by $\mathfrak{q}(R)$ the modulus q of an arithmetic progression $a + q\mathbb{Z}$.

Proposition 2 *Let* $\mathscr{Q}$ *be a finite collection of distinct arithmetic progressions in* $\mathbb{Z}$ *with square-free moduli. Let* $\mathfrak{D}$ *be a non-empty subset of* $\mathscr{Q}^\cap = \{\bigcap\mathscr{S} : \mathscr{S} \subset \mathscr{Q}\}$ *with* $\emptyset \notin \mathfrak{D}$. *Assume* $\mathfrak{D}$ *is closed under containment, i.e., if* $S \in \mathfrak{D}$, *then every superset* $S' \supset S$ *in* $\mathscr{Q}^\cap$ *is also in* $\mathfrak{D}$. *Let* $F_{\mathfrak{D}}$ *be as above. Then*

$$\begin{aligned} 1_{n\notin P\forall P\in\mathscr{Q}}(n) &= F_{\mathfrak{D}}(n) + O^*\left(\sum_{R\in\partial\mathfrak{D}} 2^{\omega(\mathfrak{q}(R))} 1_{n\in R}\right) \\ &= F_{\mathfrak{D}}(n) + O^*\left(\sum_{R\in\partial_{\text{out}}\mathfrak{D}} 3^{\omega(\mathfrak{q}(R))} 1_{n\in R}\right) \end{aligned}$$

where

$$\partial\mathfrak{D} = \{R \in \mathfrak{D} : \exists P \in \mathscr{Q} \text{ s.t. } P \cap R \notin \mathfrak{D}\},$$

$$\partial_{\text{out}}\mathfrak{D} = \{D \in \mathcal{Q}^{\cap} \setminus \mathfrak{D} : \exists P \in \mathcal{Q}, R \in \mathfrak{D} \text{ s.t. } D = P \cap R\}.$$

Moreover, we can write $F_{\mathfrak{D}}(n)$ in the form

$$F_{\mathfrak{D}}(n) = \sum_{R \in \mathfrak{D}} c_R 1_{n \in R}$$

with $|c_R| \leq 2^{\omega(\mathfrak{q}(R))}$.

We can of course think of $\partial\mathfrak{D}$ and $\partial_{\text{out}}\mathfrak{D}$ as the boundary and the outer boundary of $\mathfrak{D}$.

Proof The proof of the Proposition starts with an application of the Lemma above. In what then follows, the important thing is to prevent a combinatorial explosion. For instance, it is not a priori clear that c_R can be bounded well: there could be very many ways to express a given $R \in \mathfrak{D}$ as an intersection $\bigcap \mathscr{S}$; in fact, the number of ways could be close to $2^{2^{\omega(\mathfrak{q}(R))}}$, that is, the number of collections of subsets of a set with $\mathfrak{q}(R)$ elements. We can give the much better bound $2^{\omega(\mathfrak{q}(R))}$ by obtaining cancellation (by $(-1)^{|\mathscr{S}|}$) among those different ways. To be more precise, we apply the following Lemma, which is an easy consequence of Rota's cross-cut theorem but can also be proved from scratch in a couple of lines. The same Lemma allows us to deal with the same combinatorial explosion in the error terms.

Lemma 6 *Let $\mathscr{C}$ be a collection of subsets of a finite set X. Then*

$$\left| \sum_{\substack{\mathscr{S} \subset \mathscr{C} \\ \bigcup \mathscr{S} = X}} (-1)^{|\mathscr{S}|} \right| \leq 2^{|X|}.$$

Proof Exercise.

How to apply the Proposition? We can define $\mathfrak{D}$ to be the set of progressions in $\mathcal{Q}^{\cap}$ with "small modulus", for some notion of "small". Then its boundary consists of progressions that are "borderline small", i.e., not really small, and so the proportion of n in each one of them will not be large; we just need to control the size of the boundary to show that the total error term is acceptable.

Sieve Graphs and Their Usage

We now come to our application of the sieve we have just developed. Our aim is to prevent our walks

$$n, n + \sigma_1 p_1, n + \sigma_1 p_1 + \sigma_2 p_2, \ldots$$

from having what we called *premature revenants*. We will do so by constraining each of $n, n + \sigma_1 p_1, n + \sigma_1 p_1 + \sigma_2 p_2 \ldots$ to lie within the set Y_ℓ of integers that *cannot* give rise to premature revenants.

To be precise: we define Y_ℓ to be the set of all integers n except for those for which there are primes $p_1, \ldots, p_l \in \mathbf{P}$ and signs $\sigma_1, \ldots, \sigma_l \in \{-1, 1\}$ with $1 \leq l < \ell$ such that

$$p_1|n,\ p_2|n + \sigma_1 p_1, \ldots, p_l|n + \sigma_1 p_1 + \ldots + \sigma_{l-1} p_{l-1}, \tag{10}$$

there are no repeated primes among $p_1, \ldots, p_l$ except perhaps for consecutive primes $p_i = p_{i+1} = \ldots = p_j$ with $\sigma_i = \sigma_{i+1} = \ldots = \sigma_j$, and one of the following two conditions holds:

- there exists a prime $p_0 \in \mathbf{P}$ distinct from $p_1, \ldots, p_l$ such that

$$p_0|n \text{ and } p_0|n + \sigma_1 p_1 + \ldots + \sigma_l p_l, \tag{11}$$

- we have

$$\sigma_1 p_1 + \ldots + \sigma_l p_l = 0. \tag{12}$$

The set of integers n that obey conditions (10) and (11) is an arithmetic progression to modulus $[p_0, p_1, \ldots, p_l]$ (that is, the lcm of $p_1, \ldots, p_l$, i.e., the product of all distinct primes among them), unless it is empty. The set of integers n obeying (10) and (12) is an arithmetic progression to modulus $[p_1, p_2, \ldots, p_l]$, unless it is empty. Let $W_{\ell,\mathbf{P}}$ denote the set of all arithmetic progressions arising in this way. Then the condition $n \in Y_\ell$ is equivalent to n not lying in any of the arithmetic progressions in $W_{\ell,\mathbf{P}}$. Likewise, for $\beta_1, \ldots, \beta_{2k} \in \mathbb{Z}$, the condition that $n + \beta_i \in Y_\ell$ for all $1 \leq i \leq 2k$ is equivalent to asking that n not be in any arithmetic progression of the form $P - \beta_i$ with $P \in W_{\ell,\mathbf{P}}$ and $1 \leq i \leq 2k$.

We are thus in the kind of situation to which our sieve for composite moduli is applicable. Applying the Proposition above, we obtain, for any $m \geq 1$,

$$1_{n+\beta_i \in Y_\ell \forall 1 \leq i \leq 2k} = \sum_{\substack{R \in \mathscr{Q}^\cap \\ \omega(\mathfrak{q}(R)) \leq m}} c_R 1_{n \in R} + O^*\left(3^{m+\ell} \sum_{\substack{R \in \mathscr{Q}^\cap \\ m < \omega(\mathfrak{q}(R)) \leq m+\ell}} 1_{n \in R}\right)$$

for $\mathscr{Q} = W_{\ell,\mathbf{P}}(\beta)$ and some $c_R \in \mathbb{R}$ with $|c_R| \leq 2^{\omega(\mathfrak{q}(R))}$. (The proof is one line: we define $\mathfrak{D}$ to be the set of all non-empty $R \in \mathscr{Q}^\cap$ such that the modulus of R has $\leq m$ prime factors.)

The more obvious issue now is how to bound the error term here. (For us, in our application, there is also the related issue of showing that the main term is well behaved—in particular, its sum over certain arithmetic progressions should not be too large.)

To keep track of the kind of conditions giving rise to a progression $R \in \mathscr{Q}^\cap$, we find it sensible to define a *sieve graph*, which is, one may say, a pictorial representation

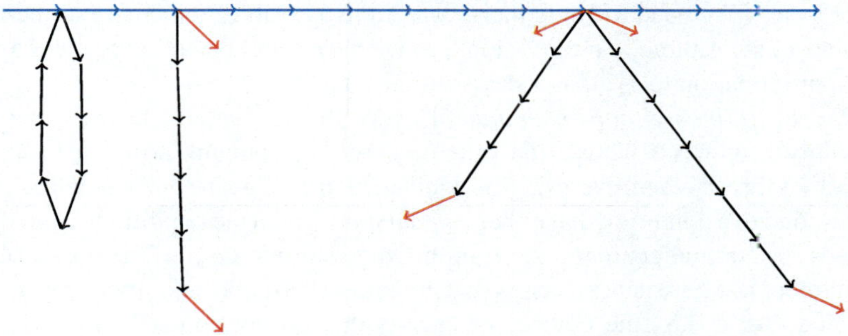

Fig. 1 A sieve graph: the blue path is the horizontal path, and the witness edges of each open path are in red. Any resemblance to a khipu is both intentional and ahistorical

of those conditions, or rather of what their general shape is and how they relate to each other.

We define a *sieve graph* to be a directed graph consisting of:

1. a path of length $2k$, called the *horizontal path*;
2. *threads* of length $< \ell$, of two kinds:
   a. a *closed thread*, which is a cycle that contains some vertex of the horizontal path, and is otherwise disjoint from it,
   b. an *open thread*, which is a path that has an endpoint at some vertex of the horizontal path, and is otherwise disjoint from it;
3. for each open thread and each of the two endpoints of that thread, an edge whose tail is that endpoint, but whose head belongs only to the edge (i.e., it is a vertex of degree 1). These two edges will be called the thread's *witnesses*; they are considered to be part of the thread (Fig. 1).

We will work with pairs $(G, \sim)$, where G is a sieve graph and $\sim$ is an equivalence relation on the edges of G such that the witnesses of a thread are equivalent to each other. We can put additional conditions, reflecting the conditions defining Y_ℓ: we require that witnesses be equivalent to no other edges in their thread, and that, in any thread, the set of edges in an equivalence class form a connected subgraph (meaning: primes do not repeat in a thread unless they are consecutive).

At any rate, it is clear what we will do: we will go over different pairs $(G, \sim)$, and, for each pair, we will consider the divisibility conditions resulting from assigning a distinct prime in $\mathbf{P}$ to each equivalence class of $\sim$. We recall that Y_ℓ is defined as the set of integers that do not satisfy any conditions of a certain kind. A pair $(G, \sim)$, together with an assignment of a prime in $\mathbf{P}$ to each equivalence class of $\sim$, corresponds to a conjunction $Q_1 \wedge Q_2 \wedge \cdots \wedge Q_j$ of some such conditions Q_i. (To be precise - the conditions are given by a pair $(G, \sim)$ together with a subset $\mathbf{l} \subset \{1, 2, \ldots, 2k\}$, corresponding to the "lit" edges: only those edges in the horizontal path whose indices are in $\mathbf{l}$ impose divisibility conditions - the unlit edges are muted, so to speak.) We need to study these conditions Q_i (all of which are of the form "n belongs to an arithmetic progression", as we have seen) because they will appear in the approximation to 1_{Y_ℓ} that a sieve will give us.

We say $(G, \sim)$ is *non-redundant* if every thread contains at least one edge x (possibly a witness) whose equivalence class $[x]$ contains no edge in any other thread. (A thread where every edge is equivalent to an edge in some other thread would correspond to a condition that either is redundant, given the conditions from the other threads, or contradicts them. As Caliph Omar did *not* say...) The *cost* $\kappa(G, \sim)$ is the number of equivalence classes that contain at least one edge (possibly a witness) in some thread, i.e., the number of classes that do not contain only edges in the horizontal path. Let $\mathbf{W}_{k,\ell,m}$ be the set of non-redundant pairs with given parameters k and ℓ and cost m. It is clear that $\mathbf{W}_{k,\ell,m}$ must be finite, since any pair of cost m contains at most m threads. It is not hard to bound the number of elements of $\mathbf{W}_{k,\ell,m}$ with a given number of threads $r \leq m$.

When we apply our sieve for composite moduli so as to approximate $1_{n+\beta_i \in Y_\ell \forall 1 \leq i \leq 2k}$, we define $\mathfrak{D}$ to be the set of conditions corresponding to non-redundant pairs $(G, \sim)$ with cost $\kappa(G, \sim) \leq m$ for some value of m we choose. Then the outer boundary $\partial_{\text{out}}\mathfrak{D}$ corresponds to non-redundant pairs $(G, \sim)$ of cost $m < \kappa(G, \sim) \leq m + \ell$. Our task is then to prove that the contribution of all $(G, \sim)$ with cost between m and $m + \ell$ is small. The crucial part is to show that, given any such $(G, \sim)$, together with a sign σ_y for each edge y and a subset $\mathbf{l} \subset \mathbf{k}$, the sum of

$$\prod_{i \in \mathbf{k} \setminus \mathbf{l}} \frac{1}{p_{[i]}} \prod_{[x] \not\subset \mathbf{k} \setminus \mathbf{l}} \frac{1}{p_{[x]}}$$

over all choices of $p_{[x]} \in \mathbf{P}$ (per equivalence class $[x]$ of $\sim$) is small: it is at most

$$\mathscr{L}^{s-r} \left(\frac{\log H}{H_0} \right)^r,$$

where r is the number of threads in G and s is the number of equivalence classes of $\sim$. (Recall that $\mathbf{P} \in [H_0, H]$ and $\mathscr{L} = \sum_{p \in \mathbf{P}} 1/p$.)

The proof is simple, and its main idea is as follows. Since $(G, \sim)$ is non-redundant, each thread contains an edge in an equivalence class that does not appear in other threads (or elsewhere in the same thread, except for consecutive appearances). For each thread, we choose one such edge x. Then the thread binds the variable $p_{[x]}$, so to speak, and so we lose one degree of freedom for each r. In detail:

1. for a closed thread, the sum $\sum_y \sigma_y p_{[y]}$ over the edges y of the thread is 0, and so $p_{[x]}$ is determined by the other $p_{[y]}$;
2. for an open thread and x a witness, $p_{[x]}$ can range only over the prime divisors of $\sum_y \sigma_y p_{[y]}$;
3. for an open thread and x not a witness, given $p_{[y]}$ for the other $[y]$ in the thread, and given $p_{[z]}$ for z the thread's witness, the class of $p_{[x]}$ modulo $p_{[z]}$ is determined.

We can give estimates on the main term of the sieve in much the same way, only keeping track of the set $\mathbf{W}'_{k,\ell,m}$ of *strongly non-redundant pairs*, meaning pairs $(G, \sim)$ such that every thread contains at least one edge x (possibly a witness) whose equivalence class $[x]$ contains no edge in any other thread and no lit edge in the horizontal path (that is, no edge in the horizontal path with index in $\mathbf{I}$).

References

1. A. C. Cojocaru and M. R. Murty. *An introduction to sieve methods and their applications*, volume 66 of *London Math. Soc. Student Texts*. Cambridge: Cambridge University Press, 2006.
2. H. Helfgott and M. Radziwiłł. Expansion, divisibility and parity. Submitted. Available at https://arxiv.org/abs/2103.06853.
3. H. Helfgott and A. Ubis. Primos, paridad y análisis. To appear in the proceedings of the AGRA III school. Available at https://arxiv.org/abs/1812.08707.
4. K. Matomäki and M. Radziwiłł. Multiplicative functions in short intervals. *Ann. of Math. (2)*, 183(3):1015–1056, 2016.
5. K. Matomäki, M. Radziwiłł, and T. Tao. An averaged form of Chowla's conjecture. *Algebra Number Theory*, 9(9):2167–2196, 2015.
6. T. Tao. The logarithmically averaged Chowla and Elliott conjectures for two-point correlations. *Forum Math. Pi*, 4:e8, 36, 2016.
7. T. Tao and J. Teräväinen. The structure of correlations of multiplicative functions at almost all scales, with applications to the Chowla and Elliott conjectures. *Algebra Number Theory*, 13(9):2103–2150, 2019.

Sums of Squares: Methods for Proving Identity Families

Russell Jay Hendel

1 Motivation

The generalized Fibonacci numbers, also known as the k-bonacci numbers, are defined by

$$F_n^{(k)} = \sum_{i=1}^{k} F_{n-i}, \qquad \text{with } F_i = 0, \text{ for } 0 \le i \le k-2, \ F_{k-1} = 1, k \ge 2. \tag{1}$$

These particular initial conditions are consistent with the Online Encyclopedia of Integer Sequences (OEIS). For $k = 2, 3, 4, 5$, we obtain the Fibonacci [9, A000045], Tribonacci [9, A000073], Tetranacci [9, A000078], and Pentenacci numbers [9, A001591], respectively.

Closed formulas for the sum of the first $m + 1$ squares of $\{F_n^{(k)}\}_{n \ge 0}$,

$$\sum_{i=0}^{m} F_i^{(k)\,2}, \tag{2}$$

are known for the cases $k = 2$, [6, pp. 77–78], $k = 3$, [7, 10, 12], $k = 4$, [7, 11, 13], $k = 5$, [11], and $k = 6$, [7].

For $k = 2$, [6, pp. 77–78], the closed formula for (2) is $\sum_{i=0}^{m} F_i^2 = F_m F_{m+1}$. [5] shows that for all k, the closed formula for (2) has a $F_m^{(k)} F_{m+1}^{(k)}$ term, generalizing the closed formula for the case $k = 2$. The algebraic methods of [13] and the generating-function methods of [11] would appear to be able to provide closed formulas for (2) for arbitrary k; however, this is not done explicitly in these papers. Moreover, neither

R. J. Hendel (✉)
Department of Mathematics, Towson University, Towson, MD 21252, USA
e-mail: RHendel@Towson.Edu

M. B. Nathanson (ed.), *Combinatorial and Additive Number Theory V*,
Springer Proceedings in Mathematics & Statistics 395,
https://doi.org/10.1007/978-3-031-10796-2_11

of these methods explicitly identify the patterns in the numerical coefficients of the closed formulas for general k.

The main contributions of this paper are (i) an explicit formula for the numerical coefficients of the closed formula for general k, (ii) the *algebraic verification* method that provides a uniform proof of the closed formula for all k, and iii) the *labeled index square* method that provides visual aids facilitating proofs.

Since the algebraic verification method is new, we should compare it to other methods. First, the algebraic verification method can simply be regarded as another tool to prove identities along with the Binet form, generating functions, and matrices. It differs from the generating-function method [11, 12] in that the proof has no algebraic prerequisites; for example, [11] succeeded in their paper by using an established theory for the Binet form of generalized Fibonacci numbers.

Both the generating functions method and algebraic verification are typically computational. A simple glance at [12] shows many lines of algebraic manipulations. In [11], these manipulations are done by software. The algebraic verification method used in this paper requires computation of the coefficients of 24 polynomials. However, while there are many computations, each of the underlying polynomials containing these coefficients are polynomials in at most three variables and at most degree 2; the computations are done by pencil and paper without a need for software.

In comparing the algebraic verification method with other algebraic-recursive methods, we see that the algebraic verification method is more direct. Once some routine simplifications are done, it is immediately transparent what has to be checked to prove the theorem. Contrastively, the methods of [5, 7, 10, 13] all require some clever algebraic manipulation to accomplish the proof.

Another important difference between the methods is that for example [11] gives methods of computing coefficients for closed formulas for sums of squares of generalized Fibonacci numbers, without showing that these coefficients have a pattern expressible as a polynomial of at most degree 2 in three variables. The identification of these patterns, as mentioned above, is one contribution of this paper.

The idea of proof methods that *uniformly* prove families of identities on recursive families has independent interest beyond the proof of the particular result in this paper. It also suggests a new trend for stating and proving identities.

These ideas motivate the following outline for this paper. Section 2 presents the closed formula for sums of the first $m+1$ squares of the $F^{(k)}$ for $k = 2, 3, 4, 5, 6$. Tables of these coefficients exhibit patterns of regularity motivating the statement of the Main Theorem in Sect. 3 which also explains the core idea of the algebraic verification method. Section 4 introduces labeled index-squares, a visual aid facilitating complex proofs. The following section presents a complete outline of the proof with an illustrative example. This is followed by two sections, which break up the sum of squares in the statement of the Main Theorem into five groups and analyze their coefficient patterns using the labeled index squares. Section 8 then combines the results of previous sections and completes the proof. Section 9 concludes the paper with a review of several recent results on families of identities and speculates on a possible new trend in approaching identities.

2 Examples

This section presents closed formulas for (2) for $k = 2, 3, 4, 5, 6$. Tables created from these formulas exhibit patterns that will enable us to formulate the Main Theorem in the next section.

First we present the examples. References for closed formulas for the cases $k = 2, 3, 4, 5, 6$ were presented in the introductory section. Note, that in these examples, and for the rest of the paper, we ease notation by letting G stand for $F^{(k)}$.

$$\textbf{For } \mathbf{G_n = F_n^{(2)}} : \sum_{i=0}^{m} G_i^2 = \frac{1}{2}\Big(2G_m G_{m+1}\Big) = G_m G_{m+1}. \tag{3}$$

$$\textbf{For } \mathbf{G_n = F_n^{(3)}} : \sum_{i=0}^{m} G_i^2 = \frac{1}{4}\Big(-G_m^2 - 4G_{m+1}^2 - G_{m+2}^2 + 2G_m G_{m+2} + 4G_{m+1}G_{m+2} + 1\Big). \tag{4}$$

$$\begin{aligned}\textbf{For } \mathbf{G_n = F_n^{(4)}} : \sum_{i=0}^{m} G_i^2 = \frac{1}{6}\Big(&-2G_m^2 - 8G_{m+1}^2 - 6G_{m+2}^2 - 2G_{m+3}^2 - 2G_m G_{m+1} \\ &+ 2G_m G_{m+3} + 4G_{m+1}G_{m+3} + 6G_{m+2}G_{m+3} + 2\Big).\end{aligned} \tag{5}$$

$$\begin{aligned}\textbf{For } \mathbf{G_n = F_n^{(5)}} : \sum_{i=0}^{m} G_i^2 = \frac{1}{8}\Big(&-3G_m^2 - 12G_{m+1}^2 - 11G_{m+2}^2 - 8G_{m-3}^2 - 3G_{m+4}^2 \\ &- 4G_m G_{m+1} - 2G_m G_{m+2} + 2G_m G_{m+4} - 4G_{m+1}G_{m+2} \\ &+ 4G_{m+1}G_{m+4} + 6G_{m+2}G_{m+4} + 8G_{m+3}G_{m+4} + 3\Big).\end{aligned} \tag{6}$$

$$\begin{aligned}\textbf{For } \mathbf{G_n = F_n^{(6)}} : \sum_{i=0}^{m} G_i^2 = \frac{1}{10}\Big(&-4G_m^2 - 16G_{m+1}^2 - 16G_{m+2}^2 - 14G_{m+3}^2 - 10G_{m+4}^2 - 4G_{m+5}^2 \\ &- 6G_m G_{m+1} - 4G_m G_{m+2} - 2G_m G_{m+3} + 2G_m G_{m+5} - 8G_{m+1}G_{m+2} - 4G_{m+1}G_{m+3} \\ &+ 4G_{m+1}G_{m+5} - 6G_{m+2}G_{m+3} + 6G_{m+2}G_{m+5} + 8G_{m+3}G_{m+5} + 10G_{m+4}G_{m+5} + 4\Big).\end{aligned} \tag{7}$$

We next discuss identifying the patterns in the coefficients in these identities and the consequent closed formulas. The general summand on the right-hand side of the above equations is

$$\frac{N_{i,j}}{D_k} G_{m+i} G_{m+j};$$

there is also a constant term, the last summand on the right-hand side. (To ease notation we omit the dependency of $N_{i,j}$ on k.) Consequently, to identify the patterns

Table 1 Coefficients, $N_{i,i}$ of G^2_{m+i}, $0 \le i \le k-1$ for $2 \le k \le 6$, based on (3)–(7)

	$i=0$	$i=1$	$i=2$	$i=3$	$i=4$	$i=5$
	G^2_m	G^2_{m+1}	G^2_{m+2}	G^2_{m+3}	G^2_{m+4}	G^2_{m+5}
$k=2$	0	0				
$k=3$	−1	−4	−1			
$k=4$	−2	−8	−6	−2		
$k=5$	−3	−12	−11	−8	−3	
$k=6$	−4	−16	−16	−14	−10	−4

in the coefficients of these identities as well as to provide closed formulas, it suffices to give explicit functional form to $N_{i,j}$, D_k, and the constant term. For $k = 2, 3, 4, 5, 6$, the denominators on the right-hand side of the examples that occur outside the big parentheses are 2,4,6,8,10 leading to the conjecture that

$$D_k = 2(k-1), \qquad k \ge 2. \tag{8}$$

The obvious approach to finding the pattern in the $N_{i,j}$, creating a table whose rows are labeled with k and whose columns are labeled with $F^{(k)}_{m+i} F^{(k)}_{m+j}$ does not immediately yield results. The correct procedure that facilitates the identification of patterns in the coefficients is to consider the diagonal elements (that is, the case $i = j$) and the non-diagonal elements separately and in different ways. The table of coefficients for the diagonal elements is presented in Table 1.

This table shows clear linear patterns in each column which naturally leads to a conjecture on the functional form of the $N_{i,i}$. The cases $i = 0$ and $i \ge 1$ are treated separately. For $i = 0$,

$$N_{0,0} = -(k-2), \tag{9}$$

while for $i \ge 1$,

$$N_{i,i} = 4 - (i+3)(k-i), \qquad \text{for } k \ge 2, 1 \le i \le k-1. \tag{10}$$

Table 1 has interesting symmetries as shown in [9, A343125].

The non-diagonal coefficients $N_{i,j}, i \ne j$, naturally form a 3-dimensional solid rather than a triangle; more specifically for each k, the coefficients $N_{i,j}$, $j \ge i+1$, form a triangle. For $k = 6$ this triangle is shown in Table 2. For $2 \le k \le 5$, the corresponding triangles are easy to construct; (they may also be found (albeit crossed out) at [9, Sequence A342955, history, item #43]).

Table 2, like Table 1, exhibits linear patterns that motivate the following conjecture on the functional form of $N_{i,j}$.

$$N_{i,j} = 2(i+1)(j-(k-2)), \qquad \text{for } 0 \le i \le k-2, i+1 \le j \le k-1. \tag{11}$$

Table 2 Coefficients $N_{i,j}$, $0 \le i \le k-2$, $i+1 \le j \le k-1$, for the case $k=6$, based on (7)

$k=6$	$j=0$	$j=1$	$j=2$	$j=3$	$j=4$	$j=5$
$i=0$		-6	-4	-2	0	2
$i=1$			-8	-4	0	4
$i=2$				-6	0	6
$i=3$					0	8
$i=4$						10

To clarify the flow of logic, we regard (8)–(11) as definitions of D_k, $N_{i,i}$, and $N_{i,j}$. The Main Theorem will then prove that these definitions correctly provide closed formulas for (2).

3 Main Theorem: Statement and Proof Outline

The previous section motivated the various components of the Main Theorem, which we now state.

Theorem 3.1 (Main Theorem) *Fix $k \ge 2$. Using (1), let $\{G_n\}_{m\ge 0} = \{F_n^{(k)}\}_{n\ge 0}$. Then, using (8)–(11),*

$$\sum_{i=0}^{m} G_i^2 = \sum_{\substack{0\le i\le k-1\\ i\le j\le k-1}} \frac{N_{i,j}}{D_k} G_{m+i}G_{m+j} - N_{k-1,k-1}. \tag{12}$$

To prove Theorem 3.1, we first make some routine simplifications.
First, clearing denominators in (12), we obtain

$$D_k \sum_{i=0}^{m} G_i^2 = \sum_{\substack{0\le i\le k-1\\ i\le j\le k-1}} N_{i,j} G_{m+i}G_{m+j} - D_k N_{k-1,k-1}. \tag{13}$$

If we replace m by $m-1$, in (13), we obtain

$$D_k \sum_{i=0}^{m-1} G_i^2 = \sum_{\substack{0\le i\le k-1\\ i\le j\le k-1}} N_{i,j} G_{m-1+i}G_{m-1+j} - D_k N_{k-1,k-1}.$$

If we now take the difference of these last two equations, we see that to prove (12), it suffices to prove

$$
\begin{aligned}
D_k G_m^2 &= \sum_{\substack{0\le i\le k-1\\ i\le j\le k-1}} N_{i,j}\Big(G_{m+i}G_{m+j} \qquad\qquad -G_{m-1+i}G_{m-1+j}\Big)\\
&= \sum_{\substack{0\le i\le k-1\\ i\le j\le k-1}} N_{i,j}G_{m+i}G_{m+j} \qquad - \sum_{\substack{-1\le i\le k-2\\ i\le j\le k-2}} N_{i+1,j+1}G_{m+i}G_{m+j}\\
&= \sum_{\substack{-1\le i\le k-2\\ i\le j\le k-2}} N'_{i,j}G_{m+i}G_{m+j}, \qquad (14)
\end{aligned}
$$

where (i) summands with a G_{m+k-1} multiplicand have been eliminated using (1) which implies that $G_{m+k-1} = \sum_{i=-1}^{k-2} G_{m+i}$, and (ii) $N'_{i,j}$ is the resulting linear combination of the $N_{i,j}$.

In the sequel, we will abuse language and refer to *the sides* of (14) that is, the first (left-most) and last (right-most) items that are set equal in (14). We similarly will refer to (i) $\sum_{\substack{0\le i\le k-1\\ i\le j\le k-1}} N_{i,j}G_{m+i}G_{m+j}$ and (ii) $-\sum_{\substack{-1\le i\le k-2\\ i\le j\le k-2}} N_{i+1,j+1}G_{m+i}G_{m+j}$, as *the two summands on the middle side of* (14). This terminology, while slightly non-standard, should cause no confusion.

We next outline the key idea in the proof of (14).

By (9)–(11), $N_{i,j}$ is a polynomial in at most three variables i, j and k, of degree at most 2, and consequently, $N'_{i,j}$ which is some linear combination of the $N_{i,j}$ is also a polynomial in at most three variables of degree at most 2.

Therefore, to prove (12), for which it suffices to prove (14), it suffices to (i) calculate the $N'_{i,j}$, $-1 \le i \le k-2$, $i \le j \le k-2$ and then (ii) *algebraically verify* that the coefficients of $G_{m+i}G_{m+j}$ on each side of (14) are identical, or equivalently, that (iii) $N'_{i,j} = D_k$ if $i = 0 = j$ and 0 otherwise.

4 The Seven Coefficient Groups

Before presenting the details of the proof, we have one subtlety to deal with. The algebraic verification method requires checking polynomial-coefficient equality over all index pairs (i, j). But these index pairs lie in the upper half of the $k \times k$ square; hence, the number of verifications could be going to infinity.

It turns out that for any $k \ge 2$, there are only seven distinct sets of index pairs that need to be considered for the proof of the Main Theorem. The polynomial form of $N'_{i,j}$ is identical for each of these seven sets. Hence, we only need algebraically verify at most seven polynomial equalities.

We will let the letters, $A, \ldots, G$ indicate these seven sets. The seven sets of index pairs are formally defined as follows.

$$A = \{i = -1, j = -1\}$$
$$B = \{i = 0, j = 0\}$$
$$C = \{1 \le i \le k-2, j = i\}$$
$$D = \{i = -1, 0 \le j \le k-3\}$$
$$E = \{i = -1, j = k-2\}$$
$$F = \{0 \le i \le k-4, i+1 \le j \le k-3\}$$
$$G = \{0 \le i \le k-3, j = k-2\}.$$

It is straightforward to check that these seven sets are mutually exclusive and completely cover the set of index pairs $\{(i, j) : -1 \le i \le k-2, i \le j \le k-2\}$. Table 3 corresponds to the formal definition just given.

In the sequel, instead of using this formal definition we will use *label indexed squares* which visually depict these seven groups. Labeled index squares are a contribution of this paper; they facilitate following the flow of a complex proof.

The following examples are illustrative of the use of the index-pair sets and labeled index squares.

Example 4.1 $\sum_A N_{i+1,j+1}G_{m+i}G_{m+j} = N_{0,0}G^2_{m-1}$. In this example, by Table 3, $A = \{i = -1, j = -1\}$. Hence $N_{i+1,j+1}G_{m+i}G_{m+j} = N_{0,0}G^2_{m-1}$ as required.

Example 4.2 A set of letters separated by commas will indicate unions of sets, for example, $\sum_{A,B,C} N_{i,j}G_{m+i}G_{m+j} = \sum_{-1 \le i \le k-2} N_{i,i}G^2_{m+i}$.

Example 4.3 $\sum_{\substack{0 \le i \le k-2 \\ 0 \le j \le k-2 \\ i \ne j}} N_{i,k-1}G_{m+i}G_{m+j} = \sum_{F,G} \left(N_{i,k-1} + N_{j,k-1}\right) G_{m+i}G_{m+j}$.

To clarify the subtleties in this sum, consider, by way of illustration two index pairs, say $i = 2, j = 4$ and $i = 4, j = 2$. For $i = 2, j = 4$ the corresponding summand is $N_{2,k-1}G_{m+2}G_{m+4} = N_{i,k-1}G_{m+i}G_{m+j}$. However, since (14) is summed over pairs of indices with $i \le j$, it follows that for $i = 4, j = 2$ the corresponding summand is

Table 3 The seven groups of $G_{m+i}G_{m+j}$

	G_{m-1}	G_m	G_{m+1}	$\cdots$	$G_{m-(k-4)}$	$G_{m-(k-3)}$	$G_{m-(k-2)}$
G_{m-1}	A	D	$\cdots$	$\cdots$	$\cdots$	D	E
G_m		B	F	$\cdots$	$\cdots$	F	G
G_{m+1}			C	$\ddots$	$\ddots$	$\vdots$	$\vdots$
$\vdots$				$\ddots$	$\ddots$	$\vdots$	$\vdots$
$G_{m-(k-4)}$					$\ddots$	F	$\vdots$
$G_{m-(k-3)}$						$\ddots$	G
$G_{m-(k-2)}$							C

$N_{4,k-1}G_{m+2}G_{m+4} = N_{j,k-1}G_{m+i}G_{m+j}$. As this example shows, the sum goes over both the index pairs with $i < j$ as well as over the index pairs with $j < i$.

5 Detailed Outline of Proof

The proof presented at the end of Sect. 3 states that the proof of the Main Theorem is accomplished through verification. The actual verification will involve distributing 24 polynomial summands over the seven index sets and then verifying the polynomial-equality corresponding to each index-set on each side of (14). To provide complete details of the flow of logic, this section organizes the proof as a series of enumerated steps with accompanying appropriate references to equations and tables.

Step i. We start with (12) the theorem statement to be proven, the Main Theorem.

Step ii. We then clear denominators, reducing proof of (12) to proof of (13)

Step iii. Equation (14) accomplishes three things. First, it eliminates the summation of the G_i^2 (and replaces it with a single G_m^2.) Additionally, this step replaces G_{m+k-1} with $\sum_{i=-1}^{k-2} G_{m+i}$ using (1).

Step iv. Equation (14) also introduces the $N'_{i,j}$, linear combinations of the $N_{i,j}$, and shows that to prove the Main Theorem it suffices to prove the equality of polynomial coefficients on *the sides* of (14).

Step v. Equations (15) and (16) break down the two summands on the middle side of (14) into three and two summands, respectively, called Summand-I through Summand-III and Summand-IV through Summand-V.

Step vi. Tables 4 through 8 analyze these fives sums in terms of the index pairs over which they are defined and the corresponding $N_{i,j}$ coefficients. This analysis is facilitated by use of labeled index squares.

Step vii. It immediately follows, that if for each index set, $A, \ldots, G$, we sum over the five summands listed in Tables 4 through 8 we will obtain the polynomial form of $N'_{i,j}$ of (14).

Step viii. Tables 9 and 10 accomplish the summation mentioned in Step vii, by first presented Table 9, a 7×5 table listing the coefficient forms of the $N_{i,j}$ over each of the seven index sets (the rows) and each of the five summands (the columns).

Step ix. Then, using (8)–(11), Table 10 evaluates each $N_{i,j}$ entry in Table 9 and additionally sums the polynomials across rows. These sums are the $N'_{i,j}$ of (14). They are linear combinations of the $N_{i,j}$. The summations are accomplished through pencil-and-paper manipulation.

Step x. A glance at this sum column in Table 10 confirms equality of the sides in (14). More specifically, for $i = 0 = j$, the polynomial coefficient is D_k and 0 otherwise.

Step xi. This completes the proof

Table 4 Summand-I $= \sum_{\substack{0 \le i \le k-2 \\ i \le j \le k-2}} N_{i,j} G_{m+i} G_{m+j} = \sum_{B,C,F,G} N_{i,j} G_{m+i} G_{m+j}$

	G_{m-1}	G_m	G_{m+1}	$\cdots$	$G_{m-(k-4)}$	$G_{m-(k-3)}$	$G_{m-(k-2)}$
G_{m-1}							
G_m		$B, N_{0,0}$	$F, N_{i,j}$	$\cdots$	$\cdots$	$F, N_{i,j}$	$G, N_{i,k-2}$
G_{m+1}			$C, N_{i,i}$	$\ddots$	$\ddots$	$\vdots$	$\vdots$
$\vdots$				$\ddots$	$\ddots$	$\vdots$	$\vdots$
$G_{m-(k-4)}$					$\ddots$	$F, N_{i,j}$	$\vdots$
$G_{m-(k-3)}$						$\ddots$	$G, N_{i,k-2}$
$G_{m-(k-2)}$							$C, N_{i,i}$

Table 5 Summand-II $= \sum_{\substack{0 \le i \le k-2 \\ j=k-1}} N_{i,j} G_{m+i} \left(\sum_{i=-1}^{k-2} G_{m+i} \right) = \sum_{B,C,F,G} N_{i,k-1} G_{m+i} G_{m+j} +$
$\sum_{D,E,F,G} N_{j,k-1} G_{m+i} G_{m+j}$

	G_{m-1}	G_m	G_{m+1}	$\cdots$	$G_{m-(k-4)}$	$G_{m-(k-3)}$	$G_{m-(k-2)}$
G_{m-1}		$D, N_{j,k-1}$	$\cdots$	$\cdots$	$\cdots$	$D, N_{j,k-1}$	$E, N_{k-2,k-1}$
G_m		$B, N_{0,k-1}$	$F, N_{i,k-1} + N_{j,k-1}$	$\cdots$	$\cdots$	$F, N_{i,k-1} + N_{j,k-1}$	$G, N_{i,k-1} + N_{k-2,k-1}$
G_{m+1}			$C, N_{i,k-1}$	$\ddots$	$\ddots$	$\vdots$	$\vdots$
$\vdots$				$\ddots$	$\ddots$	$\vdots$	$\vdots$
$G_{m-(k-4)}$					$\ddots$	$F, N_{i,k-1} + N_{j,k-1}$	$\vdots$
$G_{m-(k-3)}$						$\ddots$	$G, N_{i,k-1} + N_{k-2,k-1}$
$G_{m-(k-2)}$							$C, N_{i,k-1}$

Illustration of the proof on one summand. To further clarify these 11 steps, we illustrate them with the proof of the equality of polynomial coefficients of $G_{m-1}G_{m-1}$ on all sides of (14).

Proof. There are two summands on the middle side of (14).

The first summand. When $i = k - 1 = j$ we obtain the summand $N_{k-1,k-1}G^2_{m+k-1}$. If we then expand G^2_{m+k-1} by applying (1) we obtain $N_{k-1,k-1}\left(\sum_{i=-1}^{k-2} G_{m+i}\right)^2$. Thus we have a contribution of $N_{k-1,k-1}G^2_{m-1}$.

The second summand. When $i = -1 = j$ we obtain the summand $-N_{0,0}G^2_{m-1}$.

Table 6 Summand-III $= N_{k-1,k-1}\left(\sum_{i=-1}^{k-2} G_{m+i}\right)^2 = \sum_{A,B,C} N_{k-1,k-1}G_{m+i}^2 + \sum_{D,E,F,G} 2N_{k-1,k-1}G_{m+i}G_{m+j}$

	G_{m-1}	G_m	G_{m+1}	$\cdots$	$G_{m-(k-4)}$	$G_{m-(k-3)}$	$G_{m-(k-2)}$
G_{m-1}	$A, N_{k-1,k-1}$	$D, 2N_{k-1,k-1}$	$\ldots$	$\ldots$	$\ldots$	$D, 2N_{k-1,k-1}$	$E, 2N_{k-1,k-1}$
G_m		$B, N_{k-1,k-1}$	$F, 2N_{k-1,k-1}$	$\cdots$	$\cdots$	$F, 2N_{k-1,k-1}$	$G, 2N_{k-1,k-1}$
G_{m+1}			$C, N_{k-1,k-1}$	$\ddots$	$\ddots$	$\vdots$	$\vdots$
$\vdots$				$\ddots$	$\ddots$	$\vdots$	$\vdots$
$G_{m-(k-4)}$					$\ddots$	$F, 2N_{k-1,k-1}$	$\vdots$
$G_{m-(k-3)}$						$\ddots$	$G, 2N_{k-1,k-1}$
$G_{m-(k-2)}$							$C, N_{k-1,k-1}$

Table 7 Summand-IV $= -\sum_{-1\le i\le k-2} N_{i+1,i+1}G_{m+i}^2 = -\sum_{A,B,C} N_{i+1,i+1}G_{m+i}^2$

	G_{m-1}	G_m	G_{m+1}	$\cdots$	$G_{m-(k-4)}$	$G_{m-(k-3)}$	$G_{m-(k-2)}$
G_{m-1}	$A, -N_{0,0}$						
G_m		$B, -N_{1,1}$					
G_{m+1}			$C, -N_{i+1,i+1}$				
$\vdots$				$\ddots$			
$G_{m-(k-4)}$					$\ddots$		
$G_{m-(k-3)}$						$\ddots$	
$G_{m-(k-2)}$							$C, -N_{i+1,i+1}$

Table 8 Summand-V $= -\sum_{\substack{-1\le i\le k-3\\ i<j\le k-2}} N_{i+1,j+1}G_{m+i}G_{m+j} = -\sum_{D,E,F,G} N_{i+1,j+1}G_{m+i}G_{m+j}$

	G_{m-1}	G_m	G_{m+1}	$\cdots$	$G_{m-(k-4)}$	$G_{m-(k-3)}$	$G_{m-(k-2)}$
G_{m-1}		$D, -N_{0,j+1}$	$\ldots$	$\ldots$	$\ldots$	$D, -N_{0,j+1}$	$E, -N_{0,k-1}$
G_m			$F, -N_{i+1,j+1}$	$\ldots$	$\ldots$	$F, -N_{i+1,j+1}$	$G, -N_{i+1,k-1}$
G_{m+1}				$\ddots$	$\ddots$	$\vdots$	$\vdots$
$\vdots$					$\ddots$	$\vdots$	$\vdots$
$G_{m-(k-4)}$						$F, -N_{i+1,j+1}$	$\vdots$
$G_{m-(k-3)}$							$G, -N_{i+1,k-1}$
$G_{m-(k-2)}$							

Table 9 Coefficient of $G_{m+i}G_{m+j}$ in Summand-I through Summand-V as given in (15)–(16)

	Summand – I	Summand – II	Summand – III	Summand – IV	Summand – V
A			$N_{k-1,k-1}$	$-N_{0,0}$	
B	$N_{0,0}$	$N_{0,k-1}$	$N_{k-1,k-1}$	$-N_{1,1}$	
C	$N_{i,i}$	$N_{i,k-1}$	$N_{k-1,k-1}$	$-N_{i+1,i+1}$	
D		$N_{j,k-1}$	$2N_{k-1,k-1}$		$-N_{0,j+1}$
E		$N_{k-2,k-1}$	$2N_{k-1,k-1}$		$-N_{0,k-1}$
F	$N_{i,j}$	$N_{i,k-1}+N_{j,k-1}$	$2N_{k-1,k-1}$		$-N_{i+1,j+1}$
G	$N_{i,k-2}$	$N_{i,k-1}+N_{k-2,k-1}$	$2N_{k-1,k-1}$		$-N_{i+1,k-1}$

Table 10 Coefficient of $G_{m+i}G_{m+j}$ in equations Summand-I through Summand-V

	Summand – I	Summand – II	Summand – III	Summand – IV	Summand – V	$Sum = N'_{i,j}$
A			$2-k$	$k-2$		0
B	$2-k$	2	$2-k$	$4(k-1)-4$		$2k-2=D_k$
C	$4-(i+3)(k-i)$	$2(i+1)$	$2-k$	$(i+4)(k-i-1)-4$		0
D		$2(j+1)$	$4-2k$		$-2(j-(k-3))$	0
E		$2(k-1)$	$4-2k$		-2	0
F	$2(i+1)(j-(k-2))$	$2(i+1)+2(j+1)$	$4-2k$		$-2(i+2)(j-(k-3))$	0
G	0	$2(i+1)+2(k-1)$	$4-2k$		$-2(i+2)$	0

Thus the total contribution of the middle side of (14) to summands involving G^2_{m-1} is $(-N_{0,0}+N_{k-1,k-1})G^2_{m-1}$.

This implies that $N'_{-1,-1}=N_{k-1,k-1}-N_{0,0}$. By (9)–(10), $N'_{-1,-1}=0$ as required. □

This completes the proof of one of the seven required verifications of equality of polynomial coefficients on all sides of (14).

A similar argument applies to the other six required verifications.

We can use this sample proof to illustrate the steps enumerated above. Step iv states the overall goal of proving the equality of coefficients of $G_{m-i}G_{m+j}$ on all sides of (14). Step v, identifying the contribution of $N_{k-1,k-1}$ to $N'_{-1\ -1}$ is found in Summand-III of (15). Step vi further analyzes Summand-III with a labeled index-square in Table 6 where we find the coefficient of $N_{k-1,k-1}$ associated with index set A as required. Step viii, summarizes this association of $N_{k-1,k-1}$ with index set A and Summand-III in Table 9 in the row labeled A and the column labeled Summand-III. Then Step ix, evaluates this $N_{k-1,k-1}$ cell entry as a polynomial using (9)–(10). This evaluation may be found in Table 10 in the row labeled A and the column labeled Summand-III.

A similar argument, or trace of logical flow, applies to the contribution of $N_{0,0}$ to $N'_{-1,-1}$.

Finally, in Step x, Table 10 evaluates and sums the polynomials in its first row showing that $N'_{-1,-1} = 0$. This provides one of the seven required verifications needed for the entire proof which when done accomplishes completion of the proof as indicated in Step xi.

In the sequel, we will refer to the steps in this section as we implement each one. Steps i–iv have already been accomplished.

6 The Five Summands

This section implements Step v of the proof, breaking down the two summands on the middle side of (14) into five summands called Summand-I through Summand-V. Corresponding to the summands on the middle side of (14) we have the following decompositions.

For the first summand on the middle side of (14), we have

$$\begin{aligned}\sum_{\substack{0\le i\le k-1\\ i\le j\le k-1}} N_{i,j}G_{m+i}G_{m+j} &= \text{Summand-I} + \text{Summand-II} + \text{Summand-III}\\ &= \sum_{\substack{0\le i\le k-2\\ i\le j\le k-2}} N_{i,j}G_{m+i}G_{m+j} + \sum_{\substack{0\le i\le k-2\\ j=k-1}} N_{i,j}G_{m+i}\left(\sum_{i=-1}^{k-2} G_{m+i}\right)\\ &\quad + N_{k-1,k-1}\left(\sum_{i=-1}^{k-2} G_{m+i}\right)^2, \end{aligned} \tag{15}$$

where in summand-II and summand-III we replace G_{m+k-1} with $\sum_{i=-1}^{k-2} G_{m+i}$ using (1).

For the second summand on the middle side of (14), we have

$$\begin{aligned}-\sum_{\substack{-1\le i\le k-2\\ i\le j\le k-2}} N_{i+1,j+1}G_{m+i}G_{m+j} &= \text{Summand-IV} + \text{Summand-V}\\ &= -\sum_{-1\le i\le k-2} N_{i+1,i+1}G_{m+i}^2 - \sum_{\substack{-1\le i\le k-3\\ i<j\le k-2}} N_{i+1,j+1}G_{m+i}G_{m+j}. \end{aligned} \tag{16}$$

7 Index Tables for Summand-I Through Summand-V

This section implements Step vi of the proof, analyzing the five summands of the last section according to the index pairs over which they are summed. For the convenience of the reader, the five summands are repeated in the header of each table. The labeled index squares provide visual aids and review the definition of the seven index sets. Each cell of these labeled index squares contains (i) the index letter for that pair of indices and (ii) the linear combination of $N_{i,j}$ for that cell. These two items are separated by a comma. The index squares should be fairly straightforward to read and check (Tables 4, 5, 6, 7, 8).

8 Distribution of $N_{i,j}$ Across Index Sets

This section implements Steps vii–xi of the proof outline. Step vii states the goal, to show that the $N'_{i,j}$ are polynomially equal on *the sides* of (14).

To accomplish this goal, first, in Step viii, Table 9 summarizes the coefficients for each of the five summands of the previous section for each index set. The coefficients are expressed in terms of $N_{i,j}$.

Next, we implement Step ix. Using (8)–(11), we evaluate each entry in Table 9 as a polynomial in the variables i, j, k.

The sum column, accomplished through paper and pencil summation of polynomials of each row, implements Step x of the proof outline and calculates the $N'_{i,j}$. The sum column confirms the algebraic verification of polynomial equality of the sides of (14) since, by (8), $N'_{i,j}$ is equal to D_k for $i = 0 = j$ and 0 otherwise as required.

As observed in Step xi, the proof of the Main Theorem is complete.

9 Conclusion

This paper has proven the Main Theorem identifying the patterns in the coefficients in the closed formulas for the sums of squares of k-bonacci numbers. Along the way, the paper presented the algebraic verification method and introduced the labeled index squares as a visual aid facilitating the proof.

This result and other recent results suggest a possible new trend in approaching identities. Reference [2] reviews several stages (not necessarily consecutive) in the history of (Fibonacci-Lucas) identities. One stage is simply a concern for punchy, cute, unexpected equalities. Another stage is a concern for methods, such as Binet form, matrices, generating functions etc. The goal of this stage is to see how existing identities can be proven with a given method as well as whether new methods provide more elegant proofs. Still another stage is generalizing known Fibonacci-Lucas identities, for example, Cassini, Catalan, d'Ocagne, to other second-order recursions.

Recently, however, a new stage has emerged, families of identities. For example, Melham [8] proved a remarkable generalization of an identity of Aurifeuille for a family of identities where the number of summands is going to infinity. He used the Dresel Verification Theorem [1] to prove several individual cases thus creating credibility to the result which is still open. Hendel has proven results about families of identities where the order of the recursions is going to infinity. In [3], the family of identities corresponds to a Taylor series with the Taylor polynomials of this Taylor series equaling the characteristic polynomials which define the recursions in this family. In [4], a result about a family of recursions whose orders are going to infinity is proven uniformly for all orders, by examining the divisibility properties of characteristic polynomials of the minimal recursions of the family members. In this paper, a result about sums of squares, with initial cases each requiring different methods and ingenuity, was proven for a family of recursions, whose orders are going to infinity by a simple algebraic verification of polynomial equality.

We believe these results and methods could point to a new trend that emphasizes studying families of identities. We encourage other researchers to use the methods mentioned to achieve new results.

References

1. L. A. G. Dresel, "Transformations of Fibonacci Lucas Identities," *Applications of Fibonacci Numbers*, **Vol. 5**, 169–184, Kluwer Acad. Publ., Dordrecht, 1993
2. Russell Jay Hendel; "Review of Subsequences of Fibonacci and Lucas polynomials with geometric subscripts, Fibonacci Quart., 50, 27–35," *Mathematical Reviews*, **#2892007**, (2012).
3. Russell Jay Hendel, "Recursive Triangles Appearing Embedded in Recursive Families," *Fibonacci Quarterly*, **58**(5), (2020), 135–144.
4. Russell Jay Hendel, " A Method for Uniformly Proving a Family of Identities," *Fibonacci Quarterly*, **60**(5), 151–157, (2022).
5. Fred T. Howard and Curtis Cooper, "Some Identities for r-Fibonacci Numbers," *The Fibonacci Quarterly*, **49(3)**, 231–242, (2011).
6. T. Koshy, *Fibonacci and Lucas Numbers with Applications,* John Wiley and Sons, New York, 2001.
7. Zbigniew Jakubczyk, "Sums of Squares of Tribonacci Numbers, (Solution of Advanced Problem H-715)," *Fibonacci Quarterly*, **51(3)**, 285–286, (2013); "Alternating Sums of High Powers of Fibonacci Numbers, (Solution of Advanced Problem H-719)," *Fibonacci Quarterly*, **51(4)**, 379–380, (2013).
8. R. S. Melham, "On a Classical Fibonacci Identity of Aurifeuille", *Fibonacci Quarterly*, **54(1)**,(2016), 19–22.
9. *The on-line Encyclopedia of Integer Sequences,* http://oeis.org
10. Hideyuki Ohtsuka, "Advanced Problem H-715," *Fibonacci Quarterly*, 50(1), 90, (2012).
11. Helmut Prodinger and Sarah J. Selkirk, "Sums of Squares of Tetranacci Numbers: A Generating Function Approach," arxiv:1906.08336, (2019).
12. Raphael Schumacher, "Explicit Formulas for Sums Involving the Squares of the First n Tribonacci Numbers", *Fibonacci Quarterly*, **58**(3), 194–202, (2020).
13. Raphael Schumacher, "How to Sum the Squares of the Tetranacci Numbers and the Fibonacci m-STEP Numbers," *Fibonacci Quarterly*, **57**(2), 168–175, (2019).

Generalized Bernoulli Numbers, Cotangent Power Sums, and Higher-Order Arctangent Numbers

Brad Isaacson

1 Introduction

Let χ be a Dirichlet character modulo m. Let a, b be nonnegative integers. In [6], we obtained a formula expressing cotangent character sums

$$c_{a,b}(h,\chi) = \sum_{j=1}^{m-1} \cot^a\left(\frac{h\pi j}{m}\right)\cot^b\left(\frac{\pi j}{m}\right)\chi(j)$$

by generalized Bernoulli numbers. As a corollary, we obtained an explicit formula for cotangent power sums $c_{a,0}(1,\chi)$ introduced by Apostol [2] and later studied by Berndt [3], Lin-Zhang [7], and others.

Theorem 1 ([6]) *Let χ be as above. Let $\mathscr{B}_{k,\chi}$ be the generalized Bernoulli numbers given by (1), and $s_{n,k}$ be the numbers given by (2). For nonnegative integer n, we have*

$$\sum_{j=1}^{m-1} cot^n\left(\frac{\pi j}{m}\right)\chi(j) = -i^n \sum_{j=0}^{\left[\frac{n-1}{2}\right]} \frac{s_{n,n-2j}}{n-2j}\mathscr{B}_{n-2j,\chi} + \begin{cases} i^n\phi(m) & \text{if } \chi \text{ is principal} \\ 0 & \text{otherwise} \end{cases},$$

where $[x]$ denotes the greatest integer $\leq x$, and ϕ denotes the Euler phi function.

Remark If $\chi(-1) \neq (-1)^n$, cotangent power sums $c_{n,0}(1,\chi)$ vanish trivially for parity reasons.

In this paper, we examine some properties of $\mathscr{B}_{k,\chi}$ and several interesting consequences of Theorem 1. Before stating our results, we fix the notation.

B. Isaacson (✉)
Mathematics Department, New York City College of Technology, 300 Jay St.,, Brooklyn, NY 11201, USA
e-mail: bisaacson@citytech.cuny.edu

M. B. Nathanson (ed.), *Combinatorial and Additive Number Theory V*,
Springer Proceedings in Mathematics & Statistics 395,
https://doi.org/10.1007/978-3-031-10796-2_12

Let χ be a Dirichlet character modulo m, and ψ be the primitive Dirichlet character modulo f which induces χ. We put $\zeta = exp(2\pi i/m)$. Let $\tau(n, \chi)$ denote the Gaussian sum $\tau(n, \chi) = \sum_{j=0}^{m-1} \chi(j)\zeta^{jn}$. We will write $\tau(\chi)$ for $\tau(1, \chi)$. We define two types of generalized Bernoulli numbers $\mathscr{B}_{k,\chi}$, $B_{k,\chi}$ by the generating functions

$$\sum_{j=0}^{m-1} \frac{\tau(j, \chi)te^{jt}}{e^{mt} - 1} = \sum_{k=0}^{\infty} \mathscr{B}_{k,\chi}\frac{t^k}{k!}, \quad \sum_{j=0}^{m-1} \frac{\chi(j)te^{jt}}{e^{mt} - 1} = \sum_{k=0}^{\infty} B_{k,\chi}\frac{t^k}{k!}. \tag{1}$$

Note that if χ is trivial, $\mathscr{B}_{k,\chi}$, $B_{k,\chi}$ both reduce to ordinary Bernoulli numbers B_k defined by the generating function $\frac{t}{e^t-1} = \sum_{k=0}^{\infty} B_k \frac{t^k}{k!}$. Also note that $\mathscr{B}_{k,\chi}$, $B_{k,\chi} = 0$ whenever $\chi(-1) \neq (-1)^k$ for parity reasons except in the exceptional case

$$\mathscr{B}_{1,\chi} = \begin{cases} \phi(m)B_1 & \text{if } \chi \text{ is principal} \\ 0 & \text{otherwise} \end{cases}, \quad B_{1,\chi} = \begin{cases} B_1 & \text{if } \chi \text{ is trivial} \\ 0 & \text{otherwise} \end{cases}.$$

These definitions lead to the formulas

$$\mathscr{B}_{k,\chi} = m^{k-1} \sum_{j=0}^{m-1} B_k \left(\frac{j}{m}\right) \tau(j, \chi),$$

$$B_{k,\chi} = m^{k-1} \sum_{j=0}^{m-1} B_k \left(\frac{j}{m}\right) \chi(j).$$

The relationship between the generalized Bernoulli numbers $\mathscr{B}_{k,\chi}$, $B_{k,\overline{\psi}}$ can be seen by the following theorem which is not at all apparent from their finite sum representations given above.

Theorem 2 *Let χ, ψ be as above. For any nonnegative integer k, we have*

$$\mathscr{B}_{k,\chi} = \tau(\psi)B_{k,\overline{\psi}} \left(\frac{m}{f}\right)^k \prod_{\substack{p|m \\ p\ prime}} \left(1 - \frac{\psi(p)}{p^k}\right).$$

These two generalized Bernoulli numbers are also related to special values of Dirichlet L-functions $L(s, \chi)$ defined by $L(s, \chi) = \sum_{n=0}^{\infty} \chi(n)/n^s$.

Theorem 3 *Let χ, ψ be as above. Let k be a positive integer. If $\chi(-1) = (-1)^k$, we have*

$$(i)\quad \mathscr{B}_{k,\chi} = \frac{(-1)^{k-1}2k!m^k}{(2\pi i)^k}L(k,\chi),$$

$$(ii)\quad B_{k,\overline{\psi}} = \frac{(-1)^{k-1}2k!f^k}{\tau(\psi)(2\pi i)^k}L(k,\psi).$$

In view of Theorems 1, 2 and 3, cotangent power sums $c_{n,0}(1,\chi)$ can also be expressed by generalized Bernoulli numbers $B_{k,\overline{\psi}}$ or Dirichlet L-functions $L(k,\chi)$ or $L(k,\overline{\psi})$. We next give three interesting consequences of Theorem 1.

For positive integers n and k with $1 \le k \le n$, we set

$$s_{n,k} = \sum_{j=k}^{n} \frac{(-2)^j \binom{n}{j} \left[{j \atop k}\right]}{(j-1)!}, \tag{2}$$

where $\left[{j \atop k}\right]$ are unsigned Stirling numbers of the first kind (Stirling's cycle numbers). We note that $s_{n,k} = 0$ unless $n-k$ is even. These numbers will appear in the results below.

Let D^n be the differential operator $D^n = \frac{d^n}{dx^n}$. As a corollary to Theorem 1, we express powers of the tangent and cotangent functions as polynomials in their derivatives with coefficients in terms of the numbers $s_{n,k}$.

Theorem 4 *Let $s_{n,k}$ be the numbers given by (2). For $n \ge 1$, we put*

$$Q_n(x) = -\frac{1}{2^n}\sum_{j=0}^{[\frac{n-1}{2}]}(-4)^j s_{n,n-2j}\, x^{n-1-2j}, \quad \alpha_n = \begin{cases} 0 & \text{if } n \text{ is odd} \\ (-1)^{n/2} & \text{if } n \text{ is even} \end{cases}.$$

Then, we have

$$\cot^n x = Q_n(D(\cot x)) + \alpha_n,$$
$$\tan^n x = (-1)^{n-1}Q_n(D(\tan x)) + \alpha_n.$$

It is known that powers of the tangent and cotangent functions can also be expressed as polynomials in their derivatives with coefficients in terms of higher-order arctangent numbers $A_{n,k}$ (see (2.14) in [4]) defined by the generating function

$$\frac{\arctan^k x}{k!} = \sum_{n=k}^{\infty}\frac{A_{n,k}}{n!}x^n, \tag{3}$$

or, equivalently, by the recurrence relation

$$A_{n+1,k} = A_{n,k-1} - n(n-1)A_{n-1,k} \quad (n \geq 0,\ k \geq 1),$$
$$A_{0,0} = 1, \quad A_{n,0} = 0 \quad (n \geq 1).$$

We note that $A_{n,k} = 0$ unless $1 \leq k \leq n$ and $n - k$ is even.

Consequently, Theorem 4 implies an explicit formula for higher-order arctangent numbers $A_{n,k}$, a result for which we are unable to find in the literature.

Theorem 5 *Let $s_{n,k}$ be the numbers given by (2), and $A_{n,k}$ be the numbers given by (3). For any integer j with $0 \leq j \leq \left[\frac{n-1}{2}\right]$, we have*

$$A_{n,n-2j} = \frac{(-1)^{n+j}(n-1)!}{2^{n-2j}} s_{n,n-2j}.$$

We give one more interesting result which follows from the results above. Let $\zeta(s)$ denote the Riemann zeta function. For $n \geq 1$, we define the coefficients $c_j^{(n)}$ by

$$\left(\sum_{j=0}^{\infty} \zeta(2j)x^{2j}\right)^n = \sum_{j=0}^{\infty} c_j^{(n)} x^{2j} \quad (|x| < 1). \tag{4}$$

Although the coefficients $c_j^{(n)}$ for $0 \leq j \leq \left[\frac{n-1}{2}\right]$ can be given explicitly by the multinomial formula, we give a much simpler formula in terms of the numbers $s_{n,k}$.

Theorem 6 *Let $s_{n,k}$ be the numbers given by (2), and $c_j^{(n)}$ be the numbers given by (4). Let n be a positive integer. For any integer j with $0 \leq j \leq \left[\frac{n-1}{2}\right]$, we have*

$$c_j^{(n)} = \frac{(-1)^j \pi^{2j}(n-2j-1)!}{4^{n-j}} s_{n,n-2j}.$$

Remark The remaining coefficients $c_j^{(n)}$ for $j > \left[\frac{n-1}{2}\right]$ can be efficiently computed by the recursive formula

$$c_j^{(n)} = \frac{-2}{j} \sum_{k=1}^{j} (k(n+1) - j)\,\zeta(2k) c_{j-k}^{(n)}$$

which follows from a known formula for powers of series (see (1.3) in [5]).

2 Proofs

In this section, we supply proofs for the results given in the Introduction which are not already in the literature. We keep the notation used previously.

Theorem 2 *Let χ, ψ be as above. For any nonnegative integer k, we have*

$$\mathscr{B}_{k,\chi} = \tau(\psi) B_{k,\overline{\psi}} \left(\frac{m}{f}\right)^k \prod_{\substack{p|m \\ p\,prime}} \left(1 - \frac{\psi(p)}{p^k}\right).$$

Proof The theorem is clear for $k = 0$ since both sides vanish unless χ is trivial, in which case both sides are equal to one. Thus we assume $k \geq 1$. The proposition is clear when $\chi(-1) \neq (-1)^k$ since $A_{k,\chi} = B_{k,\overline{\psi}} = 0$ for parity reasons unless $k = 1$ and χ is principal, in which case $A_{1,\chi} = \phi(m) B_1$. When $\chi(-1) = (-1)^k$, the assertion of the proposition follows from Theorem 3, noting that

$$L(k,\chi) = L(k,\psi) \prod_{\substack{p|m \\ p\,\text{prime}}} \left(1 - \frac{\psi(p)}{p^k}\right).$$

□

The results in Theorem 3 already appear in the literature (see [1]), so we omit the proof.

Before proving Theorem 4, we prepare a lemma. We consider the sequence of functions $\{c_k(x)\}_{k=0}^{\infty}$ given in [8]. Set

$$c_k(x) = \sum_{n\in\mathbb{Z}}{}' \frac{1}{(n+x)^k} \quad (k \geq 1),$$

where x is a complex number and the prime of the summation sign indicates that meaningless terms are to be excluded. If $k = 1$, then the infinite sum on the right-hand side is understood to be

$$\lim_{t\to\infty} \left(\sum_{\substack{n\in\mathbb{Z} \\ |n+x|<t}}{}' \frac{1}{n+x} \right).$$

We also define a function $c_0(x)$ by $c_0(x) = 1$. Observe that $c_1(x) = \pi \cot \pi x$ if $x \notin \mathbb{Z}$ and 0 otherwise.

Lemma *Let k be a positive integer. We have*

$$\sum_{a=0}^{m-1} c_k\left(\frac{a}{m}\right) \chi(a) = \frac{(-1)^{k-1}(2\pi i)^k}{k!} \mathscr{B}_{k,\chi} + \begin{cases} \pi i \phi(m) & \text{if } k = 1 \text{ and } \chi \text{ is principal} \\ 0 & \text{otherwise} \end{cases}.$$

Proof We have

$$\sum_{a=0}^{m-1} c_k\left(\frac{a}{m}\right)\chi(a) = \sum_{a=0}^{m-1}{}' \sum_{n\in\mathbb{Z}} \frac{\chi(a)}{(n+a/m)^k} = m^k \sum_{a=0}^{m-1}{}' \sum_{n\in\mathbb{Z}} \frac{\chi(mn+a)}{(mn+a)^k}$$

$$= m^k \sum_{l\in\mathbb{Z}}{}' \frac{\chi(l)}{l^k} = \begin{cases} 2m^k L(k,\chi) & \text{if } \chi(-1) = (-1)^k \\ 0 & \text{if } \chi(-1) \neq (-1)^k \end{cases}.$$

Thus, in the case of $\chi(-1) = (-1)^k$, the lemma follows from Theorem 3. In the case of $\chi(-1) \neq (-1)^k$, the lemma follows from the fact that $\mathscr{B}_{k,\chi} = 0$ unless $k = 1$ and χ is principal, in which case $\mathscr{B}_{k,\chi} = -\phi(m)/2$. □

Theorem 4 *Let $s_{n,k}$ be the numbers given by (2). For $n \geq 1$, we put*

$$Q_n(x) = -\frac{1}{2^n} \sum_{j=0}^{[\frac{n-1}{2}]} (-4)^j s_{n,n-2j}\, x^{n-1-2j}, \quad \alpha_n = \begin{cases} 0 & \textit{if } n \textit{ is odd} \\ (-1)^{n/2} & \textit{if } n \textit{ is even} \end{cases}.$$

Then, we have

$$cot^n\, x = Q_n(D(cot\, x)) + \alpha_n,$$
$$tan^n\, x = (-1)^{n-1} Q_n(D(tan\, x)) + \alpha_n.$$

Proof For $x \notin \mathbb{Z}$, since $\frac{d}{dx} c_k(x) = -k c_{k+1}(x)$ for $k \geq 1$, we have $c_k(x) = P_k(c_1(x))$ with a polynomial $P_k(t)$ of degree k given recursively by

$$P_1(t) = t \quad \text{and} \quad k P_{k+1}(t) = (t^2 + \pi^2) P_k(t) \quad (k \geq 1).$$

From this, it follows that there is a polynomial $U_k(t)$ of degree k whose parity is that of k such that $c_1^k(x) = U_k(c(x))$ where we use the symbolic power notation $(c(x))^j$ to denote $c_j(x)$. Thus, we have

$$c_1^n(x) = \sum_{j=0}^{[n/2]} a_{n-2j}^{(n)}\, c_{n-2j}(x) = \sum_{j=0}^{[n/2]} a_{n-2j}^{(n)} \frac{(-1)^n}{(n-2j-1)!} D^{n-2j-1}(c_1(x)) \tag{5}$$

for some coefficients $a_{n-2j}^{(n)}$ $(0 \leq j \leq [n/2])$ which we will now determine. Suppose χ is a nonprincipal Dirichlet character. Plugging the first equality of (5) into the left-hand side of Theorem 1 and applying the Lemma, we get

$$\frac{1}{\pi^n} \sum_{j=0}^{[\frac{n-1}{2}]} a_{n-2j}^{(n)} \left(\frac{(-1)^{n-1}(2\pi i)^{n-2j}}{(n-2j)!} \right) \mathscr{B}_{n-2j,\chi} = -i^n \sum_{j=0}^{[\frac{n-1}{2}]} \frac{s_{n,n-2j}}{n-2j} \mathscr{B}_{n-2j,\chi}.$$

Hence, it follows that

$$a_{n-2j}^{(n)} = \frac{(-1)^{n+j}\pi^{2j}(n-2j-1)!}{2^{n-2j}} s_{n,n-2j} \quad (0 \le j \le [(n-1)/2]).$$

It remains to determine $a_0^{(n)}$ in the case of n even. Taking the limit of both sides of the first equality in (5) as x goes to $i\infty$, noting that $c_1(x) \to \pi i$ and $c_k(x) \to 0$ $(k \ge 2)$ as $x \to i\infty$, we get $a_0^{(2j)} = (\pi i)^{2j}$. Plugging these values of $a_{n-2j}^{(n)}$ $(0 \le j \le [n/2])$ into the second equality of (5) and simplifying, we get

$$\cot^n x = -\frac{1}{2^n} \sum_{j=0}^{[\frac{n-1}{2}]} (-4)^j s_{n,n-2j} D^{n-2j-1}(\cot x) + \begin{cases} 0 & \text{if } n \text{ is odd} \\ (-1)^{n/2} & \text{if } n \text{ is even} \end{cases},$$

as desired. Since $\cot(\pi/2 - x) = \tan x$ and $D^{n-2j-1}(\cot(\pi/2 - x)) = (-1)^{n-1} D^{n-2j-1}(\tan x)$, the analogous assertion for the tangent function readily follows. □

Theorem 5 *Let $s_{n,k}$ be the numbers given by (2), and $A_{n,k}$ be the numbers given by (3). For any integer j with $0 \le j \le \left[\frac{n-1}{2}\right]$, we have*

$$A_{n,n-2j} = \frac{(-1)^{n+j}(n-1)!}{2^{n-2j}} s_{n,n-2j}.$$

Proof From (2.14) in [4], we have

$$\tan^n x = \frac{1}{(n-1)!} \sum_{j=0}^{[\frac{n-1}{2}]} A_{n,n-2j}\, D^{n-1-2j}(\tan x) + \begin{cases} 0 & n \text{ odd} \\ (-1)^{\frac{n}{2}} & n \text{ even} \end{cases},$$

$$\cot^n x = \frac{(-1)^{n-1}}{(n-1)!} \sum_{j=0}^{[\frac{n-1}{2}]} A_{n,n-2j}\, D^{n-1-2j}(\cot x) + \begin{cases} 0 & n \text{ odd} \\ (-1)^{\frac{n}{2}} & n \text{even} \end{cases}.$$

Comparing the expressions above with those given by Theorem 4 and equating coefficients, we get the assertion of the theorem. □

Theorem 6 *Let $s_{n,k}$ be the numbers given by (2), and $c_j^{(n)}$ be the numbers given by (4). Let n be a positive integer. For any integer j with $0 \le j \le \left[\frac{n-1}{2}\right]$, we have*

$$c_j^{(n)} = \frac{(-1)^j \pi^{2j}(n-2j-1)!}{4^{n-j}} s_{n,n-2j}.$$

Proof From the Taylor expansion at $x = 0$ of

$$x\,c_1(x) = \sum\nolimits_{n\in\mathbb{Z}}' \frac{x}{n+x} = 1 + \sum_{j=0}^{\infty}\left(\sum_{\substack{n\in\mathbb{Z}\\ n\neq 0}} \frac{1}{n^{j+1}}\right)(-1)^j x^{j+1} = 1 - 2\sum_{j=1}^{\infty}\zeta(2j)x^{2j}$$

$$= -2\sum_{j=0}^{\infty}\zeta(2j)x^{2j} \quad (|x|<1),$$

we get the Laurent expansion at $x = 0$ of

$$c_1^n(x) = \left(\frac{-2}{x}\right)^n \left(\sum_{j=0}^{\infty}\zeta(2j)x^{2j}\right)^n = (-2)^n \sum_{j=0}^{\infty} c_j^{(n)} x^{2j-n} \quad (0<|x|<1,\ n\geq 1).$$

On the other hand, from (5), for $x \notin \mathbb{Z}$, we have

$$c_1^n(x) = \sum_{j=0}^{[\frac{n-1}{2}]} a_{n-2j}^{(n)}\, c_{n-2j}(x) + \begin{cases} 0 & \text{if } n \text{ is odd} \\ (\pi i)^n & \text{if } n \text{ is even} \end{cases},$$

where

$$a_{n-2j}^{(n)} = \frac{(-1)^{n+j}\pi^{2j}(n-2j-1)!}{2^{n-2j}}\, s_{n,n-2j} \quad (0 \leq j \leq [(n-1)/2]).$$

Comparing the principal parts of both expressions for $c_1^n(x)$ given above, we see that

$$(-2)^n c_j^{(n)} = a_{n-2j}^{(n)} \quad \text{for } 0 \leq j \leq \left[\frac{n-1}{2}\right].$$

Thus we get the assertion of the theorem. □

Acknowledgements Support for this project was provided by a PSC-CUNY Award, jointly funded by The Professional Staff Congress and The City University of New York. This paper is dedicated to Robert Sczech on the occasion of his 75th birthday.

References

1. E. ALKAN, *Values of Dirichlet L-functions, Gauss sums and trigonometric sums*, Ramanujan J., **26** (2011), pp. 375–398.
2. T. APOSTOL, *Dirichlet L-functions and character power sums*, J. Number Theory, **2** (1970), pp. 223–234.
3. B. C. BERNDT, *An elementary proof of some character sum identities of Apostol*, Glasgow Math. J., **14**(1) (1973), pp. 50–53.
4. D. CVIJOVIĆ, *Higher-order tangent and secant numbers*, Comput. Math. Appl., **62**(4) (2011), pp. 1879–1886.

5. H. W. GOULD, *Coefficient identities for powers of Taylor and Dirichlet series*, Am. Math. Mon., **81**(1) (1974), pp. 3–14.
6. B. ISAACSON, *Three imprimitive character sums*, Integers, **21** (2021). #A103.
7. X. LIN AND W. ZHANG, *Identities involving trigonometric functions and Bernoulli numbers*, Appl. Math. Comput., **331** (2018), pp. 228–294.
8. A. WEIL, *Elliptic Functions According to Eisenstein and Kronecker*, Springer, 1976.

A New Class of Minimal Asymptotic Bases

Melvyn B. Nathanson

2020 Mathematics Subject Classification: 11B13, 11B05, 11B34, 11B75

1 $\mathcal{G}$-Adic Asymptotic Bases

Let $\mathbf{N}_0 = \{0, 1, 2, 3, \ldots\}$ be the set of nonnegative integers and let h be a positive integer. Let $A, A_1, \ldots, A_h$ be subsets of $\mathbf{N}_0$. We define the *sumset*

$$A_1 + \cdots + A_h = \{a_1 + \cdots + a_h : a_i \in A_i \text{ for all } i = 1, 2, \ldots, h\}$$

and the *h-fold sumset*

$$hA = \underbrace{A + \cdots + A}_{h\text{summands}} = \{a_1 + \cdots + a_h : a_i \in A \text{ for all } i = 1, 2, \ldots, h\}.$$

The set A is a *basis of order h* if every nonnegative integer can be represented as the sum of h not necessarily distinct elements of A, that is, if $hA = \mathbf{N}_0$. The set A is an *asymptotic basis of order h* if hA contains all sufficiently large integers. An asymptotic basis of order h is *minimal* if no proper subset of A is an asymptotic basis of order h. Thus, if A is a minimal asymptotic basis of order h, then, for all $a \in A$, there are infinitely many integers n such that removing a from A destroys every representation of n as a sum of h elements of A.

Supported in part by a grant from the PSC-CUNY Research Award Program.

M. B. Nathanson (✉)
Lehman College (CUNY), Bronx, NY 10468, USA
e-mail: melvyn.nathanson@lehman.cuny.edu

M. B. Nathanson (ed.), *Combinatorial and Additive Number Theory V*,
Springer Proceedings in Mathematics & Statistics 395,
https://doi.org/10.1007/978-3-031-10796-2_13

Minimal asymptotic bases are extremal objects in additive number theory and are related to the conjecture of Erdős and Turán [4] that the representation function of an asymptotic basis of order h must be unbounded.

At this time, there are few explicit constructions of minimal asymptotic bases. Nathanson [11, 12] used a 2-adic construction to produce the first examples of minimal asymptotic bases. This method was extended to g-adically defined sets by Chen [2], Chen and Chen [1], Chen and Tang [3], Jia [5], Jia and Nathanson [6], Lee [7], Li and Li [8], Ling and Tang [9, 10], Sun [15, 16], and Sun and Tao [17]. This paper constructs a new class of minimal asymptotic bases.

An *interval of integers of length* t is a set of t consecutive integers. For $u, v \in \mathbf{N}_0$ with $u \leq v$, the set

$$[u, v] = \{x \in \mathbf{N}_0 : u \leq x \leq v\}$$

is an interval of integers of length $v - u + 1$. Thus, $[1, h] = \{1, 2, \ldots, h\}$.

A $\mathcal{G}$*-adic sequence* is a strictly increasing sequence of positive integers $\mathcal{G} = (g_i)_{i=0}^{\infty}$ such that $g_0 = 1$ and g_{i-1} divides g_i for all $i \geq 1$. Let $(d_i)_{i=1}^{\infty}$ be the sequence of positive integers defined by

$$d_i = \frac{g_i}{g_{i-1}}.$$

For all $i \geq 1$, we have

$$d_i \geq 2 \tag{1}$$

and

$$g_i = d_1 d_2 \cdots d_i. \tag{2}$$

For all i and j with $0 \leq i < j$, we have

$$\frac{g_{i+j}}{g_i} = d_{i+1} d_{i+2} \cdots d_{i+j}. \tag{3}$$

Every positive integer n has a unique $\mathcal{G}$*-adic representation*

$$n = \sum_{j=0}^{\infty} x_j g_j$$

, where

$$x_j \in [0, d_{j+1} - 1]$$

for all $j \in \mathbf{N}_0$ and $x_j = 0$ for all sufficiently large j (Nathanson [13, 14]). Equivalently, for every positive integer n, there is a unique nonempty finite set $F \subseteq \mathbf{N}_0$ and a unique set $\{x_j : j \in F\}$ such that

$$n = \sum_{j \in F} x_j g_j, \tag{4}$$

where $x_j \in [1, d_{j+1} - 1]$ for all $j \in F$.

For every integer $g \geq 2$, the usual g-adic representation uses the $\mathcal{G}$-adic sequence $\mathcal{G} = \left(g^i\right)_{i=0}^{\infty}$ with quotients $d_i = g$ for all $i \geq 1$.

Lemma 1 *Let $\mathcal{G} = (g_i)_{i=0}^{\infty}$ be a $\mathcal{G}$-adic system. If $n = \sum_{j \in F} x_j g_j$ is a positive integer with $x_j \in [1, d_{j+1} - 1]$ for all $j \in F$, then*

$$g_M \leq n < g_{M+1}$$

if and only if

$$\max(F) = M.$$

Proof If $\max(F) = M$, then $F \subseteq [0, M]$ and

$$\begin{aligned} g_M \leq x_M g_M \leq n &\leq \sum_{j \in F} (d_{j+1} - 1) g_j \\ &\leq \sum_{j=0}^{M} (d_{j+1} - 1) g_j = \sum_{j=0}^{M} (g_{j+1} - g_j) \\ &= g_{M+1} - 1 < g_{M+1}. \end{aligned}$$

Conversely, if $g_M \leq n < g_{M+1}$ and $\max(F) = M'$, then the inequality

$$g_{M'} \leq x_{M'} g_{M'} \leq n < g_{M+1}$$

implies $M' \leq M$. If $M' \leq M - 1$, then

$$n \leq \sum_{j=0}^{M'} (d_{j+1} - 1) g_j < g_{M'+1} \leq g_M$$

which is absurd. Therefore, $g_M \leq n < g_{M+1}$ implies $\max(F) = M$. This completes the proof.

Let W be a nonempty set of nonnegative integers, and let $\mathcal{F}^*(W)$ be the set of all nonempty finite subsets of W. Let $\mathcal{G} = (g_i)_{i=0}^{\infty}$ be a $\mathcal{G}$-adic sequence. We define the set of positive integers

$$A_{\mathcal{G}}(W) = \left\{ \sum_{j \in F} x_j g_j : F \in \mathcal{F}^*(W) \text{ and } x_j \in [1, d_{j+1} - 1] \right\}.$$

Note that $0 \notin A_{\mathcal{G}}(W)$ because $\emptyset \notin \mathcal{F}^*(W)$.

Let $h \geq 2$. A *partition* of $\mathbf{N}_0$ is a sequence $\mathcal{W} = (W_i)_{i=0}^{h-1}$ of nonempty pairwise disjoint sets such that

$$\mathbf{N}_0 = W_0 \cup W_1 \cup \cdots \cup W_{h-1}.$$

Theorem 1 *Let $h \geq 2$ and let $\mathcal{W} = (W_i)_{i=0}^{h-1}$ be a partition of $\mathbf{N}_0$. Let $\mathcal{G} = (g_i)_{i=0}^{\infty}$ be a $\mathcal{G}$-adic sequence. The set*

$$A_{\mathcal{G}}(\mathcal{W}) = \bigcup_{i=0}^{h-1} A_{\mathcal{G}}(W_i)$$

is an asymptotic basis of order h with h-fold sumset

$$hA_{\mathcal{G}}(\mathcal{W}) = \{n \in \mathbf{N}_0 : n \geq h\}.$$

Proof The smallest integer in the set $A_{\mathcal{G}}(\mathcal{W})$ is $1 = 1 \cdot g_0$. It follows that $h \in hA_{\mathcal{G}}(\mathcal{W})$ but $[0, h-1] \cap hA_{\mathcal{G}}(\mathcal{W}) = \emptyset$.

Every positive integer n has a unique $\mathcal{G}$-adic representation $n = \sum_{j \in F} x_j g_j$, where F is a nonempty finite set of nonnegative integers and $x_j \in [0, d_{j+1} - 1]$. For $i \in [0, h-1]$, let

$$F_i = F \cap W_i$$

and

$$n_i = \sum_{j \in F_i} x_j g_j.$$

If $F_i = \emptyset$, then $n_i = 0 \notin A_{\mathcal{G}}(\mathcal{W})$. If $F_i \neq \emptyset$, then

$$n_i = \sum_{j \in F_i} x_j g_j \in A_{\mathcal{G}}(W_i) \subseteq A_{\mathcal{G}}(\mathcal{W}).$$

Let

$$L = \{i \in [0, h-1] : F_i \neq \emptyset\} = \{i \in [0, h-1] : n_i \geq 1\}$$

and

$$|L| = \ell_0.$$

We have $\ell_0 \in [1, h]$ and

$$n = \sum_{i \in L} n_i \in \ell_0 A_{\mathcal{G}}(\mathcal{W}).$$

Let ℓ be the largest integer such that $\ell \leq h$ and $n \in \ell A_{\mathcal{G}}(\mathcal{W})$. We must prove that $\ell = h$.

If $n \in \ell A_{\mathcal{G}}(\mathcal{W})$, then there exist integers $n_1, \ldots, n_\ell \in A_{\mathcal{G}}(\mathcal{W})$ such that

$$n = n_1 + \cdots + n_{k-1} + n_k + n_{k+1} + \cdots + n_\ell.$$

For each $i \in [1, \ell]$ there is an integer $s_i \in [0, h-1]$ and a set $F_{s_i} \in \mathcal{F}^*(W_{s_i})$ such that n_i has the $\mathcal{G}$-adic representation

$$n_i = \sum_{j \in F_{s_i}} x_j g_j \in A_{\mathcal{G}}(W_{s_i}).$$

Suppose that $\ell < h$. If $|F_{s_k}| \geq 2$ for some $k \in [1, \ell]$, then there are nonempty sets F'_{s_i} and F''_{s_i} such that

$$F_{s_k} = F'_{s_k} \cup F''_{s_k} \quad \text{and} \quad F'_{s_k} \cap F''_{s_k} = \emptyset.$$

The integers

$$n'_k = \sum_{j \in F'_{s_k}} x_j g_j \in A_{\mathcal{G}}(W_{s_k}) \quad \text{and} \quad n''_k = \sum_{j \in F''_{s_k}} x_j g_j \in A_{\mathcal{G}}(W_{s_k})$$

satisfy

$$n_k = n'_k + n''_k$$

and so

$$n = n_1 + \cdots n_{k-1} + n'_k + n''_k + n_{k+1} + \cdots + n_\ell \in (\ell+1) A_{\mathcal{G}}(\mathcal{W}).$$

This contradicts the maximality of ℓ, and so $|F_{s_i}| = 1$ for all $i \in [1, \ell]$ and

$$n_i = x_{j_i} g_{j_i}$$

for some $g_{j_i} \in W_{s_i}$ and $x_{j_i} \in [0, d_{j_i+1} - 1]$.

If $x_{j_k} \geq 2$ for some $k \in [1, \ell]$, then

$$g_{j_k} \in A_{\mathcal{G}}(W_{s_k}) \quad \text{and} \quad (x_{j_k} - 1) g_{j_k} \in A_{\mathcal{G}}(W_{s_k})$$

and

$$n_k = g_{j_k} + (x_{j_k} - 1) g_{j_k}.$$

It follows that

$$n = n_1 + \cdots n_{k-1} + g_{j_k} + (x_{j_k} - 1) g_{j_k} + n_{k+1} + \cdots + n_\ell \in (\ell+1) A_{\mathcal{G}}(\mathcal{W}),$$

which again contradicts the maximality of ℓ. Therefore, $x_{j_i} = 1$ for all $i \in [1, \ell]$, and

$$n = g_{j_1} + \cdots + g_{j_k} + \cdots + g_{j_\ell}.$$

If $j_k \geq 1$ for some $k \in [1, \ell]$, then

$$g_{j_k} = g_{j_k - 1} + (d_{j_k} - 1) g_{j_k - 1}$$

and

$$n = g_{j_1} + \cdots + g_{j_k - 1} + (d_{j_k} - 1) g_{j_k - 1} + \cdots + g_{j_\ell} \in (\ell + 1) A_{\mathcal{G}}(\mathcal{W}),$$

which also contradicts the maximality of ℓ. Therefore, $j_k = 0$ and $g_{j_k} = g_0 = 1$ for all $k \in [1, \ell]$, and so

$$n = \underbrace{g_0 + \cdots + g_0}_{\ell \text{ summands}} = \underbrace{1 + \cdots + 1 + \cdots + 1}_{\ell \text{ summands}} = \ell \leq h - 1,$$

which is absurd. This completes the proof.

Theorem 2 *Let $h \geq 2$ and let $\mathcal{W} = (W_i)_{i=0}^{h-1}$ be a partition of $\mathbf{N}_0$. The set*

$$A = \{0\} \cup A_{\mathcal{G}}(\mathcal{W})$$

is a basis of order h but not a minimal asymptotic basis of order h.

Proof Because $0 \in A$ and $1 \in A_{\mathcal{G}}(\mathcal{W}) \subseteq A$, we have

$$\ell = (h - \ell) \cdot 0 + \ell \cdot 1 \in hA$$

for all $\ell \in [0, h - 1]$. By Theorem 1,

$$\{n \in \mathbf{N}_0 : n \geq h\} = hA_{\mathcal{G}}(\mathcal{W}) \subseteq hA$$

and so A is a basis of order h.

The set A is not a minimal asymptotic basis of order h because $0 \in A$ and the removal of 0 from A gives the set $A \setminus \{0\} = A_{\mathcal{G}}(\mathcal{W})$, which is still an asymptotic basis of order h. This completes the proof.

2 Minimal Asymptotic Bases

The following lemma generalizes a result of Jia [5].

Lemma 2 *Let $\mathcal{G} = (g_i)_{i=0}^{\infty}$ be a $\mathcal{G}$-adic sequence. Let $(u_i)_{i=1}^{p}$ be a strictly increasing finite sequence of nonnegative integers, and let $(v_j)_{j=1}^{q}$ be a finite sequence of not necessarily distinct nonnegative integers. Let*

$$x_i \in [1, d_{u_i + 1} - 1]$$

for all $i \in [1, p]$ *and*

$$y_j \in [1, d_{v_j+1} - 1]$$

for all $j \in [1, q]$. *If*

$$n = \sum_{i=1}^{p} x_i g_{u_i} = \sum_{j=1}^{q} y_j g_{v_j} \tag{5}$$

then

$$\sum_{u_i \le u_k} x_i g_{u_i} \le \sum_{v_j \le u_k} y_j g_{v_j} \tag{6}$$

for all $k \in [1, p]$.

Proof Because the sequence $(u_i)_{i=1}^{p}$ is strictly increasing, Lemma 1 implies

$$\sum_{u_i \le u_k} x_i g_{u_i} = \sum_{i=1}^{k} x_i g_{u_i} < g_{u_{k+1}}$$

for all $k \in [1, p]$. Choosing $k = p$ gives

$$n = \sum_{i=1}^{p} x_i g_{u_i} < g_{u_{p+1}}.$$

Relation (5) implies

$$g_{v_j} \le n < g_{u_p+1}$$

and so $v_j \le u_p$ for all $j \in [1, q]$. This implies (6) for $k = p$.

In the sequence $\mathcal{G} = (g_i)_{i=0}^{\infty}$, the integer g_i divides g_j for all $i \le j$. Let $k \in [1, p - 1]$. Because the sequence $(u_i)_{i=1}^{p}$ is strictly increasing, for all $i \in [k + 1, p]$, we have

$$u_k < u_k + 1 \le u_{k+1} \le u_i$$

and g_{u_k+1} divides g_{u_i}, that is,

$$g_{u_i} \equiv 0 \pmod{g_{u_k+1}}.$$

If $v_j \ge u_k + 1$, then

$$g_{v_j} \equiv 0 \pmod{g_{u_k+1}}.$$

Rearranging (5), we obtain

$$\sum_{u_i \leq u_k} x_i g_{u_i} - \sum_{v_j \leq u_k} y_j g_{v_j} = \sum_{v_j \geq u_k+1} y_j g_{v_j} - \sum_{u_i > u_k} x_i g_{u_i} \tag{7}$$
$$\equiv 0 \pmod{g_{u_k+1}}.$$

If

$$\sum_{u_i \leq u_k} x_i g_{u_i} > \sum_{v_j \leq u_k} y_j g_{v_j}$$

then Lemma 1 gives

$$0 < \sum_{i=1}^{k} x_i g_{u_i} - \sum_{v_j \leq u_k} y_j g_{v_j} \leq \sum_{i=1}^{k} x_i g_{u_i} < g_{u_{k+1}}.$$

This inequality contradicts congruence (7). This completes the proof.

Theorem 3 *Let $h \geq 2$ and let t be an integer such that*

$$t \geq 1 + \frac{\log h}{\log 2}. \tag{8}$$

Let $\mathcal{W} = (W_i)_{i=0}^{h-1}$ be a partition of $\mathbf{N}_0$ such that, for all $i \in [0, h-1]$, there is an infinite set $\mathcal{M}_i$ of positive integers such that

$$[M_i - t + 1, M_i] \subseteq W_i \tag{9}$$

for all $M_i \in \mathcal{M}_i$. Let $\mathcal{G} = (g_i)_{i=0}^{\infty}$ be a $\mathcal{G}$-adic sequence. The set

$$A_{\mathcal{G}}(\mathcal{W}) = \bigcup_{i=0}^{h-1} A_{\mathcal{G}}(W_i)$$

is a minimal asymptotic basis of order h.

Proof By Theorem 1, the set $A_{\mathcal{G}}(\mathcal{W})$ is an asymptotic basis of order h.

Let $a \in A_{\mathcal{G}}(\mathcal{W})$. Without loss of generality, we can assume that $a = a_0 \in A_{\mathcal{G}}(W_0)$ and

$$a_0 = \sum_{j \in F_0} x_{0,j} g_j, \tag{10}$$

where $F_0 \in \mathcal{F}^*(W_0)$ and $x_{0,j} \in [1, d_{j+1} - 1]$ for all $j \in F_0$. Let $M_0 = \max(F_0)$. By Lemma 1,

$$g_{M_0} \leq a_0 < g_{M_0+1}.$$

For all $i \in [1, h-1]$, choose an integer M_i in the infinite set $\mathcal{M}_i$ such that

$$M_i \geq M_0 + t \tag{11}$$

and let

$$a_i = \sum_{\substack{j \in W_i \\ j < M_0}} (d_{j+1} - 1)g_j + g_{M_i} \in A_{\mathcal{G}}(W_i). \tag{12}$$

This is the $\mathcal{G}$-adic representation of a_i. Let

$$n = \sum_{i=0}^{h-1} a_i = a_0 + \sum_{i=1}^{h-1} \sum_{\substack{j \in W_i \\ j < M_0}} (d_{j+1} - 1)g_j + \sum_{i=1}^{h-1} g_{M_i}. \tag{13}$$

This is the $\mathcal{G}$-adic representation of n.

Let

$$n = \sum_{i=0}^{h-1} b_{k_i}$$

be any representation of n as the sum of h elements of $A_{\mathcal{G}}(\mathcal{W})$, where $k_i \in [0, h-1]$ for all $i \in [0, h-1]$ and $b_{k_i} \in A_{\mathcal{G}}(W_{k_i})$. We must prove that $b_{k_i} = a_0$ for some $i \in [0, h-1]$.

Each integer b_{k_i} is of the form

$$b_{k_i} = \sum_{j \in E_{k_i}} y_{i,j} g_j \in A_{\mathcal{G}}(W_{k_i})$$

, where $E_{k_i} \in \mathcal{F}^*(W_{k_i})$ and $y_{i,j} \in [1, g_{j+1} - 1]$ for all $j \in E_{k_i}$. The uniqueness of the $\mathcal{G}$-adic representation implies that if $\{k_0, k_1, \ldots, k_{h-1}\} = [0, h-1]$, then, after rearrangement, $k_i = i$ and $a_i = b_i$ for all $i \in [0, h-1]$.

If $\{k_0, k_1, \ldots, k_{h-1}\} \neq [0, h-1]$, then there exists $s \in [0, h-1]$ such that $s \notin \{k_0, k_1, \ldots, k_{h-1}\}$. Suppose that $s \neq 0$. Recall that

$$M_s \geq M_0 + t$$

and

$$[M_s - t + 1, M_s] \subseteq W_s.$$

Because $k_i \neq s$ for all $i \in [0, h-1]$, we have

$$[M_s - t + 1, M_s] \cap E_{k_i} \subseteq W_s \cap W_{k_i} = \emptyset.$$

We construct the partition

$$E_{k_i} = E'_{k_i} \cup E''_{k_i}$$

with

$$E'_{k_i} = \{j \in E_{k_i} : j \leq M_s - t\}$$

and

$$E''_{k_i} = \{j \in E_{k_i} : j \geq M_s + 1\}.$$

The sets E'_{k_i} and E''_{k_i} are not necessarily nonempty. Let

$$b_{k_i} = b'_{k_i} + b''_{k_i}$$

, where

$$b'_{k_i} = \sum_{j \in E'_{k_i}} y_{i,j} g_j \quad \text{and} \quad b''_{k_i} = \sum_{j \in E''_{k_i}} y_{i,j} g_j.$$

Note that $b'_{k_i} = 0$ if $E'_{k_i} = \emptyset$ and $b''_{k_i} = 0$ if $E''_{k_i} = \emptyset$.

By Lemma 1,

$$b'_{k_i} < g_{M_s - t + 1} \tag{14}$$

and

$$b''_{k_i} = 0 \quad \text{or} \quad b''_{k_i} \geq g_{M_s + 1}.$$

Let

$$n' = \sum_{i=0}^{h-1} b'_{k_i} \quad \text{and} \quad n'' = \sum_{i=0}^{h-1} b''_{k_i}. \tag{15}$$

Recall inequalities (1) and (8):

$$d_i \geq 2 \quad \text{and} \quad h \leq 2^{t-1}.$$

From (15), (14), and (3), we obtain

$$n' < h g_{M_s - t + 1} \leq 2^{t-1} g_{M_s - t + 1} \leq g_{M_s - t + 1} \prod_{i=1}^{t-1} d_{M_s - t + i} = g_{M_s}.$$

Therefore, the $\mathcal{G}$-adic representation of n' is of the form

$$n' = \sum_{j=0}^{M_s - 1} z_j g_j$$

with $z_j \in [0, d_{j+1} - 1]$. Because

$$n'' = 0 \quad \text{or} \quad n'' \geq g_{M_s + 1},$$

the $\mathcal{G}$-adic representation of n'' is of the form

$$n'' = \sum_{j=M_s+1}^{\infty} z_j g_j$$

with $z_j \in [0, d_{j+1} - 1]$ and $z_j \geq 1$ for only finitely many j. Therefore,

$$n = n' + n'' = \sum_{i=0}^{M_s-1} z_i g_i + \sum_{i=M_s+1}^{\infty} z_i g_i$$

is the $\mathcal{G}$-adic representation of n. In this representation, the coefficient of g_{M_s} is 0, which contradicts the construction of n. It follows that

$$[1, h-1] \subseteq \{k_0, k_1, \ldots, k_{h-1}\}.$$

Renumbering the integers b_i, we can assume that $k_i = i$ and $b_i \in A_{\mathcal{G}}(W_i)$ for all $i \in [1, h-1]$.

We must prove that $k_0 = 0$, or, equivalently, that $b_0 \in A_{\mathcal{G}}(W_0)$. If not, then $b_0 \in A_{\mathcal{G}}(W_r)$ for some $r \in [1, h-1]$. Because

$$M_0 = \max(F_0) \in F_0 \subseteq W_0$$

we have

$$M_0 \notin \bigcup_{i=0}^{h-1} E_{k_i} \subseteq \bigcup_{i=1}^{h-1} W_i.$$

Identity (10) and Lemma 1 give

$$g_{M_0} \leq a_0 = \sum_{j \in F_0} x_{0,j} g_j < g_{M_0+1}.$$

From (12), we have

$$\begin{aligned} n &= a_0 + \sum_{i=1}^{h-1} \sum_{\substack{j \in W_i \\ j < M_0}} (d_{j+1} - 1) g_j + \sum_{i=1}^{h-1} g_{M_i} \\ &= \sum_{i=0}^{h-1} \sum_{j \in E_i} y_{i,j} g_j. \end{aligned}$$

Summing only over terms g_j with $j \leq M_0$ and applying Lemma 2, we obtain

$$
\begin{aligned}
a_0 + \sum_{i=1}^{h-1} \sum_{\substack{j \in W_i \\ j < M_0}} (d_{j+1} - 1) g_j &\leq \sum_{i=0}^{h-1} \sum_{\substack{j \in E_i \\ j \leq M_0}} y_{i,j} g_j = \sum_{i=0}^{h-1} \sum_{\substack{j \in E_i \\ j < M_0}} y_{i,j} g_j \\
&= \sum_{\substack{j \in E_0 \\ j < M_0}} y_{0,j} g_j + \sum_{i=1}^{h-1} \sum_{\substack{j \in E_i \\ j < M_0}} y_{i,j} g_j \\
&< g_{M_0} + \sum_{i=1}^{h-1} \sum_{\substack{j \in W_i \\ j < M_0}} (d_{j+1} - 1) g_j \\
&\leq a_0 + \sum_{i=1}^{h-1} \sum_{\substack{j \in W_i \\ j < M_0}} (d_{j+1} - 1) g_j .
\end{aligned}
$$

This is absurd, and so $b_0 \in A_{\mathcal{G}}(W_0)$. It follows that the integer n defined by (13) has a unique representation as the sum of h elements of $A_{\mathcal{G}}(\mathcal{W})$. Therefore,

$$
n \notin h\left(A_{\mathcal{G}}(\mathcal{W}) \setminus \{a\}\right). \tag{16}
$$

For all $i \in [1, h-1]$, there are infinitely many integers $M_i \in \mathcal{M}_i$ with $M_i \geq M_0 + t$, and so infinitely many positive integers n satisfying (16). It follows that $A_{\mathcal{G}}(\mathcal{W}) \setminus \{a\}$ is not an asymptotic basis of order h for all $a \in A_{\mathcal{G}}(\mathcal{W})$. Equivalently, $A_{\mathcal{G}}(\mathcal{W})$ is a minimal asymptotic basis of order h. This completes the proof.

Corollary 1 *Let $\mathcal{W} = W_0 \cup W_1$ be a partition of $\mathbf{N}_0$ such that both W_0 and W_1 contain infinitely many pairs of consecutive integers. Let $\mathcal{G} = (g_i)_{i=0}^{\infty}$ be a $\mathcal{G}$-adic sequence. The set*

$$
A_{\mathcal{G}}(\mathcal{W}) = A_{\mathcal{G}}(W_1) \cup A_{\mathcal{G}}(W_2)
$$

is a minimal asymptotic basis of order 2.

Proof This is the case $h = 2$ of Theorem 3.

3 Open Problems

1. Let $h \geq 2$. By Theorem 1, for every partition $\mathcal{W} = (W_i)_{i=0}^{h-1}$ of $\mathbf{N}_0$ and every $\mathcal{G}$-adic sequence, the set

$$
A = \{0\} \cup A_{\mathcal{G}}(\mathcal{W})
$$

is a basis of order h that is not a minimal asymptotic basis of order h.

   a. Determine the set of all integers $a \in A$ such that $A \setminus \{a\}$ is an asymptotic basis of order h.

b. Determine the set of all integers $a \in A$ such that $A \setminus \{a\}$ is a minimal asymptotic basis.

2. Let $\mathcal{G} = (g_i)_{i=0}^{\infty}$ be a $\mathcal{G}$-adic sequence. Let $h \geq 2$.
   a. Construct partitions $\mathcal{W} = (W_i)_{i=0}^{h-1}$ of $\mathbf{N}_0$ such that the set $A_{\mathcal{G}}(\mathcal{W})$ is a minimal asymptotic basis of order h.
   b. Construct partitions $\mathcal{W} = (W_i)_{i=0}^{h-1}$ of $\mathbf{N}_0$ such that the set $A_{\mathcal{G}}(\mathcal{W})$ is not a minimal asymptotic basis of order h.
3. Let $(d_i)_{i=1}^{\infty}$ be a sequence of 2s and 3s. Let $\mathcal{G} = (g_i)_{i=0}^{\infty}$ be the $\mathcal{G}$-adic sequence defined by $g_0 = 1$ and $g_i = \prod_{j=1}^{i} d_i$ for $i \geq 1$. Consider problem (2) with respect to this $\mathcal{G}$-adic sequence. Of particular interest are the infinitely many $\mathcal{G}$-adic sequences $\mathcal{G} = (g_i)_{i=0}^{\infty}$ with quotients $\{d_{2i-1}, d_{2i}\} = \{2, 3\}$ for all $i = 1, 2, 3, \ldots$. In this case, $g_{2i} = 6^i$ for all i.

References

1. F.-J. Chen and Y.-G. Chen, *On minimal asymptotic bases*, European J. Combin. **32** (2011), no. 8, 1329–1335.
2. Y.-G. Chen, *On the Erdős-Turán conjecture*, C. R. Math. Acad. Sci. Paris **350** (2012), no. 21-22, 933–935.
3. Y.-G. Chen and M. Tang, *On a problem of Nathanson*, Acta Arith. **185** (2018), no. 3, 275–280.
4. P. Erdős and P. Turán, *On a problem of Sidon in additive number theory, and on some related problems*, J. London Math. Soc. **16** (1941), 212–215.
5. X.-D. Jia, *Minimal bases and g-adic representations of integers*, Number Theory: New York Seminar 1991-1995 (New York), Springer-Verlag, 1996, pp. 201–209.
6. X.-D. Jia and M. B. Nathanson, *A simple construction of minimal asymptotic bases*, Acta Arith. **52** (1989), no. 2, 95–101.
7. J. B. Lee, *A construction of minimal asymptotic bases*, Periodica Math. Hungar. **26** (1993), 211–218.
8. J. Li and J. Li, *On minimal asymptotic basis of order 4*, J. Math. Res. Appl. **36** (2016), no. 6,16, 651–658.
9. D. Ling and M. Tang, *On minimal asymptotic g-adic bases*, Bull. Aust. Math. Soc. **92** (2015), no. 3, 374–379.
10. D. Ling and M. Tang, *On minimal asymptotic bases of order three*, Colloq. Math. **151** (2018), no. 1, 9–18.
11. M. B. Nathanson, *Minimal bases and maximal nonbases in additive number theory*, J. Number Theory **6** (1974), 324–333.
12. M. B. Nathanson, *Minimal bases and powers of* 2, Acta Arith. **49** (1988), no. 5, 525–532.
13. M. B. Nathanson, *Additive systems and a theorem of de Bruijn*, Amer. Math. Monthly **121** (2014), no. 1, 5–17.
14. M. B. Nathanson, *Limits and decomposition of de Bruijn's additive systems*, Combinatorial and additive number theory. II, Springer Proc. Math. Stat., vol. 220, Springer, Cham, 2017, pp. 255–267.
15. C.-F. Sun, *On minimal asymptotic basis of order $g - 1$*, Indag. Math. (N.S.) **30** (2019), no. 1, 128–135.

16. C.-F. Sun, *On a problem of Nathanson on minimal asymptotic bases*, Journal of Number Theory **218** (2021), 152–160.
17. C.-F. Sun and T.-T. Tao, *On g-adic minimal asymptotic bases of order h*, Int. J. Number Theory **15** (2019), no. 2, 389–406.

An Inverse Problem for Finite Sidon Sets

Melvyn B. Nathanson

2020 Mathematics Subject Classification: 11B13, 11B24, 11B75, 11P99

1 Sidon Sets for Linear Forms

A classical Sidon set is a set A of positive integers such that every integer has at most one representation as the sum of two elements of A. Equivalently, A has the property that if $a_1, a_2, a_1', a_2' \in A$ and $a_1 + a_2 = a_1' + a_2'$, then $\{a_1, a_2\} = \{a_1', a_2'\}$. In this paper, we consider an analog of Sidon sets for linear forms.

Let $\mathbf{F}$ be a field, let $\mathbf{F}^\times = \mathbf{F} \setminus \{0\}$, and let h be a positive integer. We consider linear forms

$$\varphi = \varphi(x_1, x_2, \ldots, x_h) = c_1x_1 + c_2x_2 + \cdots + c_hx_h$$

where $c_i \in \mathbf{F}^\times$ for all $i \in \{1, \ldots, h\}$. The linear form φ is *monic* if $c_1 = 1$. A linear form with two variables is *binary*. A linear form with h variables is called *h-ary*.

Let A be a nonempty subset $\mathbf{F}$. For $c, t \in \mathbf{F}$, the *c-dilate* of A is the set

$$c * A = \{ca : a \in A\}$$

and the *t-translate* of A by t is the set

$$A + t = \{a + t : a \in A\}.$$

Supported in part by a grant from the PSC-CUNY Research Award Program.

M. B. Nathanson (✉)
Lehman College (CUNY), Bronx, NY 10468, USA
e-mail: melvyn.nathanson@lehman.cuny.edu

M. B. Nathanson (ed.), *Combinatorial and Additive Number Theory V*,
Springer Proceedings in Mathematics & Statistics 395,
https://doi.org/10.1007/978-3-031-10796-2_14

Let

$$A^h = \{(a_1, \ldots, a_h) : a_i \in A \text{ for all } i \in \{1, \ldots, h\}\}$$

be the set of all h-tuples of elements of A. The *φ-image of* A is the set

$$\begin{aligned}\varphi(A) &= \left\{\varphi(a_1, \ldots, a_h) : (a_1, \ldots, a_h) \in A^h\right\} \\ &= \left\{c_1 a_1 + \cdots + c_h a_h : (a_1, \ldots, a_h) \in A^h\right\} \\ &= c_1 * A + \cdots + c_h * A.\end{aligned}$$

Thus, $\varphi(A)$ is a sum of dilates. We define $\varphi(\emptyset) = \{0\}$.

A nonempty subset A of $\mathbf{F}$ is a *Sidon set for the linear form* φ or, simply, a *φ-Sidon set* if it satisfies the following property: For all h-tuples $(a_1, \ldots, a_h) \in A^h$ and $(a'_1, \ldots, a'_h) \in A^h$, if

$$\varphi(a_1, \ldots, a_h) = \varphi(a'_1, \ldots, a'_h)$$

then $(a_1, \ldots, a_h) = (a'_1, \ldots, a'_h)$, that is, $a_i = a'_i$ for all $i \in \{1, \ldots, h\}$. Thus, A is a φ-Sidon set if the linear form φ is one-to-one on A^h. If A is a finite set with k elements, then A is a φ-Sidon set for the h-ary linear form φ if and only if $|\varphi(A)| = k^h$.

Of special interest is the case $\mathbf{F} = \mathbf{Q}$ with φ-Sidon sets of integers or positive integers.

A *direct problem* for Sidon sets for linear forms is the following: Given a linear form φ, describe the subsets of $\mathbf{F}$ that are φ-Sidon sets. In this paper, we consider an *inverse problem*: Given a subset A of $\mathbf{F}$, describe the linear forms φ such that A is a φ-Sidon set.

Sidon sets for linear forms were introduced by Nathanson [1, 2].

2 Monic Forms and Normalized Sets

Associated with the linear form

$$\varphi = c_1 x_1 + c_2 x_2 + \cdots + c_h x_h$$

is the monic linear form

$$\psi = x_1 + c'_2 x_2 + \cdots + c'_h x_h$$

where

$$c'_i = \frac{c_i}{c_1} \qquad \text{for } i = 2, 3, \ldots, h.$$

Let A be a nonempty subset of V. For all $(a_1, \ldots, a_h) \in A^h$, we have

$$\begin{aligned}\varphi(a_1, \dots, a_h) &= c_1 a_1 + c_2 a_2 + \cdots + c_h a_h \\ &= c_1 \left(a_1 + \frac{c_2}{c_1} a_2 + \cdots + \frac{c_h}{c_1} a_h \right) \\ &= c_1 \psi(a_1, \dots, a_h)\end{aligned}$$

and so

$$\varphi(A) = c_1 * \psi(A).$$

Because $c_1 \neq 0$, the set A is a φ-Sidon set if and only if A is a ψ-Sidon set. Thus, it suffices to consider only monic linear forms.

A subset A of $\mathbf{F}$ is *normalized* if $\{0, 1\} \subseteq A$.

Let A be a subset of $\mathbf{F}$ of cardinality $|A| \geq 2$. Let $a_0, a_1 \in A$ with $a_0 \neq a_1$. For all $a \in A$, let

$$a' = \frac{a - a_0}{a_1 - a_0}$$

and

$$A' = \left\{ \frac{a - a_0}{a_1 - a_0} : a \in A \right\}.$$

We have

$$a = (a_1 - a_0)a' + a_0$$

and

$$A = (a_1 - a_0) * A' + a_0.$$

Note that $a'_0 = 0$ and $a'_1 = 1$, and so $\{0, 1\} \subseteq A'$, that is, A' is normalized. For all $(a_1, \dots, a_h) \in A^h$ we have

$$\begin{aligned}\varphi(a_1, \dots, a_h) &= \sum_{i=1}^{h} c_i a_i = \sum_{i=1}^{h} c_i \left((a_i - a_0)a'_i + a_0 \right) \\ &= (a_1 - a_0) \sum_{i=1}^{h} c_i a'_i + a_0 \sum_{i=1}^{h} c_i \\ &= (a_1 - a_0)\varphi\left(a'_1, \dots, a'_h\right) + a_0 \sum_{i=1}^{h} c_i.\end{aligned}$$

Thus,

$$\varphi(A) = (a_1 - a_0) * \varphi(A') + a_0 \sum_{i=1}^{h} c_i.$$

Because $a_1 - a_0 \neq 0$, the set A is a φ-Sidon set if and only if A' is a φ-Sidon set. Thus, it suffices to consider only normalized subsets of $\mathbf{F}$.

3 Binary Linear Forms

Consider monic binary linear forms $\varphi = x_1 + cx_2$ with $c \in \mathbf{F}^\times$. For every nonempty subset A of $\mathbf{F}$, let

$$\mathcal{E}(A) = \left\{ c \in \mathbf{F}^\times : A \text{ is not a } \varphi - \text{ Sidon set for the form } \varphi = x_1 + cx_2 \right\}.$$

We shall prove that $\mathcal{E}(A)$ is finite if A is finite. Equivalently, every nonempty finite subset A of $\mathbf{F}$ is a φ-Sidon set for all but finitely many monic binary linear forms φ. Moreover, we compute $\mathcal{E}(A)$ explicitly for sets with at most 4 elements.

Let A be a nonempty subset of $\mathbf{F}$. Define the *difference set*

$$D(A) = A - A = \{a' - a : a, a' \in A\}$$

and the *set of difference quotients*

$$D^*(A) = \left\{ \frac{d'}{d} : d, d' \in D(A) \text{ and } d \neq 0 \right\}.$$

Theorem 1 *Let A be a nonempty subset of the field $\mathbf{F}$. Then*

$$\mathcal{E}(A) \subseteq D^*(A).$$

Proof Let $c \in \mathbf{F}^\times$ and $\varphi = x_1 + cx_2$. If $c \in \mathcal{E}(A)$, then there exist $(a_1, a_2), (a_1', a_2') \in A^2$ with $(a_1, a_2) \neq (a_1', a_2')$ such that

$$a_1 + ca_2 = \varphi(a_1, a_2) = \varphi(a_1', a_2') = a_1' + ca_2'.$$

Equivalently,

$$c(a_2' - a_2) = a_1 - a_1'.$$

If $a_2 = a_2'$, then $a_1 = a_1'$ and $(a_1, a_2) = (a_1', a_2')$, which is absurd. Therefore, $a_2 \neq a_2'$ and

$$c = \frac{a_1 - a_1'}{a_2' - a_2} \in D^*(A).$$

This completes the proof.

Corollary 1 *Every nonempty finite subset of a field is a φ-Sidon set for all but finitely many monic binary forms.*

Proof If A is finite, then the sets $D(A)$ and $D^*(A)$ are also finite, and so $\mathcal{E}(A)$ is finite.

We shall explicitly compute $\mathcal{E}(A)$ for sets with 2, 3, or 4 elements.

Lemma 1 *If A and B are nonempty subsets of a field and $A \subseteq B$, then $\mathcal{E}(A) \subseteq \mathcal{E}(B)$.*

Proof Let $c \in \mathcal{E}(A)$. There exist distinct pairs $(a_1, a_2), (a_1', a_2') \in A^2$ such that $a_1 + ca_2 = a_1' + ca_2'$. If $A \subseteq B$, then $(a_1, a_2), (a_1', a_2')$ are also distinct pairs in B^2 and so $c \in \mathcal{E}(B)$. This completes the proof.

For $c \in \mathbf{F}^\times$, let

$$E(c) = \left\{ \pm c, \pm \frac{1}{c} \right\}.$$

Lemma 2 *Let A be a nonempty subset of the field $\mathbf{F}$. If $c \in \mathcal{E}(A)$, then $E(c) \subseteq \mathcal{E}(A)$.*

Proof We associate to every $c \in \mathbf{F}^\times$ the monic binary linear forms

$$\begin{aligned}
\varphi_1 &= x_1 + cx_2 \\
\varphi_2 &= x_1 - cx_2 \\
\varphi_3 &= x_1 + \frac{1}{c}x_2 \\
\varphi_4 &= x_1 - \frac{1}{c}x_2.
\end{aligned}$$

Let $(a_1, a_2), (a_1', a_2') \in A^2$. We have

$$\varphi_1(a_1, a_2) = a_1 + ca_2 = a_1' + ca_2' = \varphi_1(a_1', a_2')$$

if and only if

$$\varphi_2(a_1, a_2') = a_1 - ca_2' = a_1' - ca_2 = \varphi_2(a_1', a_2)$$

if and only if

$$\varphi_3(a_2, a_1) = a_2 + \frac{1}{c}a_1 = a_2' + \frac{1}{c}a_1' = \varphi_3(a_2', a_1')$$

if and only if

$$\varphi_4(a_2, a_1') = a_2 - \frac{1}{c}a_1' = a_2' - \frac{1}{c}a_1 = \varphi_4(a_2', a_1).$$

Also,

$$(a_1, a_2) \neq (a_1', a_2')$$

if and only if

$$(a_1, a_2') \neq (a_1', a_2)$$

if and only if

$$(a_2, a_1) \neq (a_2', a_1')$$

if and only if

$$(a_2, a_1') \neq (a_2', a_1).$$

Thus, the set A is a φ-Sidon set for some linear form $\varphi \in \{\varphi_1, \varphi_2, \varphi_3, \varphi_4\}$ if and only if A is a φ-Sidon set for all $\varphi \in \{\varphi_1, \varphi_2, \varphi_3, \varphi_4\}$. Equivalently, $c \in \mathcal{E}(A)$ implies $E(c) \subseteq \mathcal{E}(A)$. This completes the proof.

Theorem 2 *Let* $\mathbf{F}$ *be a field and let* $A = \{0, 1\}$ *be the normalized subset of* $\mathbf{F}$ *of cardinality 2. Then*

$$\mathcal{E}(0, 1) = \{0, \pm 1\} = \{0\} \cup E(1).$$

Proof For $c \in \mathbf{F}$ and $\varphi = x_1 + cx_2$, we have $\varphi(A) = \{0, 1, c, 1 + c\}$, and $|\varphi(A)| = 4$ if and only if $c \neq 0, \pm 1$. Therefore, $\mathcal{E}(A) = \{0, \pm 1\}$. This completes the proof.

Theorem 3 *Let* $\mathbf{F}$ *be a field and let* $A = \{0, 1, a\}$ *be a normalized subset of* $\mathbf{F}$ *of cardinality 3. Then*

$$\begin{aligned} \mathcal{E}(0, 1, a) &= \{0, \pm 1\} \cup \left\{\pm a, \pm \frac{1}{a}\right\}, \left\{\pm (a-1), \pm \frac{1}{a-1}\right\}, \left\{\pm \frac{a-1}{a}, \pm \frac{a}{a-1}\right\} \\ &= \mathcal{E}(0, 1) \cup E(a) \cup E(a-1) \cup E\left(\frac{a-1}{a}\right) \end{aligned}$$

Proof For the monic binary form $\varphi = x_1 + cx_2$, the set $A = \{0, 1, a\}$ is a φ-Sidon set if and only if the set

$$\varphi(A) = \{0, 1, c, 1 + c, a, ac, a + c, 1 + ac, a + ac\}$$

has cardinality 9. There are $\binom{9}{2} = 36$ pairs of distinct elements of $\varphi(A)$. We obtain the set $\mathcal{E}(A)$ by equating pairs of distinct elements of $\varphi(A)$ and solving for c. For example, for the elements $c \in \varphi(A)$ and $1 + ac \in \varphi(A)$, we have $c = 1 + ac$ if and only if $c = -1/(a - 1)$, and so $c = -1/(a - 1) \in \mathcal{E}(A)$. Lemma 2 implies that $E(a - 1) \subseteq \mathcal{E}(A)$. Computing c for all 36 pairs determines $\mathcal{E}(A)$. This completes the proof.

Note that the set

$$E(a) \cup E(a-1) \cup E\left(\frac{a-1}{a}\right) =$$
$$\left\{\pm a, \pm \frac{1}{a}, \pm (a-1), \pm \frac{1}{a-1}, \pm \frac{a-1}{a}, \pm \frac{a}{a-1}\right\}$$

is the set of linear fractional transformations associated with the following set of matrices in $SL_2(\mathbf{Z})$:

$$\pm\left\{\begin{pmatrix}1&0\\0&1\end{pmatrix},\begin{pmatrix}0&-1\\1&0\end{pmatrix},\begin{pmatrix}1&-1\\0&1\end{pmatrix},\begin{pmatrix}0&-1\\1&-1\end{pmatrix},\begin{pmatrix}1&-1\\1&0\end{pmatrix},\begin{pmatrix}-1&0\\1&-1\end{pmatrix}\right\}.$$

Theorem 4 *Let* $\mathbf{F}$ *be a field and let* $A = \{0, 1, a, b\}$ *be a normalized subset of* $\mathbf{F}$ *of cardinality 4. Then*

$$\begin{aligned}\mathcal{E}(0,1,a,b) =& \mathcal{E}(0,1,a) \cup \mathcal{E}(0,1,b)\\ &\cup E(a-b) \cup E\left(\frac{a-b}{a}\right) \cup E\left(\frac{a-b}{b}\right)\\ &\cup E\left(\frac{a-1}{b}\right) \cup E\left(\frac{b-1}{a}\right) \cup E\left(\frac{a-1}{b-1}\right)\\ &\cup E\left(\frac{a-1}{a-b}\right) \cup E\left(\frac{b-1}{a-b}\right) \cup E\left(\frac{a}{b}\right).\end{aligned}$$

Proof The set A is a φ-Sidon set if and only if the set

$$\begin{aligned}\varphi(A) =& \{0, 1, a, b, c, ca, cb, 1+c, a+c, b+c,\\ & ac+1, ac+a, bc+1, bc+b, bc+a, ac+b\}\end{aligned}$$

has cardinality 16. As in the proof of Theorem 3, we determine the elements $c \in \mathcal{E}(A)$ by solving the $\binom{16}{2} = 120$ equations for c. Maple does the calculation. Note that Lemma 1 implies that $\mathcal{E}(0, 1, a) \cup \mathcal{E}(0, 1, b) \subseteq \mathcal{E}(0, 1, a, b)$.

4 Sidon Sets for Linear Forms in $h \geq 2$ Variables

Here are two simple constructions of h-ary linear forms φ for which a finite set A is a φ-Sidon set.

Theorem 5 *Let* A *be a finite set of complex numbers with* $|A| \geq 2$. *Let*

$$\delta = \min\{|a - a'| : a, a' \in A \text{ and } a \neq a'\}$$

and

$$\Delta = \max\{|a - a'| : a, a' \in A \text{ and } a \neq a'\}.$$

Let $(c_i)_{i=1}^{\infty}$ *be sequence of complex numbers such that*

$$|c_j| > \frac{\Delta}{\delta} \sum_{i=1}^{j-1} |c_i|$$

for all $j \geq 2$. The set A is a φ_h-Sidon set for the h-ary linear form

$$\varphi_h = \sum_{i=1}^{h} c_i x_i$$

for all $h \geq 2$.

Proof If A is not a φ_h-Sidon set, then there exist distinct h-tuples $(a_1, \ldots, a_h)$ and $(a'_1, \ldots, a'_h)$ in A^h such that

$$\varphi_h(a_1, \ldots, a_h) = \varphi_h(a'_1, \ldots, a'_h).$$

There is a largest integer $j \in \{1, \ldots, h\}$ such that $a_j \neq a'_j$. It follows that

$$\sum_{i=1}^{j} c_i a_i = \sum_{i=1}^{j} c_i a'_i$$

and so

$$c_j \left(a'_j - a_j\right) = \sum_{i=1}^{j-1} c_i (a_i - a'_i).$$

Therefore,

$$|c_j|\delta \leq \left|c_j\right| \left|a'_j - a_j\right| \leq \sum_{i=1}^{j-1} |c_i| \left|a_i - a'_i\right| \leq \Delta \sum_{i=1}^{j-1} |c_i| < \delta |c_j|$$

which is absurd. This completes the proof.

The following construction appeared in [1].

Theorem 6 *Let A be a nonempty finite set of positive integers and let $a^* = \max(A)$. For all integers $g > a^*$ and for all $h \geq 2$, the set A is a φ-Sidon set for the h-ary linear form*

$$\varphi = x_1 + g x_2 + g^2 x_3 + \cdots + g^{h-1} x_h.$$

Proof This follows immediately from the uniqueness of the g-adic representation of a positive integer.

5 Open Problems

1. Let Φ_h be the set of all monic h-ary linear forms $\varphi = x_1 + cx_2 + \cdots + c_h x_h$ with nonzero coefficients c_i in the field $\mathbf{F}$. For every nonempty subset A of $\mathbf{F}$, let

$$\mathcal{E}_h(A) = \{\varphi \in \Phi_h : A \text{ is not a } \varphi - \text{ Sidon set for some } h - \text{ ary form } \varphi\}.$$

   Compute the sets $\mathcal{E}_h(A)$ for small subsets A of $\mathbf{F}$.
2. The *representation function* of the set A with respect to an h-ary linear form φ with coefficients in the field $\mathbf{F}$ is

$$r_{A,\varphi}(b) = \text{card}\left\{(a_1, \ldots, a_h) \in A^h : \varphi(a_1, \ldots, a_h) = b\right\}$$

   for $b \in \mathbf{F}$. Let $g \geq 1$. The set A is a *φ-Sidon set of order g* if $r_{A,\varphi}(b) \leq g$ for all $b \in \mathbf{F}$. The set A is a φ-Sidon set if A is a φ-Sidon set of order 1. It is of interest to investigate inverse problems for φ-Sidon sets of orders $g \geq 2$.
3. Consider the relationship between the exceptions sets $\mathcal{E}_h(A)$ and subsets of the special linear group $SL_h(\mathbf{F})$ both for $h = 2$ and for $h \geq 3$.
4. Consider the inverse Sidon problem for infinite sets. Let $h \geq 2$. Does there exist an infinite set A such that A is not a φ-Sidon set for only finitely many h-ary linear forms φ?

References

1. M. B. Nathanson, *Representation functions of bases for binary linear forms*, Funct. Approx. Comment. Math. **37** (2007), 341–350.
2. M. B. Nathanson, Sidon sets for linear forms. J. Number Theory **239**, 207–227 (2022)